ELEMENTARY CALCULUS

VOLUME II

ELEMENTARY CALCULUS

VOLUME II

BY

C. V. DURELL, M.A.

AND

A. ROBSON, M.A.

Joint Authors of "Advanced Algebra", "Advanced Trigonometry", etc.

LONDON

G. BELL AND SONS, LTD

1977

First published 1934
Reprinted 1936, 1938, 1941, 1942, 1944, 1946 (*twice*), 1948, 1953, 1954, 1956, 1958, 1960, 1961, 1963, 1965, 1968, 1977

PRINTED IN GREAT BRITAIN BY
THE CAMELOT PRESS LTD, SOUTHAMPTON

PREFACE

THE first volume of this text-book was designed to meet the needs of the ordinary student and to supply a preliminary course for those who might afterwards become mathematical specialists. It includes all that is necessary for what is usually known as "additional mathematics" or "mathematics more advanced" in school certificate and matriculation examinations.

This second volume deals comprehensively with higher certificate work and covers many of the special requirements of distinction papers. It also contains the extensions required by physics and engineering students. Certain matters, such as the validity of some results in Chapter XIV, the transformation of integrals in Chapter XV, and the use of sign-conventions in Chapters XVI, XVII have been treated more carefully than has been usual in elementary text-books, and there is greater precision in the statements of some of the results, both in the text and the answers, than is usually demanded of an average student.

The exercises are necessarily longer than those in Volume I. For the convenience of those teachers who require a short preliminary course a list of selected examples has been compiled. This will be found after the Table of Contents; it is printed in two columns: the first contains a minimum course and some teachers will probably use the second column to supplement it; alternatively the second column may be used for revision. It is hoped that this innovation may be of assistance in securing that the work of the student is distributed fairly over the various parts of the subject. For the sake of those teachers who prefer to have a much larger number of questions at their disposal, an **Appendix** of supplementary examples has been drawn up on the same lines as that in Volume I. The exercises are given the same headings as the corresponding exercises in the main body of the text so that the teacher can easily pass on to the supplementary questions; at the head of each appendix exercise a list of selected questions for a short course is given.

For the convenience of students whose work is restricted to

the easier mathematical papers in the various higher certificate examinations, a suitable selection of the material in Volume II is published separately under the title **Higher Certificate Calculus,** in two parts.

The authors have also in preparation an *Advanced Calculus* which will form a sequel to the *Elementary Calculus*. This will include a more rigorous treatment of the fundamentals, will provide for the extra practice necessary to develop the technique of the mathematician, and will deal with a few more advanced topics.

The authors have again to thank Mr. J. C. Manisty for his valuable assistance in the correction of the proofs.

C. V. D.

A. R.

November 1933.

CONTENTS

VOLUME II

SELECTED EXAMPLES

The following lists are provided for those who prefer a short preliminary course. The list on the left is a *minimum* selection for a first reading; that on the right is intended for a revision course.

	First Reading		Revision Course
XI a.	1, 3, 5, 7, 9, 11, 14, 15, 16.	**XI** a.	13, 17.
XI b.	1, 4, 5, 6, 9, 10, 11, 12.	**XI** b.	2, 8, 13, 15.
XI c.	1, 3, 5, 7, 14.	**XI** c.	4, 6, 8, 12.
XII a.	1, 2, 4, 5, 8, 11, 13, 17.	**XII** a.	10, 15, 16, 18.
XII b.	1, 2, 5, 7, 8, 10, 12, 14, 17.	**XII** b.	6, 9, 13, 15.
XII c.	1, 3, 5, 7, 11, 13, 16, 17, 19, 23.	**XII** c.	6, 12, 18, 20, 24.
XII d.	1, 4, 7, 11, 13, 14, 18.	**XII** d.	2, 5, 9, 12, 17.
XII e.	1, 2, 3, 5, 6, 9, 16, 19, 20.	**XII** e.	10, 11, 18, 21, 22.
XIII a.	1, 2, 3, 4, 6, 7, 8, 9, 11.	**XIII** a.	5, 12.
XIII b.	1, 2, 3, 4, 5, 6, 7, 9, 11, 14, 16, 18, 19, 24, 25, 27.	**XIII** b.	8, 12, 13, 22, 23, 26.
XIII c.	1, 2, 3, 6, 7, 9, 13, 14, 15, 16, 19, 20, 26, 27, 29.	**XIII** c.	8, 11, 12, 18, 21, 25, 30.
XIII d.	1, 2, 3, 5, 9, 10.	**XIII** d.	4, 7, 8.
XIII e.	1, 3, 5, 7, 8, 9, 17.	**XIII** e.	2, 10, 11, 12, 18.
XIII f.	1, 3, 5, 8, 9, 12, 13, 15, 18, 19, 20, 21, 28.	**XIII** f.	6, 17, 22, 23, 27.
XIV a.	(i), (ii) of 1, 2, 3, 4, 5, 6, 7, 8; 9, 12, 13, 14, 15, 16, 17, 19, 20, 21, 24, 25, 30, 31, 34, 37, 39, 40, 42.	**XIV** a.	(iii) of 1, 2, 3, 4, 5, 6, 7, 8; 18, 23, 27, 35, 38, 41, 44.
XIV b.	1, 2, 3, 5, 6, 8, 10, 11, 13, 17.	**XIV** b.	4, 9, 15, 16, 18.
XIV c.	1, 2, 3, 7, 8, 12, 14, 17, 19, 23.	**XIV** c.	9, 15, 21, 24, 27.
XIV d.	1, 2, 6, 7, 8, 9, 13, 14, 15, 19, 21, 24, 28, 29, 30.	**XIV** d.	5, 12, 16, 20, 25, 32, 33, 34.
XIV e.	1, 2, 4, 5, 7, 8, 10, 11, 14, 20.	**XIV** e.	3, 9, 12, 16, 18, 19.
XIV f.	1, 2, 4, 6, 9, 10.	**XIV** f.	3, 7, 12, 15.
XIV g.	1, 2, 3, 4, 6, 7, 8, 11.	**XIV** g.	5, 9, 10, 14, 15.
XV a.	1, 2, 3, 4, 7, 8, 11, 13, 17, 19, 25, 27, 30, 32.	**XV** a.	5, 6, 10, 12, 14, 16, 18, 20, 31, 34.
XV b.	1, 2, 3, 6, 7, 9.	**XV** b.	4, 5, 10, 13.
XV c.	1, 2, 3, 4, 5, 16, 20.	**XV** c.	6, 7, 10, 14, 18, 21, 22.
XV d.	1, 2, 4, 6, 12, 13.	**XV** d.	5, 8, 14, 15.

First Reading		Revision Course	
XV e.	1, 2, 3, 6, 9, 10, 11, 15, 16, 18.	**XV** e.	4, 7, 12, 13, 17, 19.
XVI a.	1, 2, 5, 6.	**XVI** a.	3, 7, 8.
XVI b.	1, 2, 3, 5, 7, 9.	**XVI** b.	4, 6, 8, 11.
XVI c.	1, 2, 4, 6, 7, 8, 10, 14.	**XVI** c.	9, 12, 15.
XVI d.	1, 2, 3, 6, 9.	**XVI** d.	4, 8, 10, 11.
XVI e.	1, 3, 6, 7, 8, 12, 14, 17.	**XVI** e.	2, 5, 10, 13, 16.
XVII a.	1, 3, 4, 5, 8, 10, 13, 14.	**XVII** a.	2, 6, 7, 9, 11.
XVII b.	1, 2, 3, 5, 6, 8.	**XVII** b.	4, 7, 9, 12.
XVII c.	1, 2, 4, 6, 9, 11.	**XVII** c.	3, 5, 7, 10.
XVII d.	1, 2, 5, 6, 8, 9, 11.	**XVII** d.	3, 10, 12.
XVIII a.	1, 2, 4, 8, 9, 11, 12, 13.	**XVIII** a.	3, 5, 14.
XVIII b.	1, 2, 5, 6, 9, 10.	**XVIII** b.	4, 7, 8, 12.
XVIII c.	1, 2, 5, 8, 9.	**XVIII** c.	4, 7, 10.
XVIII d.	1, 2, 3, 4, 6, 8.	**XVIII** d.	5, 7, 10.
XVIII e.	1, 2, 3, 5, 7, 9, 11, 12.	**XVIII** e.	4, 6, 8, 10, 13, 16.
XIX a.	1, 2, 3, 7, 8, 11, 14, 15, 18, 20.	**XIX** a.	6, 12, 16, 17, 19, 21.
XIX b.	1, 2(i), 3, 4, 8, 11(i), 12, 13.	**XIX** b.	5, 10, 11(ii), 14.
XIX c.	1, 3, 4, 7, 9, 10, 13, 15.	**XIX** c.	2, 5, 8, 14, 16.
XX a.	1, 2, 3, 4, 7(i), 8.	**XX** a.	6, 7(ii), 9.
XX b.	1, 2, 4, 5, 7, 9, 11, 18, 19, 23.	**XX** b.	3, 6, 8, 10, 12, 13, 16, 17, 22.
XX c.	1, 2, 3, 5, 7, 10.	**XX** c.	4, 6, 8, 9, 11.
XX d.	1, 2, 3, 7, 8.	**XX** d.	5, 6, 9, 10, 11.
XX e.	1, 2, 5, 7, 10, 11, 15, 16.	**XX** e.	3, 6, 8, 12, 17.
XX f.	1, 2, 3, 4, 5, 6, 7, 8, 9, 10, 11, 12, 13, 14, 15, 16, 17, 18, 19, 20, 21, 23, 25, 26, 27, 29, 30, 32.	**XX** f.	2, 4, 5, 7, 22, 24, 28, 33(i), 34, 35.
XXI a.	1, 2, 4, 5, 7, 8, 11, 15, 16, 18, 19.	**XXI** a.	3, 6, 9, 10, 13, 14, 20, 22.
XXI b.	1, 2, 3, 5, 7, 10, 11, 15, 19, 21, 23.	**XXI** b.	6, 12, 16, 18, 22, 24.
XXI c.	1, 2, 3, 4, 8, 9, 10, 11, 12, 14, 15, 19, 21, 23.	**XXI** c.	6, 7, 13, 16, 18, 20, 24.
XXI d.	1, 2, 3, 4, 5, 7, 9, 11, 13, 17, 20, 22, 24, 26.	**XXI** d.	6, 8, 10, 14, 18, 19, 21, 23, 27, 28, 29.
XXII a.	1, 3, 4, 5, 6, 8.	**XXII** a.	2, 7, 9, 10.
XXII b.	1, 2, 3, 4, 6, 7, 8, 10, 12, 16.	**XXII** b.	5, 9, 11, 14, 15.
XXII c.	1, 2, 3, 4, 5, 8.	**XXII** c.	6, 7, 9.

NATURAL LOGARITHMS

	·0	·1	·2	·3	·4	·5	·6	·7	·8	·9
1	0·000	0·095	0·182	0·262	0·336	0·405	0·470	0·531	0·588	0·642
2	0·693	0·742	0·788	0·833	0·875	0·916	0·956	0·993	1·030	1·065
3	1·099	1·131	1·163	1·194	1·224	1·253	1·281	1·308	1·335	1·361
4	1·386	1·411	1·435	1·459	1·482	1·504	1·526	1·548	1·569	1·589
5	1·609	1·629	1·649	1·668	1·686	1·705	1·723	1·740	1·758	1·775
6	1·792	1·808	1·825	1·841	1·856	1·872	1·887	1·902	1·917	1·932
7	1·946	1·960	1·974	1·988	2·001	2·015	2·028	2·041	2·054	2·067
8	2·079	2·092	2·104	2·116	2·128	2·140	2·152	2·163	2·175	2·186
9	2·197	2·208	2·219	2·230	2·241	2·251	2·262	2·272	2·282	2·293

log 10 = 2·302585 . . .

x	e^x	e^{-x}	ch x	sh x	th x
0·0	1·00	1·00	1·00	0·00	0·00
0·2	1·22	0·82	1·02	0·20	0·20
0·4	1·49	0·67	1·08	0·41	0·38
0·6	1·82	0·55	1·19	0·64	0·54
0·8	2·23	0·45	1·34	0·89	0·66
1·0	2·72	0·37	1·54	1·18	0·76
1·2	3·32	0·30	1·81	1·51	0·83
1·4	4·06	0·25	2·15	1·90	0·89
1·6	4·95	0·20	2·58	2·38	0·92
1·8	6·05	0·17	3·11	2·94	0·95
2·0	7·39	0·14	3·76	3·63	0·96
2·2	9·03	0·11	4·57	4·46	0·98
2·4	11·02	0·09	5·56	5·47	0·98
2·6	13·46	0·07	6·77	6·69	0·99
2·8	16·44	0·06	8·25	8·19	0·99
3·0	20·09	0·05	10·07	10·02	1·00

CHAPTER XI

TWO METHODS OF INTEGRATION

HITHERTO we have dealt only with integrals whose values can be written down in virtue of previous experience of differentiation. It is not possible to give a complete set of rules for integration like those given for differentiation in Chapter V. Before discussing what functions can be integrated it is convenient to explain two methods which are often used to reduce a given integral to another which is already known.

Integration by Substitution.

Example 1. Evaluate $\int x\sqrt{(x+1)}\,dx$.

Denoting the integral by y, $\dfrac{dy}{dx}=x\sqrt{(x+1)}$.

$$\text{Put } x+1=z^2,\ (z>0),\ \text{then } \frac{dx}{dz}=2z\,;$$

$$\therefore \frac{dy}{dz}\equiv\frac{dy}{dx}\frac{dx}{dz}=x\sqrt{(x+1)}\,.\,2z\,;$$

$$\therefore \frac{dy}{dz}=(z^2-1)z\,.\,2z=2z^4-2z^2.$$

$$\therefore y=\int(2z^4-2z^2)\,dz=\tfrac{2}{5}z^5-\tfrac{2}{3}z^3+c\,;$$

$$\therefore \int x\sqrt{(x+1)}\,dx=\tfrac{2}{5}\sqrt{(x+1)^5}-\tfrac{2}{3}\sqrt{(x+1)^3}+c.$$

Example 2. Evaluate $\displaystyle\int\frac{x}{(3x-2)^3}\,dx$.

Denoting the integral by y, $dy=x(3x-2)^{-3}dx$.

$$\text{Put } 3x-2=z,\ \text{then } 3dx=dz\,;$$

$$\therefore 3dy=x(3x-2)^{-3}dz\,;$$

$$\therefore 9dy=(2+z)z^{-3}\,dz=(2z^{-3}+z^{-2})dz\,;$$

$$\therefore 9y=\int(2z^{-3}+z^{-2})dz=-z^{-2}-z^{-1}+c\,;$$

$$\therefore y=-\frac{1}{9(3x-2)^2}-\frac{1}{9(3x-2)}+c'.$$

The method used in these examples may be expressed in general terms as follows :

$$\text{If } y=\int f(x)\,dx, \quad dy=f(x)dx.$$

$$\text{Put } x=\phi(z), \text{ then } dx=\phi'(z)dz.$$

$$\therefore\ dy=f(x)\phi'(z)dz=f\{\phi(z)\}\phi'(z)dz.$$

$$\therefore\ \int f(x)\,dx=y=\int f\{\phi(z)\}\phi'(z)\,dz.$$

This proves that the integral of $f(x)$ with respect to x is equal to that of another function $f\{\phi(z)\}\phi'(z)$ with respect to z, where $x=\phi(z)$; but it is conveniently remembered as follows :

the indefinite integral $\int f(x)\,dx$ can be transformed by writing $\phi(z)$ for x in $f(x)$ and **replacing dx by φ′(z)dz.**

The convenience is due to the similarity in form between this transformation and the properties of differentials, but it must be realised that, in $\int \dots dx$ and $\int \dots dz$, dx and dz are not differentials and that these symbols are merely shorthand expressions for the phrases, integral with respect to x and integral with respect to z.

Experience is necessary to decide what substitution will lead to a function $f\{\phi(z)\}\phi'(z)$ which can be integrated. Example 1 illustrates the fact that the substitution $ax+b=z^2$ is useful when $\sqrt{(ax+b)}$ occurs in the integrand. On the other hand, it is not usually advisable to put $ax^2+bx+c=z^2$ when $\sqrt{(ax^2+bx+c)}$ occurs.

Integrals involving $\sqrt{(a^2-x^2)}$ are often evaluated by trigonometrical substitutions. See also Example 4, p. 244.

Example 3. Evaluate $\int\sqrt{(1-x^2)}\,dx$.

The form of the integrand implies that $-1\leqslant x\leqslant 1$, and hence an angle θ, between $-\frac{1}{2}\pi$ and $+\frac{1}{2}\pi$, exists such that $x=\sin\theta$. Using this as the substitution, $dx=\cos\theta\,d\theta$;
also, $\sqrt{(1-x^2)}=\sqrt{(1-\sin^2\theta)}=\cos\theta$, since $\cos\theta$ is positive for $-\frac{1}{2}\pi<\theta<\frac{1}{2}\pi$. Hence

$$\begin{aligned}\int\sqrt{(1-x^2)}\,dx&=\int\cos\theta\cos\theta\,d\theta=\int\cos^2\theta\,d\theta.\\&=\tfrac{1}{2}\int(1+\cos 2\theta)\,d\theta\\&=\tfrac{1}{2}(\theta+\tfrac{1}{2}\sin 2\theta)+c\\&=\tfrac{1}{2}(\theta+\sin\theta\cos\theta)+c.\end{aligned}$$

But θ is the principal value of $\mathrm{Sin}^{-1}x$ (i.e. the value between $-\frac{1}{2}\pi$ and $+\frac{1}{2}\pi$), which we denote by $\sin^{-1}x$. Hence

$$\int\sqrt{(1-x^2)}\,dx=\tfrac{1}{2}\{\sin^{-1}x+x\sqrt{(1-x^2)}\}+c.$$

EXERCISE XI a

Use the given substitutions to find the integrals in Nos. 1-10:

1. $\int x(x-3)^5\,dx$; $x-3=z$.
2. $\int \frac{x\,dx}{(2x+1)^3}$; $2x+1=z$.
3. $\int \frac{x^2\,dx}{(x+2)^4}$; $x+2=z$.
4. $\int \frac{x\,dx}{\sqrt{(x-2)}}$; $x-2=z^2$.
5. $\int x\sqrt{(x+1)}\,dx$; $x+1=z^2$.
6. $\int x(1+x^2)^3\,dx$; $x^2=z$.
7. $\int \frac{x\,dx}{(4x^2+9)^3}$; $x^2=u$, $4u+9=z$.
8. $\int \frac{x^2\,dx}{\sqrt{(1-x^3)}}$; $x^3=z$.
9. $\int \sin x \cos^3 x\,dx$; $\cos x=z$.
10. $\int \tan^3 x \sec^2 x\,dx$; $\tan x=z$.

Evaluate the integrals in Nos. 11-14:

11. $\int x(x+2)^4\,dx$.
12. $\int (x+1)(x-1)^5\,dx$.
13. $\int \frac{x^2\,dx}{(x+3)^5}$.
14. $\int \frac{x\,dx}{\sqrt{(4-x)}}$.

Use the given substitutions to find the integrals in Nos. 15-18:

15. $\int \frac{dx}{\sqrt{(4-9x^2)}}$; $x=\frac{2}{3}\sin\theta$.
16. $\int \frac{dx}{9+4x^2}$; $x=\frac{3}{2}\tan\theta$.
17. $\int \sqrt{\left(\frac{x}{1-x}\right)}\,dx$; $x=\sin^2\theta$.
18. $\int \frac{\sin^3 x}{\sqrt{(\cos x)}}\,dx$; $\cos x=z^2$.

The method of substitution may also be applied to *definite* integrals. If, in Example 3, p. 242, the limits of $\int \sqrt{(1-x^2)}\,dx$ had been 0 and 1, we could still have used the substitution $x=\sin\theta$ to show that

$$\int \sqrt{(1-x^2)}\,dx = \tfrac{1}{2}\{\sin^{-1} x + x\sqrt{(1-x^2)}\} + c,$$

and then

$$\int_0^1 \sqrt{(1-x^2)}\,dx = \tfrac{1}{2}\Big[\sin^{-1} x + x\sqrt{(1-x^2)}\Big]_0^1$$

$$= \tfrac{1}{2}(\sin^{-1} 1 - \sin^{-1} 0) = \tfrac{1}{2}\cdot\tfrac{1}{2}\pi = \tfrac{1}{4}\pi.$$

It is, however, unnecessary to change back the primitive into a function of x. When $x=\sin\theta$, the values of $\sin^{-1} x + x\sqrt{(1-x^2)}$ for $x=0, 1$ are the same as the values of $\theta+\sin\theta\cos\theta$ for $\theta=0, \frac{1}{2}\pi$ (given by $\sin\theta=0, 1$) respectively, and therefore

$$\Big[\sin^{-1} x + x\sqrt{(1-x^2)}\Big]_{x=0}^{x=1} = \Big[\theta + \sin\theta\cos\theta\Big]_{\theta=0}^{\theta=\pi/2}.$$

The work is therefore set out as follows :

$$\text{Put } x=\sin\theta, \text{ then } dx=\cos\theta\,d\theta;$$

$$\therefore \int_0^1 \surd(1-x^2)\,dx=\int_0^{\pi/2}\cos\theta\,.\cos\theta\,d\theta=\tfrac{1}{2}\int_0^{\pi/2}(1+\cos 2\theta)\,d\theta$$

$$=\tfrac{1}{2}\Big[\theta+\sin\theta\cos\theta\Big]_0^{\pi/2}$$

$$=\tfrac{1}{2}(\tfrac{1}{2}\pi-0)=\tfrac{1}{4}\pi.$$

Note. The equation of the circle centre the origin and radius 1 is $x^2+y^2=1$, that is $y=\pm\surd(1-x^2)$; the upper half of it is given by $y=\surd(1-x^2)$. Hence $\int_0^1\surd(1-x^2)\,dx$ is the measure of the area of the quarter of this circle which is in the first quadrant. This verifies the value $\frac{1}{4}\pi$.

Example 4. Evaluate $\int_0^1 \frac{dx}{\surd(1+x^2)^3}$.

Let θ be the acute angle such that $x=\tan\theta$; then as x increases from 0 to 1, θ increases from 0 to $\frac{1}{4}\pi$.

Also, $dx=\sec^2\theta\,d\theta$ and $1+x^2=1+\tan^2\theta=\sec^2\theta$,

$$\therefore \int_0^1\frac{dx}{\surd(1+x^2)^3}=\int_0^{\pi/4}\frac{\sec^2\theta\,d\theta}{\sec^3\theta}$$

$$=\int_0^{\pi/4}\cos\theta\,d\theta$$

$$=\Big[\sin\theta\Big]_0^{\pi/4}=\sin\tfrac{1}{4}\pi=\tfrac{1}{2}\surd 2.$$

If the substitution $x=\phi(z)$ is used to transform the definite integral $\int_a^b f(x)\,dx$, and if z takes the values α, β when x takes the values a, b respectively, the formula used in Example 4 may be written

$$\int_a^b f(x)\,dx=\int_\alpha^\beta f\{\phi(z)\}\phi'(z)\,dz.$$

A general consideration of the conditions for the validity of this transformation is reserved for the advanced volume. For practical purposes it is sufficient for the reader to satisfy himself that x is an increasing (or decreasing) function of z for the interval from α to β; this covers most ordinary cases, but sometimes it may be necessary to express the given definite integral as the sum of two or more integrals by dividing up the range of integration.

For example, the substitution $x=\frac{1}{z}$ does not prove that $\int_{-1}^{+2} x^2\,dx$ is equal to $\int_{-1}^{1/2} \frac{1}{z^2}\cdot\frac{-1}{z^2}\,dz$. The first integral is 3 and the second integral does not exist. The transformation is invalid because $1/z$ does not change continuously from -1 to $+2$ as z increases from -1 to $\frac{1}{2}$. In this example $\phi(z)=1/z$ and $\phi'(z)$ does not exist for $z=0$, nor even does $\phi(z)$.

It is also sometimes necessary to exercise care in forming the transformed integrand.

Thus using the substitution $x=\tan\theta$, in Example 4, x would increase from 0 to 1 if θ increased from $-\pi$ to $-3\pi/4$ instead of from 0 to $\pi/4$; but

$$\int_0^1 \frac{dx}{\sqrt{(1+x^2)^3}} \text{ is not equal to } \int_{-\pi}^{-3\pi/4} \cos\theta\,d\theta.$$

Here $\cos\theta$ is negative when θ increases from $-\pi$ to $-3\pi/4$,

$$\therefore\ \sqrt{(1+x^2)}=\sqrt{(\sec^2\theta)}=-\sec\theta;$$

$$\therefore\ \int_0^1 \frac{dx}{\sqrt{(1+x^2)^3}}=\int_{-\pi}^{-3\pi/4}(-\cos\theta)\,d\theta.$$

For examples where careful examination of the transformation is required see Exercise XI b, Nos. 16-20.

Sometimes a substitution which fails to evaluate the integral directly leads to a form which suggests a second substitution.

Example 5. Evaluate $\int \frac{x^3}{\sqrt{(1+x^2)}}\,dx.$

Put $x^2=u$, then $2x\,dx=du$;

$$\therefore\ \int\frac{x^3}{\sqrt{(1+x^2)}}\,dx=\int\frac{u}{\sqrt{(1+u)}}\,\tfrac{1}{2}du.$$

Put $1+u=z^2$, then $du=2z\,dz$;

$$\therefore\ \int\frac{x^3}{\sqrt{(1+x^2)}}\,dx=\int\frac{z^2-1}{z}\,z\,dz.$$

$$=\int(z^2-1)\,dz=\tfrac{1}{3}z^3-z+c.$$

But $z^2=1+u=1+x^2$, $\quad\therefore\ z=\sqrt{(1+x^2)}$;

$$\therefore\ \int\frac{x^3}{\sqrt{(1+x^2)}}\,dx=\tfrac{1}{3}\sqrt{(1+x^2)}\{(1+x^2)-3\}+c$$

$$=\tfrac{1}{3}(x^2-2)\sqrt{(1+x^2)}+c.$$

The substitutions in Example 5 are equivalent to the single one $x^2+1=z^2$ or, more precisely, to $x=\sqrt{(z^2-1)}$. The reader should verify that this leads directly to the same result.

If successive substitutions are made in the reduction of a definite integral, it is only necessary to use the limits of the last variable and to verify that the substitutions are legitimate.

Thus to evaluate $\int_0^2 \frac{x^3}{\sqrt{(1+x^2)}}\,dx$ we proceed as in Example 5 and note that x increases from 0 to 2 when u increases from 0 to 4, and that u increases from 0 to 4 when z increases from 1 to $\sqrt{5}$.

Hence
$$\int_0^2 \frac{x^3\,dx}{\sqrt{(1+x^2)}}=\int_1^{\sqrt{5}} (z^2-1)\,dz.$$

EXERCISE XI b

Evaluate the integrals in Nos. 1-4:

1. $\int_0^1 \frac{dx}{\sqrt{(4+5x)}}$. **2.** $\int_4^6 \frac{x\,dx}{\sqrt{(25-4x)}}$.

3. $\int_0^4 x\sqrt{(9+x^2)}\,dx$. **4.** $\int_0^{\pi/2} \sin^4 x\cos x\,dx$.

Use the given substitutions to find the integrals in Nos. 5-9:

5. $\int \frac{\sqrt{(1+x^2)}}{x^4}\,dx$; $x=\tan\theta$, $\sin\theta=z$.

6. $\int \frac{dx}{x\sqrt{(x^2-1)}}$; $x=\frac{1}{u}$, $u=\sin\theta$.

7. $\int \frac{x^2\,dx}{1+x^6}$; $x^3=u$, $u=\tan\theta$.

8. $\int_1^2 \frac{dx}{x^2\sqrt{(x^2+1)}}$; $x=\frac{1}{u}$, $1+u^2=z$.

9. $\int_3^{9/2} \frac{dx}{\sqrt{(6x-x^2)}}$; $x-3=u$, $u=3\sin\theta$.

Evaluate the integrals in Nos. 10-15 :

10. $\int_0^1 \frac{dx}{\sqrt{(4-x^2)}}$. 11. $\int_0^3 \frac{dx}{9+x^2}$.

12. $\int_0^{\pi/2} \cos^3 x \sin^8 x \, dx$. 13. $\int_0^{1/4} \sqrt{\frac{x}{1-x}} \, dx$.

14. $\int_0^1 x^3 \sqrt{(1-x^2)} \, dx$. 15. $\int_0^{1/2} \sqrt{\frac{1+x}{1-x}} \, dx$.

16. If $x = \tan\theta$, prove that $\int \frac{2dx}{(1+x^2)^2} = \int (1+\cos 2\theta) \, d\theta$.

Is $\int_0^1 \frac{2dx}{(1+x^2)^2}$ equal to

$$\int_{\pi}^{\pi/4} (1+\cos 2\theta) \, d\theta \quad \text{or} \quad \int_{\pi}^{5\pi/4} (1+\cos 2\theta) \, d\theta \, ?$$

What are the values of these definite integrals ?

17. What is the integral obtained from $\int_{-1}^{+1} x^2 \, dx$ by substituting $x^2 = z$? What is its value ?

[The integral must be regarded as $\int_{-1}^{0} x^2 \, dx + \int_{0}^{+1} x^2 \, dx$.]

18. If $x = 1/z$, prove that $\int \frac{dx}{1+x^2} = -\int \frac{dz}{1+z^2}$. Is this substitution applicable to $\int_1^2 \frac{dx}{1+x^2}$ and $\int_{-1}^{+1} \frac{dx}{1+x^2}$? What are the values of these integrals ?

19. What integrals are obtained from $\int_0^{\pi/2} 1 \, dx$ and $\int_{\pi/2}^{\pi} 1 \, dx$ by making the substitution $\sin x = z$? Verify that each of them equals $\frac{1}{2}\pi$.

20. Verify that the value of $\int_0^{\pi} \cos^2 x \, dx$ is $\frac{1}{2}\pi$ by making the substitution $\sin x = z$. [Use the note on p. 244.]

Integration by Parts. The formula for differentiating a product is

$$\frac{d}{dx}(uv) = v\frac{du}{dx} + u\frac{dv}{dx};$$

∴ by integration,

$$uv = \int v\frac{du}{dx}\,dx + \int u\frac{dv}{dx}\,dx \quad . \quad . \quad . \quad \text{(i)}$$

If either of the integrals in (i) is known, the other can be found.

For example, if $u = x$, $v = \sin x$,

$$\frac{d}{dx}(x \sin x) = \sin x + x \cos x;$$

∴ by integration,

$$x \sin x = \int \sin x\,dx + \int x \cos x\,dx$$

$$= -\cos x + c + \int x \cos x\,dx;$$

$$\therefore \int x \cos x\,dx = x \sin x + \cos x - c.$$

Denoting $\frac{du}{dx}, \frac{dv}{dx}$ by u', v' respectively in (i),

$$uv = \int vu'\,dx + \int uv'\,dx;$$

$$\therefore \int uv'\,dx = uv - \int vu'\,dx \quad . \quad . \quad . \quad \text{(ii)}$$

This is called the formula for **Integration by Parts.**

It is a formula for the integration of the product of the two functions u, v'; but it is only successful in effecting the integration if v, i.e. $\int v'\,dx$, and $\int vu'\,dx$ can both be evaluated.

Example 6. Evaluate $\int x^2 \sin x\,dx$.

Put $u = x^2$, $v' = \sin x$; hence $v = -\cos x$.

∴ the formula gives

$$\int x^2 \sin x\,dx = x^2(-\cos x) - \int(-\cos x)\frac{d(x^2)}{dx}\,dx$$

$$= -x^2 \cos x + \int 2x \cos x\,dx.$$

But by repeating the process, or by the method above,

$$\int x \cos x\,dx = x \sin x + \cos x + c_1,$$

$$\therefore \int x^2 \sin x\,dx = -x^2 \cos x + 2(x \sin x + \cos x) + c_2.$$

By using the formula for the change of the variable of an integral we may replace, in equation (i), p. 248,

$$\int v\frac{du}{dx}\,dx \text{ by } \int v\,du \quad\text{and}\quad \int u\frac{dv}{dx}\,dx \text{ by } \int u\,dv.$$

Hence
$$uv = \int v\,du + \int u\,dv;$$

$$\therefore \int \mathbf{u}\,\mathbf{dv} = \mathbf{uv} - \int \mathbf{v}\,\mathbf{du} \qquad . \qquad . \qquad . \qquad \text{(iii)}$$

which is another way of expressing the formula for integration by parts.

Many students find form (iii) easier to remember and apply than form (ii).

Using it in Example 6, we write:

$$\begin{aligned}\int x^2 \sin x\,dx &= \int x^2\,d(-\cos x)\\ &= x^2(-\cos x) - \int(-\cos x)\,d(x^2)\\ &= -x^2\cos x + 2\int x\cos x\,dx.\end{aligned}$$

Example 7. Evaluate $\int \sin^{-1} x\,dx$.

From (iii), putting $u = \sin^{-1} x$, $v = x$,

$$\int \sin^{-1} x\,dx = x\sin^{-1} x - \int x\,d(\sin^{-1} x);$$

$$\text{but } d(\sin^{-1} x) = \frac{1}{\sqrt{(1-x^2)}}\,dx,$$

$$\therefore \int \sin^{-1} x\,dx = x\sin^{-1} x - \int\frac{x\,dx}{\sqrt{(1-x^2)}};$$

$$\text{but } \frac{d}{dx}\{\sqrt{(1-x^2)}\} = \tfrac{1}{2}(1-x^2)^{-\frac{1}{2}}(-2x) = -\frac{x}{\sqrt{(1-x^2)}},$$

$$\therefore \int \sin^{-1} x\,dx = x\sin^{-1} x + \sqrt{(1-x^2)} + c.$$

If formula (ii) is used for Example 7, we write:

$$u = \sin^{-1} x,\ v' = 1;\ \text{hence } v = x.$$

$$\therefore \int \sin^{-1} x\,dx = x\sin^{-1} x - \int x\frac{d}{dx}(\sin^{-1} x)\,dx.$$

We then continue as before.

Those who use formula (ii) may find it easier to remember it in words as follows:

The integral of a product of two functions
= (1st function)($\int$ 2nd) − $\int$($\int$ 2nd)(differential of 1st).

The method of integration by parts may also be used to calculate definite integrals.

Example 8. Evaluate $\int_0^{\pi/2} x \sin x \, dx$.

$$\int x \sin x \, dx = \int x \, d(-\cos x) = x(-\cos x) - \int (-\cos x) \, dx;$$

$$\therefore \int_0^{\pi/2} x \sin x \, dx = \Big[-x \cos x + \int \cos x \, dx \Big]_0^{\pi/2}$$

$$= \Big[-x \cos x \Big]_0^{\pi/2} + \int_0^{\pi/2} \cos x \, dx$$

$$= 0 + \Big[\sin x \Big]_0^{\pi/2} = 1.$$

EXERCISE XI c

Evaluate the integrals in Nos. 1-9:

1. $\int x \sin 3x \, dx$.
2. $\int x \cos 2x \, dx$.
3. $\int x^2 \cos x \, dx$.
4. $\int x \sin^2 x \, dx$.
5. $\int \cos^{-1} x \, dx$.
6. $\int x \tan^{-1} x \, dx$.
7. $\int_0^{\pi} x \cos \tfrac{1}{2}x \, dx$.
8. $\int_0^{\pi/2} x^2 \sin x \, dx$.
9. $\int_0^{\pi} x^3 \cos x \, dx$.
10. By writing $\int \cos^3 \theta \, d\theta$ in the form $\int \cos^2 \theta \cos \theta \, d\theta$ or $\int \cos^2 \theta \, d(\sin \theta)$, prove that it equals
$$\cos^2 \theta \sin \theta + 2\int (\cos \theta - \cos^3 \theta) \, d\theta$$
and hence find its value.
11. Prove by the method of No. 10 that $\int_0^{\pi/2} \sin^3 \theta \, d\theta = \frac{2}{3}$.
12. Evaluate $\int x \sin x \cos 3x \, dx$.
13. Evaluate $\int_0^{\pi/4} x \cos x \cos 3x \, dx$.
14. Draw the graph of $y = x \sin 2x$ from $x = 0$ to $x = \pi$, and prove that one of the areas bounded by the curve and Ox is 3 times the other.

[*Additional examples of "integration by parts" are given in Exercises XII e, XIII c, and XIII f.*]

CHAPTER XII

LOGARITHMIC AND EXPONENTIAL FUNCTIONS

THE first step in a systematic investigation of what functions can be integrated is to evaluate $\int x^n\,dx$.

Since $\qquad \dfrac{d}{dx}(x^{n+1}) = (n+1)x^n, \quad \int (n+1)x^n\,dx = x^{n+1} + c\,;$

$$\therefore \int x^n\,dx = \frac{x^{n+1}}{n+1} + c', \textit{ unless } n = -1.$$

In the present chapter we prepare the way for Chapter XIV by investigating the integral

$$\int x^n\,dx, \text{ when } n = -1.$$

It is convenient to begin with the definite integral

$$\int_1^t \frac{1}{x}\,dx.$$

For given positive values of t, approximate values of this integral can be obtained and these values are the measures of the areas bounded by the curve $y = \dfrac{1}{x}$, the x-axis, and the ordinates $x=1$, $x=t$; in particular from p. 109, $\displaystyle\int_1^2 \frac{1}{x}\,dx \simeq 0{\cdot}6931$.

$\displaystyle\int_1^t \frac{1}{x}\,dx$ cannot be expressed as an *algebraic* function of t. If, however, we write

$$\int_1^t \frac{1}{x}\,dx \equiv \mathsf{F}(t), \text{ where } t > 0,$$

we can establish properties of $\mathsf{F}(t)$ which supply all the information that we need about this function.

Fig. 108 represents part of the graph of $y = \frac{1}{x}$.

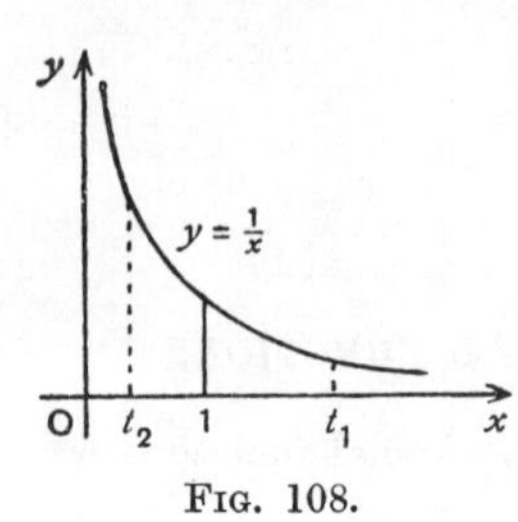

Fig. 108.

If $t_1 > 1$, the area between the curve $y = \frac{1}{x}$, the x-axis, and the ordinates $x = 1$, $x = t_1$ is measured by

$$\int_1^{t_1} \frac{1}{x}\,dx, \text{ that is by } \mathsf{F}(t_1).$$

If $t_2 < 1$, the area between the curve, the x-axis, and the ordinates $x = t_2$, $x = 1$ is measured by

$$\int_{t_2}^{1} \frac{1}{x}\,dx, \text{ which is } -\int_1^{t_2} \frac{1}{x}\,dx, \text{ that is } -\mathsf{F}(t_2).$$

Thus $\mathsf{F}(t)$ is positive for $t > 1$, zero for $t = 1$, and negative for $0 < t < 1$; also, $\mathsf{F}(t)$ increases steadily as t increases through positive values. And, by p. 109 (Vol. I), $\mathsf{F}(2) \simeq 0{\cdot}6931$.

$\mathsf{F}(t)$ is not defined for $t \leqslant 0$.

If a is positive and n is any rational number,

$$\mathbf{F(a^n) = nF(a),}$$

where a^n is supposed to have its *positive* value. For example, if $n = \frac{1}{2}$, a^n is to be interpreted as $\sqrt{a}$, not $-\sqrt{a}$.

By definition, $$\mathsf{F}(a^n) = \int_1^{a^n} \frac{1}{x}\,dx.$$

Let z be the positive value of $x^{1/n}$;

$$\text{then } x = z^n \quad \text{and} \quad dx = nz^{n-1}\,dz.$$

Also, as x varies continuously from 1 to a^n, z varies continuously from 1 to a.

$$\therefore\ \mathsf{F}(a^n) = \int_1^a \frac{1}{z^n} nz^{n-1}\,dz = n\int_1^a \frac{1}{z}\,dz;$$

$$\therefore\ \mathsf{F}(a^n) = n\mathsf{F}(a).$$

[*The student should now work Exercise XII a, Nos. 1-3.*]

Values of F(x) when x is large or small.

Putting $a=2$, $n=100$ in the relation $F(a^n)=nF(a)$, and using the approximate value 0·6931 of F(2), we have

$$F(2^{100}) \simeq 69{\cdot}31.$$

Similarly, $\quad F(2^{1000}) \simeq 693{\cdot}1, \quad F(2^{10000}) \simeq 6931$, etc.,

and by giving n a large enough value we can arrive at a value of $F(x)$ which is as large as we please ; and we see that $F(x)$ *increases steadily* as x increases.

Again, $\quad F(2^{-100}) \simeq -69{\cdot}31,$
$\quad F(2^{-1000}) \simeq -693{\cdot}1$, etc.,

and by making x sufficiently small, but positive, values of $F(x)$ can be found which are negative and numerically as large as we please.

We may express these results shortly by saying that $F(x)$ increases steadily from $-\infty$ to $+\infty$ as x increases from 0 to $+\infty$. Consequently $F(x)$ assumes any given positive, zero, or negative value for one and only one value of x.

The graph of $F(x)$ is shown in Fig. 109.

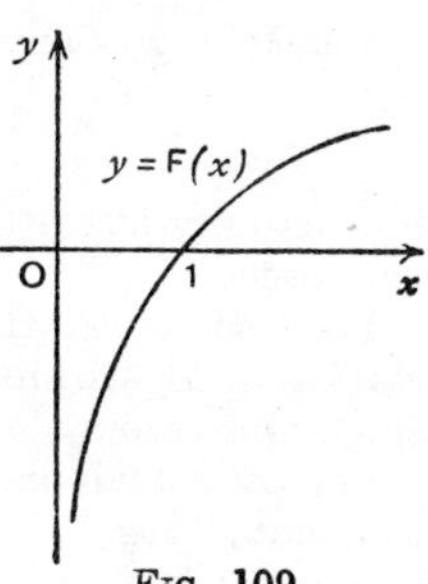

FIG. 109.

The Definition of e. We have proved that

$$F(2^n) = nF(2) \simeq n \times 0{\cdot}6931 ;$$

$$\therefore F(2^n) \simeq 1 \text{ if } n = \frac{1}{0{\cdot}6931} \simeq 1{\cdot}443 ;$$

$$\therefore F(x) \simeq 1 \text{ if } x \simeq 2^{1{\cdot}443} \simeq 2{\cdot}718.$$

The value of x for which $F(x)$ is equal to 1 is denoted by ***e***.

Thus $$F(e) = \int_1^e \frac{1}{x}\,dx = 1,$$

and $e \simeq 2{\cdot}718$. The actual value of e is not rational, but approximations can be calculated to as many decimal places as desired. A closer approximation than that given above is **2·7182818285.**

Logarithms. If any positive number t is expressed in the form 10^n, n is called the logarithm of t to the *base* 10 and is denoted by $\log_{10} t$; thus

$$n = \log_{10} t \text{ means the same as } 10^n = t.$$

Logarithms to other bases are defined in a similar way:

If $t = e^n$, n is called the logarithm of t to the base e and is denoted by $\log_e t$; thus

$$n = \log_e t \text{ means the same as } e^n = t.$$

Similarly, if p is any positive number,

$$n = \log_p t \text{ means the same as } p^n = t,$$

but actually logarithms to bases other than 10 and e are hardly ever used.

Logarithms to the base e are called *natural logarithms* or Napierian logarithms; they are of great theoretical importance chiefly on account of the results proved in this chapter.

For the remainder of this book the symbol "log" will be used to denote "$\log_e$", logarithms to base 10 or base p being denoted by $\log_{10}$ or $\log_p$.

$$\text{If } t = e^n, \quad n = \log t;$$

but

$$\mathsf{F}(t) = \mathsf{F}(e^n) = n\mathsf{F}(e) = n;$$

$$\therefore \mathsf{F}(t) \equiv \log t,$$

that is

$$\int_1^t \frac{1}{x}\,dx = \log t.$$

We have thus found the meaning of the definite integral introduced on p. 251, and shall now use the notation $\log t$ instead of $\mathsf{F}(t)$. The function $\log t$ is defined for *positive* values of t only. When therefore a function $\log f(x)$ occurs, it is implied that the values of x are restricted to those for which $f(x)$ is positive; usually this limitation will not be stated explicitly. We proceed to find the meaning of the indefinite integral $\int \frac{1}{x}\,dx$.

Derivative of log x.

If t and $t+h$ are positive,

$$\log(t+h)-\log t=\int_1^{t+h}\frac{1}{x}\,dx-\int_1^{t}\frac{1}{x}\,dx$$
$$=\int_t^{t+h}\frac{1}{x}\,dx.$$

But when x lies between the positive values t and $t+h$, $\frac{1}{x}$ lies between $\frac{1}{t}$ and $\frac{1}{t+h}$;

$$\therefore\ \int_t^{t+h}\frac{1}{x}\,dx \text{ lies between } \int_t^{t+h}\frac{1}{t}\,dx \text{ and } \int_t^{t+h}\frac{1}{t+h}\,dx,$$

$$\text{that is, between } \frac{h}{t} \text{ and } \frac{h}{t+h};$$

$$\therefore\ \frac{\log(t+h)-\log t}{h} \text{ lies between } \frac{1}{t} \text{ and } \frac{1}{t+h}.$$

But when $h\to 0$, $\frac{1}{t+h}\to\frac{1}{t}$,

$$\therefore \text{ when } h\to 0,\quad \frac{\log(t+h)-\log t}{h}\to\frac{1}{t};$$

that is
$$\frac{d}{dt}(\log t)=\frac{1}{t}.$$

With a change of notation this gives

$$\frac{\mathbf{d}}{\mathbf{dx}}(\mathbf{\log x})=\frac{\mathbf{1}}{\mathbf{x}} \quad\text{and}\quad \mathbf{d}(\mathbf{\log x})=\frac{\mathbf{dx}}{\mathbf{x}};$$

and this is equivalent to the statement

$$\int\frac{\mathbf{1}}{\mathbf{x}}\,\mathbf{dx}=\mathbf{\log x+C}$$

for positive values of x.

Also,
$$\frac{d}{dx}\log(ax+b)=\frac{d\log u}{du}\,\frac{du}{dx}, \text{ where } u=ax+b,$$
$$=\frac{1}{u}\frac{du}{dx};$$
$$\therefore\ \frac{d}{dx}\log(ax+b)=\frac{a}{ax+b},$$

and this is equivalent to the statement

$$\int\frac{\mathbf{dx}}{\mathbf{ax+b}}=\frac{\mathbf{1}}{\mathbf{a}}\mathbf{\log(ax+b)+C}$$

for positive values of $ax+b$, where $a\neq 0$.

If we put $a = -1$, $b = 0$ in the formula for $\frac{d}{dx}\log(ax+b)$, we obtain $\frac{d}{dx}\log(-x) = \frac{1}{x}$; this holds only if x is negative, since otherwise $\log(-x)$ has no meaning. Similarly, the formula $\frac{d}{dx}\log x = \frac{1}{x}$ holds only if x is positive.

The symbol $|z|$, called mod z, is used to denote z if $z > 0$ and to denote $-z$ if $z < 0$. With this notation

$$\frac{d}{dx}\log|x| = \frac{1}{x}, \text{ unless } x \text{ is zero.}$$

General Properties of Logarithms.

If $\log_p a = u$ and $\log_p b = v$, where a, b, p are positive,

then $$a = p^u \quad \text{and} \quad b = p^v;$$

$$\therefore ab = p^u p^v = p^{u+v};$$

$$\therefore \text{by definition, } \log_p(ab) = u + v = \log_p a + \log_p b;$$

and similarly,

$$\log_p(a/b) = \log_p a - \log_p b;$$

also, $$\log_p(a^b) = b\log_p a.$$

These general properties are the same for all positive bases. In dealing with natural logarithms, we can deduce them directly from the corresponding definite integrals. For example,

$$\log(ab) = \int_1^{ab} \frac{1}{x}\,dx = \int_1^a \frac{1}{x}\,dx + \int_a^{ab} \frac{1}{x}\,dx;$$

in the second integral put $x = az$; then when x varies from a to ab, z varies from 1 to b;

$$\therefore \int_a^{ab} \frac{1}{x}\,dx = \int_1^b \frac{1}{az}a\,dz = \int_1^b \frac{1}{z}\,dz;$$

$$\therefore \log(ab) = \int_1^a \frac{1}{x}\,dx + \int_1^b \frac{1}{x}\,dx = \log a + \log b.$$

Example 1. Find $\frac{d}{dx}\log\frac{1+x}{1-x}$, $(-1 < x < +1)$.

$$\log\frac{1+x}{1-x} = \log(1+x) - \log(1-x);$$

$$\therefore \frac{d}{dx}\log\frac{1+x}{1-x} = \frac{1}{1+x} - \frac{1}{1-x}(-1) = \frac{2}{1-x^2}.$$

EXERCISE XII a

1. Verify that $F(2) \simeq 0{\cdot}7$ by means of the graph of $1/x$, and then use the formula $F(2^n) = nF(2)$ to draw a rough graph of $F(x)$. [Take $n = 0, \pm\frac{1}{4}, \pm\frac{1}{2}, \pm 1, \pm 2$.]

2. Give the approximate values of $F(8)$, $F(16)$, $F(32)$, $F(1/8)$, $F(1/16)$, $F(1/32)$.

3. Obtain graphically an approximate value of $F(3)$.

Differentiate with respect to x the functions in Nos. 4-15:

4. $\log 3x$.
5. $\log x^3$.
6. $\log(1/x)$.
7. $\log \sqrt{x}$.
8. $\log \sin x$.
9. $\log(1 - x)$.
10. $\log \cos^2 x$.
11. $\log(4x + 3)$.
12. $\log \tan x$.
13. $x \log x$.
14. $(\log x)/x$.
15. $\log \sqrt{\{(1 + x)/(1 - x)\}}$.

16. If $y = x^2 \log x$, prove that $\dfrac{d^2y}{dx^2} = \dfrac{2y}{x^2} + 3$.

17. Find the maximum value of $x^2 \log(1/x)$.

18. Find the values of θ between 0 and $\frac{1}{2}\pi$ for which $\tan\theta + 3\log\cos\theta + \theta$ is stationary, and distinguish between them.

Integration.

$$\text{If } u \equiv \phi(x), > 0,$$

$$\frac{d}{dx}(\log u) = \frac{1}{u}\frac{du}{dx} = \frac{\phi'(x)}{\phi(x)};$$

$$\therefore \int \frac{\boldsymbol{\varphi'(x)}}{\boldsymbol{\varphi(x)}}\,\mathbf{dx} = \mathbf{\log \varphi(x) + C}.$$

This result is a generalisation of the formula for $\displaystyle\int \frac{dx}{ax + b}$ on p. 255.

Example 2. Find $\displaystyle\int \frac{x}{1 + x^2}\,dx$.

$$\text{Put } 1 + x^2 = z, \text{ then } 2x\,dx = dz;$$

$$\therefore \int \frac{x}{1 + x^2}\,dx = \int \frac{\frac{1}{2}dz}{z} = \tfrac{1}{2}\log z + c$$

$$= \tfrac{1}{2}\log(1 + x^2) + c.$$

Example 3. Find $\int_0^{\pi/4} \tan x\,dx$.

Since $\tan x = \dfrac{\sin x}{\cos x}$, we put $\cos x = z$,

$$\therefore\ -\sin x\,dx = dz.$$

$$\therefore \int \tan x\,dx = \int \frac{\sin x\,dx}{\cos x} = -\int \frac{dz}{z}$$
$$= -\log z + c$$

where z is positive.

$$\therefore \int_0^{\pi/4} \tan x\,dx = \Big[-\log\cos x\Big]_0^{\pi/4}$$
$$= -\log(1/\sqrt{2}) + \log 1 = \tfrac{1}{2}\log 2.$$

Note. For an integral such as $\int_{3\pi/4}^{\pi} \tan x\,dx$, $\cos x$ is negative throughout the range of integration, and

$$-\int \frac{dz}{z} = -\log(-z), \text{ where } z = \cos x.$$

Using the modulus notation we may then write

$$\int \tan x\,dx = -\log|\cos x| + c.$$

EXERCISE XII b

Integrate with respect to x the functions in Nos. 1-9:

1. $\dfrac{1}{5x}$. 2. $\dfrac{1}{1+x}$. 3. $\dfrac{x}{1-x^2}$.

4. $\dfrac{x^2}{1+x^3}$. 5. $\dfrac{6x+5}{3x^2+5x+7}$. 6. $\dfrac{x-2}{x^2-4x+3}$.

7. $\dfrac{\cos x}{\sin x}$. 8. $\tan 2x$. 9. $\cot 3x$.

Evaluate the integrals in Nos. 10-15:

10. $\displaystyle\int_4^{12} \frac{dx}{1+2x}$. 11. $\displaystyle\int_3^4 \frac{x\,dx}{25-x^2}$. 12. $\displaystyle\int_0^{\pi/3} \tan x\,dx$.

13. $\displaystyle\int_1^e \frac{\log x}{x}\,dx$. 14. $\displaystyle\int_0^1 \frac{x\,dx}{1+x}$. 15. $\displaystyle\int_2^3 \frac{3-2x}{1-x}\,dx$.

16. Prove directly from the definition (see pp. 251, 254) that
$$\log(a/b) = \log a - \log b.$$

17. Find the area bounded by the x-axis, the ordinates $x = a$, $x = 2a$, and the arc of the curve $xy = x^2 + a^2$.

The Function e^x.

If $y = e^x$, then $x = \log y$.

Hence the graph of $y = e^x$ can be deduced from the graph of $y = \log x$ (see Fig. 109) by interchanging the axes of x and y, or by taking the image of the curve in the line $y = x$. The graph of e^x therefore takes the form shown in Fig. 110, and we see that e^x increases steadily from 0 to $+\infty$ as x increases from $-\infty$ to $+\infty$.

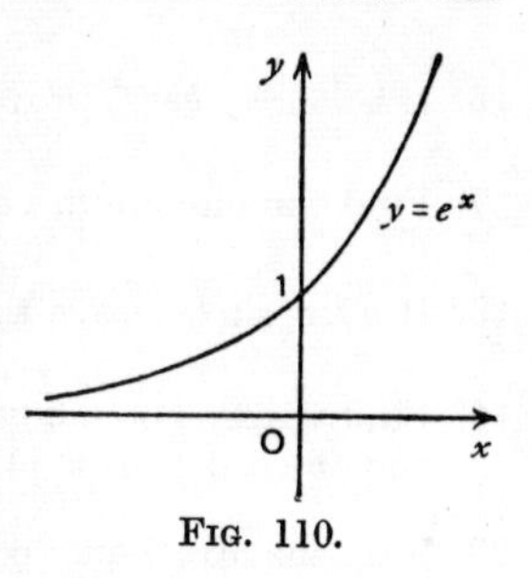

FIG. 110.

The function e^x is called the **exponential function of x.**

Derivative and Integral of the Exponential Function.

$$\text{If } y = e^x, \quad \log y = x;$$

$$\therefore \frac{1}{y}\,dy = dx; \quad \therefore \frac{dy}{dx} = y;$$

that is
$$\frac{d}{dx}(e^x) = e^x.$$

Hence also
$$\int e^x\,dx = e^x + C.$$

More generally,

$$\frac{d}{dx}(e^{ax}) = ae^{ax} \quad \text{and} \quad \int e^{ax}\,dx = \frac{1}{a}e^{ax} + c, \ (a \neq 0).$$

EXERCISE XII c

Differentiate with respect to x the functions in Nos. 1-6:

1. e^{4x}. 2. e^{-2x}. 3. xe^x.

4. e^{x^2}. 5. $e^{2x}\sin 3x$. 6. $(x^2 - x)/e^x$.

Integrate with respect to x the functions in Nos. 7-12:

7. e^{3x}. 8. $1/e^x$. 9. $\sqrt{e^x}$.

10. xe^{x^2}. 11. $e^x/(e^x + 1)$. 12. $1/(e^x + 1)$.

Evaluate the integrals in Nos. 13-15:

13. $\int_0^4 e^{\frac{1}{2}x}\,dx$. 14. $\int_{-1}^{+1} \frac{dx}{e^{2x}}$. 15. $\int_0^{\frac{1}{2}\pi} e^{\sin x}\cos x\,dx$.

16. If $y = ae^{cx} + be^{-cx}$, prove that $\dfrac{d^2y}{dx^2} = c^2y$.

17. Find the maximum value of x/e^x.

18. If $y = e^x \sin x$, prove that $\dfrac{d^4y}{dx^4} = -4y$.

19. Find the area bounded by the axes, the catenary $y = \frac{1}{2}(e^x + e^{-x})$, and the ordinate $x = 1$.

20. Find the area bounded by the axes, the curve $ye^x = c$, and the ordinate $x = a$. What happens when a increases indefinitely?

21. Find the values of x which give a maximum point and points of inflexion on $y = x^3e^{-x}$ and sketch the curve.

22. Find n if $y = e^{nx}$ satisfies the equation $\dfrac{d^2y}{dx^2} - 4\dfrac{dy}{dx} + 3y = 0$.

23. If $f(x) \equiv e^x$ when $0 \leqslant x < 1$ and $f(x) \equiv ax + b$ when $1 \leqslant x \leqslant 2$, find for what values of a and b, $f(x)$ has a derivative at $x = 1$.

24. Find values of the constants p, q in terms of a, b such that $\dfrac{d}{dx}\{e^{ax}(p \sin bx + q \cos bx)\}$ is equal to $e^{ax} \sin bx$. Hence find $\int e^{ax} \sin bx\, dx$.

Logarithmic Differentiation. It is often convenient, especially when differentiating a continued product, to take logarithms before differentiating.

If
$$y = uvw$$
where u, v, w are functions of x, then
$$\log y = \log u + \log v + \log w;$$
$$\therefore \frac{1}{y}dy = \frac{1}{u}du + \frac{1}{v}dv + \frac{1}{w}dw;$$
or
$$\frac{1}{uvw}\frac{d}{dx}(uvw) = \frac{1}{u}\frac{du}{dx} + \frac{1}{v}\frac{dv}{dx} + \frac{1}{w}\frac{dw}{dx}.$$

This result was obtained on p. 61 (Vol. I) in another way, and it can be extended to any number of factors.

Another type for which it is useful to take logarithms is illustrated in Examples 6, 7.

Example 4. Differentiate $e^x x^3 \sin x \log x$ with respect to x.

$$\text{If } y = e^x x^3 \sin x \log x,$$

$$\log y = \log e^x + \log x^3 + \log \sin x + \log \log x$$
$$= x + 3 \log x + \log \sin x + \log \log x;$$

$$\therefore \frac{1}{y}\frac{dy}{dx} = 1 + \frac{3}{x} + \frac{\cos x}{\sin x} + \frac{1}{\log x} \cdot \frac{1}{x};$$

$$\therefore \frac{dy}{dx} = e^x x^3 \sin x \log x \left(1 + \frac{3}{x} + \cot x + \frac{1}{x \log x}\right).$$

Example 5. If $y = ae^{-mx} \sin nx$, prove that

$$\frac{d^2y}{dx^2} + 2m\frac{dy}{dx} + (m^2 + n^2)y = 0.$$

$$\log y = \log a - mx + \log \sin nx;$$

$$\therefore \frac{1}{y}\frac{dy}{dx} = -m + n \cot nx;$$

$$\therefore \frac{dy}{dx} + my = ny \cot nx = nae^{-mx} \cos nx.$$

Differentiate again with respect to x, then

$$\frac{d^2y}{dx^2} + m\frac{dy}{dx} = -mnae^{-mx} \cos nx - n^2ae^{-mx} \sin nx$$
$$= -m\left(\frac{dy}{dx} + my\right) - n^2y;$$

$$\therefore \frac{d^2y}{dx^2} + 2m\frac{dy}{dx} + (m^2 + n^2)y = 0.$$

Example 6. Find (i) $\frac{d}{dx}a^x$ and (ii) $\int a^x\,dx$; $(a > 0)$.

$$\text{If } y = a^x, \quad \log y = x \log a;$$

$$\therefore \frac{1}{y}dy = dx \,.\, \log a; \quad \therefore \frac{dy}{dx} = y \log a$$

that is $$\frac{d}{dx}(a^x) = a^x \log a.$$

Hence $$\int a^x\,dx = \frac{a^x}{\log a} + c.$$

Example 7. Find the derivative of x^x, $(x>0)$.

$$\text{If } y=x^x, \quad \log y = x\log x,$$

$$\therefore \frac{1}{y}dy = dx \,.\, \log x + x\frac{1}{x}dx$$

$$= (1+\log x)\,dx;$$

$$\therefore \frac{dy}{dx} = y(1+\log x);$$

that is
$$\frac{d}{dx}(x^x) = x^x(1+\log x).$$

EXERCISE XII d

Differentiate with respect to x the functions in Nos. 1-9.

1. $(x+1)^2(x+2)^3(x+3)^4$.
2. $(x-1)^5(3x+4)^2/(5x-2)^3$.
3. $(x-3)^{\frac{1}{2}}(2x+1)^{\frac{1}{3}}(x+4)^3$,
4. $e^x \sin x \log x$.
5. $x^3e^x \sin 2x$.
6. $e^{2x}\sin 3x \cos 4x$.
7. 10^x.
8. $x^2 10^x$.
9. $\log_{10} x$.

Integrate with respect to x the functions in Nos. 10-12.

10. $e^{\cos x}\sin x$.
11. 2^x.
12. $\log(x^{1/x})$, $(x>0)$.

13. Find the maximum value of $x^{1/x}$, $(x>0)$.

14. If $y = a\cos(\log x) + b\sin(\log x)$, prove that

$$x^2\frac{d^2y}{dx^2} + x\frac{dy}{dx} + y = 0.$$

15. If $xy = \log x$, find $\dfrac{dy}{dx}$ and $\dfrac{d^2y}{dx^2}$.

16. If $y = e^{xy}$, find $\dfrac{dy}{dx}$ in terms of x and y.

17. If $x = \log\tan\frac{1}{2}t$, $y = \cos 3t$, find $\dfrac{dy}{dx}$ in terms of t.

18. The space-time relation for the motion of a particle along Ox is $x = ae^{-kt}\cos nt$. Prove that its velocity v and acceleration α in any position are such that $\alpha + 2kv + (k^2+n^2)x = 0$.

We shall conclude this chapter with some additional examples of the use of integration by parts.

Example 8. Find (i) $\int \log x\,dx$; (ii) $\int xe^x\,dx$.

(i) Using formula (iii) on p. 249,

$$\int \log x\,dx = (\log x)x - \int x\,d\log x$$
$$= x\log x - \int x\frac{1}{x}\,dx = x\log x - \int 1\,dx$$
$$= x\log x - x + c.$$

(ii) Using the rule given in words on p. 249,

$$\int xe^x\,dx = xe^x - \int e^x \,.\, 1\,dx$$
$$= xe^x - e^x + c.$$

Example 9. Find $\int e^{ax}\sin bx\,dx$ and $\int e^{ax}\cos bx\,dx$.

If $y = \int e^{ax}\sin bx\,dx$ and $z = \int e^{ax}\cos bx\,dx$,

$$y = \frac{1}{a}\int \sin bx\,d(e^{ax})$$
$$= \frac{1}{a}e^{ax}\sin bx - \frac{1}{a}\int e^{ax}\,d(\sin bx)$$
$$= \frac{1}{a}e^{ax}\sin bx - \frac{b}{a}\int e^{ax}\cos bx\,dx$$
$$\therefore\; y = \frac{1}{a}e^{ax}\sin bx - \frac{b}{a}z.$$

Similarly it may be proved that

$$z = \frac{1}{a}e^{ax}\cos bx + \frac{b}{a}y.$$

The student should work this out and then solve the equations,

$$y + \frac{b}{a}z = \frac{1}{a}e^{ax}\sin bx, \quad z - \frac{b}{a}y = \frac{1}{a}e^{ax}\cos bx.$$

The results are

$$(a^2+b^2)y = e^{ax}(a\sin bx - b\cos bx), \quad (a^2+b^2)z = e^{ax}(a\cos bx + b\sin bx).$$

Note. For an alternative method see Exercise XII c, No. 24.

EXERCISE XII e

Integrate with respect to x the functions in Nos. 1-15:

1. $x \log x$.
2. $\tan^{-1} x$.
3. xe^{3x}.
4. x^2e^x.
5. $e^x \sin x$.
6. $e^{-x} \cos x$.
7. $(\log x)/x^2$.
8. $(\sin 2x)/e^x$.
9. $x \sec^2 x$.
10. $x \tan^2 x$.
11. $e^x \sin x \cos x$.
12. $e^x \cos^2 x$.
13. $e^x \sin 3x \cos x$.
14. $(x \log x)^2$.
15. $x^5 \log x$.

Find the values of the integrals in Nos. 16-21:

16. $\int_1^e \log x \, dx$.
17. $\int_1^{\sqrt{e}} x \log x \, dx$.
18. $\int_0^1 \cot^{-1} x \, dx$.
19. $\int_1^e x^n \log x \, dx$.
20. $\int_0^1 \sin^{-1} x \, dx$.
21. $\int_0^\pi x \sin^2 x \, dx$.

22. Find the value of $\int_\epsilon^1 x^4 \log x \, dx$, $(\epsilon > 0)$, and assuming that $\epsilon \log \epsilon \to 0$ when $\epsilon \to 0$, find the limit when $\epsilon \to 0$ of the integral.

CHAPTER XIII

HYPERBOLIC FUNCTIONS

THE functions $\frac{1}{2}(e^x - e^{-x})$ and $\frac{1}{2}(e^x + e^{-x})$ have properties analogous to those of $\sin x$ and $\cos x$, and are called respectively the "hyperbolic sine" and the "hyperbolic cosine" of x, and we write

$$\text{sh}\, x = \tfrac{1}{2}(e^x - e^{-x})\ ;\quad \text{ch}\, x = \tfrac{1}{2}(e^x + e^{-x}).$$

The hyperbolic sine, sh, is pronounced "shine" or "sinsh", and ch is pronounced "cosh". They are often written "sinh" and "cosh".

EXERCISE XIII a

1. What are the numerical values of sh 0 and ch 0 ?

2. Verify that (i) $\text{sh}(-x) = -\text{sh}\, x$; (ii) $\text{ch}(-x) = \text{ch}\, x$.

3. Prove that the sign of $\text{sh}\, x$ is the same as the sign of x.

4. Prove that $\text{ch}\, x \geqslant 1$.

5. Prove that $\text{ch}\, x > \text{sh}\, x$ and that $\text{ch}\, x > -\text{sh}\, x$.

6. The following table of values can be obtained from the approximation $e \simeq 2{\cdot}7183$ or from a book of tables :

x	-2	$-1{\cdot}5$	-1	$-0{\cdot}5$	$0{\cdot}5$	1	$1{\cdot}5$	2
e^x	0·14	0·22	0·37	0·61	1·65	2·72	4·48	7·39

Make a corresponding table for the values of $\text{sh}\, x$ and $\text{ch}\, x$ and draw their graphs.

Use the definitions of $\text{sh}\, x$ and $\text{ch}\, x$ to simplify the expressions in Nos. 7-12 :

7. $\text{ch}\, x + \text{sh}\, x$.   **8.** $\text{ch}\, x - \text{sh}\, x$.   **9.** $\text{ch}^2 x - \text{sh}^2 x$.

10. $\text{ch}^2 x + \text{sh}^2 x$.   **11.** $2\,\text{sh}\, x\, \text{ch}\, x$.   **12.** $\sqrt{\{\frac{1}{2}(\text{ch}\, 2x + 1)\}}$.

From the definitions it follows that

$$\operatorname{sh}(-x) = -\operatorname{sh} x \quad \text{and} \quad \operatorname{ch}(-x) = \operatorname{ch} x;$$

therefore the graph of sh x is symmetrical through the origin, and the graph of ch x is symmetrical about Oy. The forms of the graphs, which can be sketched by the method indicated in Exercise XIII a, No. 6, or by using the tables on p. xii, are shown in Fig. 111.

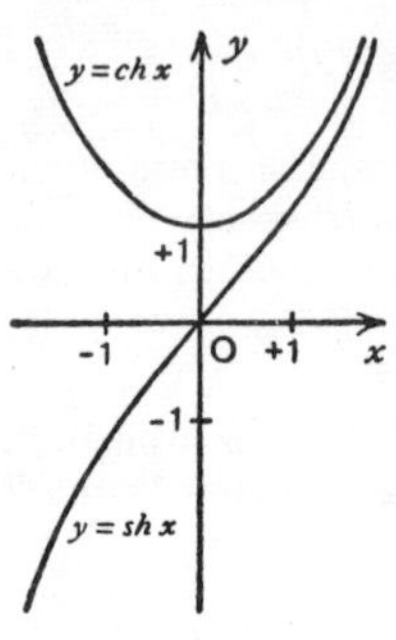

FIG. 111.

It is easy to prove that the gradient of sh x at the origin is 1 and that the origin is a point of inflexion. It can also be proved that the graph of ch x is the form assumed by a uniform flexible chain with fixed ends, and for this reason the curve is called a *catenary* (Latin, *catena* = a chain).

Continuing the analogy with the circular functions, the function (sh x)/(ch x) is called the "hyperbolic tangent" of x, and we write

$$\operatorname{th} x = \frac{\operatorname{sh} x}{\operatorname{ch} x} = \frac{e^x - e^{-x}}{e^x + e^{-x}}.$$

"th" is pronounced "than" or "tansh" and is often written "tanh".

Since sh x has the same sign as x, and since ch x is always positive, th x, $= \dfrac{\operatorname{sh} x}{\operatorname{ch} x}$, has the same sign as x; also, th 0 = 0. Further,

$$\operatorname{th} x = \frac{e^x - e^{-x}}{e^x + e^{-x}} = \frac{e^{2x} - 1}{e^{2x} + 1} = 1 - \frac{2}{e^{2x} + 1}.$$

But e^{2x} increases steadily from 0 to ∞ when x increases from $-\infty$ to $+\infty$; therefore th x increases steadily from -1 to $+1$ when x increases from $-\infty$ to $+\infty$.

The form of the graph of th x, which can be sketched by using the tables on p. xii, is shown in Fig. 112; since th $(-x) = -$th x, the graph is symmetrical through the origin.

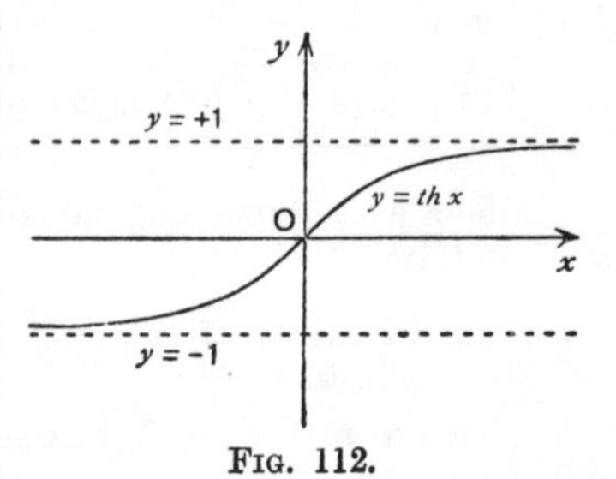

FIG. 112.

The remaining hyperbolic functions may be defined as follows:

$$\text{cosech}\, x = \frac{1}{\text{sh}\, x}; \ \text{sech}\, x = \frac{1}{\text{ch}\, x}; \ \text{coth}\, x = \frac{\text{ch}\, x}{\text{sh}\, x}.$$

To illustrate the analogy between the circular and the hyperbolic functions we shall prove, in Examples 1, 2, two of the fundamental formulae; others are given in Exercise XIII b. For further details the reader is referred to Durell and Robson, *Advanced Trigonometry*, ch. vi. p. 104.

Example 1. Prove that $\text{ch}\, 2x = \text{ch}^2 x + \text{sh}^2 x$.

$$\begin{aligned}\text{ch}^2 x + \text{sh}^2 x &= \tfrac{1}{4}(e^x + e^{-x})^2 + \tfrac{1}{4}(e^x - e^{-x})^2 \\ &= \tfrac{1}{2}(e^{2x} + e^{-2x}) = \text{ch}\, 2x.\end{aligned}$$

Example 2. Prove that $\text{sh}\,(\mathsf{A} + \mathsf{B}) = \text{sh}\,\mathsf{A}\,\text{ch}\,\mathsf{B} + \text{ch}\,\mathsf{A}\,\text{sh}\,\mathsf{B}$.

Put $e^{\mathsf{A}} = a$ and $e^{\mathsf{B}} = b$; then

$$\begin{aligned}\text{sh}\,\mathsf{A}\,\text{ch}\,\mathsf{B} + \text{ch}\,\mathsf{A}\,\text{sh}\,\mathsf{B} &= \tfrac{1}{4}\left\{\left(a - \frac{1}{a}\right)\left(b + \frac{1}{b}\right) + \left(a + \frac{1}{a}\right)\left(b - \frac{1}{b}\right)\right\} \\ &= \tfrac{1}{4}\left(2ab - \frac{2}{ab}\right) \\ &= \tfrac{1}{2}(e^{\mathsf{A}+\mathsf{B}} - e^{-\mathsf{A}-\mathsf{B}}) = \text{sh}\,(\mathsf{A} + \mathsf{B}).\end{aligned}$$

The results already proved, including those of Examples 1, 2, suggest that the relations connecting hyperbolic functions are closely analogous to those connecting circular functions. They may in fact be written down by *Osborn's* rule:

In any formula connecting the circular functions of **general** *angles, replace each circular function by the corresponding hyperbolic function and change the sign of every product* (*or implied product*) *of two sines.*

This gives the corresponding formula for hyperbolic functions.

Thus, from $\tan(\mathsf{A} + \mathsf{B}) = \dfrac{\tan\mathsf{A} + \tan\mathsf{B}}{1 - \tan\mathsf{A}\tan\mathsf{B}}$, we may infer that

$$\text{th}\,(\mathsf{A} + \mathsf{B}) = \frac{\text{th}\,\mathsf{A} + \text{th}\,\mathsf{B}}{1 + \text{th}\,\mathsf{A}\,\text{th}\,\mathsf{B}},$$

since $\tan\mathsf{A}\tan\mathsf{B}$, which is $(\sin\mathsf{A}\sin\mathsf{B})/(\cos\mathsf{A}\cos\mathsf{B})$, contains a product of two sines.

EXERCISE XIII b

[Some of the results in Nos. 1-12 have already been proved ; the others may be verified by Osborn's rule or proved independently. Nos. 1-6 are of great importance.]

1. $\text{ch}^2 x - \text{sh}^2 x = 1$.
2. $\text{ch}^2 x + \text{sh}^2 x = \text{ch } 2x$.
3. $1 - \text{th}^2 x = \text{sech}^2 x$.
4. $\text{ch } 2x + 1 = 2 \text{ ch}^2 x$.
5. $\text{sh } 2x = 2 \text{ sh } x \text{ ch } x$.
6. $\text{ch } 2x - 1 = 2 \text{ sh}^2 x$.
7. $\text{sh }(A+B) = \text{sh } A \text{ ch } B + \text{ch } A \text{ sh } B$.
8. $\text{sh } A + \text{sh } B = 2 \text{ sh } \frac{1}{2}(A+B) \text{ ch } \frac{1}{2}(A-B)$.
9. $\text{ch }(A+B) = \text{ch } A \text{ ch } B + \text{sh } A \text{ sh } B$.
10. $\text{ch } A + \text{ch } B = 2 \text{ ch } \frac{1}{2}(A+B) \text{ ch } \frac{1}{2}(A-B)$.
11. $2 \text{ ch } A \text{ sh } B = \text{sh }(A+B) - \text{sh }(A-B)$.
12. $\text{ch } A - \text{ch } B = 2 \text{ sh } \frac{1}{2}(A+B) \text{ sh } \frac{1}{2}(A-B)$.
13. Deduce formulae from Nos. 7, 8, 9 by changing B into $-B$.

In Nos. 14-17 write down the corresponding formulae for the hyperbolic functions :

14. $\text{cosec}^2 \theta = 1 + \cot^2 \theta$.
15. $2 \sin \theta \cos \phi = \sin(\theta + \phi) + \sin(\theta - \phi)$.
16. $\tan(\theta - \phi) = \dfrac{\tan \theta - \tan \phi}{1 + \tan \theta \tan \phi}$.
17. $\cot(\theta + \phi) = \dfrac{\cot \theta \cot \phi - 1}{\cot \theta + \cot \phi}$.

Write down simple alternative forms of the expressions in Nos. 18-23 :

18. $1 - \text{ch}^2 x$.
19. $1 - \text{coth}^2 x$.
20. $\text{ch}^2 2x + \text{sh}^2 2x$.
21. $\text{th}^2 x + \text{sech}^2 x$.
22. $\dfrac{\text{sh } 2x}{1 + \text{ch } 2x}$.
23. $\dfrac{\text{ch } x + 1}{\text{ch } x - 1}$.
24. If $\text{sh } x = \frac{3}{4}$, find the value of $\text{ch } x$.
25. If $\text{ch } x = 3$, find the values of $\text{sh } x$.
26. If $\text{th } x = \frac{1}{3}$, find the value of $\text{sech } x$.

27. (i) Prove that $\operatorname{ch} x + \operatorname{sh} x = e^x$ and that $\operatorname{ch} x - \operatorname{sh} x = e^{-x}$.

(ii) If $\operatorname{ch} x = 5/3$, prove that $\operatorname{sh} x = \pm 4/3$ and hence find the values of e^x and x.

28. Sketch the graph of $\coth x$.

Differentials and Integrals of sh x and ch x.

Since $d(e^x) = e^x dx$ and $d(e^{-x}) = -e^{-x}dx$,

$$d(\operatorname{sh} x) = d\{\tfrac{1}{2}(e^x - e^{-x})\} = \tfrac{1}{2}(e^x + e^{-x})dx;$$

$$\therefore \mathbf{d(sh\,x) = ch\,x\,dx.}$$

Also,
$$d(\operatorname{ch} x) = d\{\tfrac{1}{2}(e^x + e^{-x})\} = \tfrac{1}{2}(e^x - e^{-x})dx;$$

$$\therefore \mathbf{d(ch\,x) = sh\,x\,dx.}$$

Hence also

$$\mathbf{\int sh\,x\,dx = ch\,x + C; \quad \int ch\,x\,dx = sh\,x + C.}$$

And more generally, if $n \neq 0$,

$$\int \operatorname{sh} nx\,dx = \frac{1}{n}\operatorname{ch} nx + c; \quad \int \operatorname{ch} nx\,dx = \frac{1}{n}\operatorname{sh} nx + c.$$

th x and coth x.

$$\frac{d}{dx}(\operatorname{th} x) = \frac{d}{dx}\left(\frac{\operatorname{sh} x}{\operatorname{ch} x}\right) = \frac{\operatorname{ch}^2 x - \operatorname{sh}^2 x}{\operatorname{ch}^2 x},$$

and
$$\operatorname{ch}^2 x - \operatorname{sh}^2 x = 1;$$

$$\therefore \mathbf{\frac{d}{dx}(th\,x) = sech^2\,x.}$$

Similarly it may be proved that

$$\mathbf{\frac{d}{dx}(coth\,x) = -cosech^2\,x.}$$

Hence also

$$\int \operatorname{sech}^2 x\,dx = \operatorname{th} x + c; \quad \int \operatorname{cosech}^2 x\,dx = -\coth x + c.$$

Further,
$$\int \operatorname{th} x\,dx = \int \frac{\operatorname{sh} x\,dx}{\operatorname{ch} x} = \int \frac{d(\operatorname{ch} x)}{\operatorname{ch} x};$$

$$\therefore \mathbf{\int th\,x\,dx = log\,(ch\,x) + C.}$$

Similarly,
$$\mathbf{\int coth\,x\,dx = log\,(sh\,x) + C}$$

provided that $\operatorname{sh} x$ is positive.

In evaluating other integrals of hyperbolic functions the formulae printed in bold type in Exercise XIII b are often required.

Example 3. Evaluate (i) $\int \text{sh}^2 x\,dx$; (ii) $\int \text{th}^2 x\,dx$.

$$\text{(i)}\ \int \text{sh}^2 x\,dx = \int \tfrac{1}{2}(\text{ch}\,2x - 1)\,dx$$
$$= \tfrac{1}{4}\,\text{sh}\,2x - \tfrac{1}{2}x + c.$$
$$\text{(ii)}\ \int \text{th}^2 x\,dx = \int (1 - \text{sech}^2 x)\,dx$$
$$= x - \text{th}\,x + c.$$

Example 4. Evaluate $\dfrac{d}{dx}\{\tan^{-1}(\text{th}\,\tfrac{1}{2}x)\}$.

$$\frac{d}{dx}\{\tan^{-1}(\text{th}\,\tfrac{1}{2}x)\} = \frac{1}{1 + \text{th}^2\,\tfrac{1}{2}x}\,\tfrac{1}{2}\,\text{sech}^2\,\tfrac{1}{2}x$$
$$= \frac{1}{2(\text{ch}^2\,\tfrac{1}{2}x + \text{sh}^2\,\tfrac{1}{2}x)}$$
$$= \frac{1}{2\,\text{ch}\,x} = \tfrac{1}{2}\,\text{sech}\,x.$$

Example 5. Evaluate $\int \text{sech}\,x\,dx$.

$$\int \text{sech}\,x\,dx = \int \frac{2}{e^x + e^{-x}}\,dx = \int \frac{2e^x}{e^{2x} + 1}\,dx.$$

Put $e^x = z$, then $e^x dx = dz$,

$$\therefore \int \text{sech}\,x\,dx = \int \frac{2dz}{z^2 + 1} = 2\tan^{-1} z + c;$$
$$\therefore \int \text{sech}\,x\,dx = 2\tan^{-1}(e^x) + c.$$

Note. From Example 4 we obtain an integral of $\text{sech}\,x$ in the form $2\tan^{-1}\text{th}\,\tfrac{1}{2}x$ apparently different from the result of Example 5. But since

$$\text{th}\,\tfrac{1}{2}x = \frac{e^x - 1}{e^x + 1} \quad \text{and} \quad \text{Tan}^{-1}u - \text{Tan}^{-1}v = \text{Tan}^{-1}\frac{u - v}{1 + uv},$$

we have $$\text{Tan}^{-1} e^x - \text{Tan}^{-1} 1 = \text{Tan}^{-1}\text{th}\,\tfrac{1}{2}x;$$

and therefore $\tan^{-1}\text{th}\,\tfrac{1}{2}x$ differs from $\tan^{-1} e^x$ only by a constant. Cf. p. 116 (Vol. I).

EXERCISE XIII c

Differentiate with respect to x the functions in Nos. 1-12:

1. $\text{sh}\,2x$. **2.** $\text{ch}\,3x$. **3.** $\text{sech}\,x$.

4. $\text{cosech}\,\tfrac{1}{2}x$. **5.** $\text{sh}^2\,3x$. **6.** $\log\text{sh}\,x$.

7. th $2x$. 8. $\text{th}^2 x$. 9. $\coth \frac{1}{2}x$.

10. $\log \text{th}\, x$. 11. $\text{sech}^2 2x$. 12. $x - \coth x$.

Integrate with respect to x the functions in Nos. 13-21 :

13. ch $2x$. 14. sh $\frac{1}{2}x$. 15. th $2x$.

16. $\text{sech}^2 2x$. 17. $\text{cosech}^2 \frac{1}{2}x$. 18. $\coth 3x$.

19. $\text{ch}^2 x$. 20. $\text{sh}^2 \frac{1}{2}x$. 21. $\coth^2 x$.

22. Evaluate $\int \text{sech}\, 2x\, dx$ by the method of Example 5.

23. Find $\dfrac{d}{dx} \log \text{th} \frac{1}{2}x$. Hence evaluate $\int \text{cosech}\, x\, dx$.

24. Use the formula in Exercise XIII b, No. 11, to evaluate $\int \text{ch}\, 3x\, \text{sh}\, x\, dx$. Also evaluate $\int \text{sh}\, 3x\, \text{ch}\, x\, dx$.

25. Evaluate (i) $\int \text{ch}\, 2x\, \text{ch}\, 3x\, dx$; (ii) $\int \text{sh}\, 2x\, \text{sh}\, 3x\, dx$.

26. Evaluate (i) $\displaystyle\int_{-1}^{+1} \text{ch}\, x\, dx$; (ii) $\displaystyle\int_0^1 \text{sh}^2 x\, dx$.

Apply the method of integration by parts to evaluate the integrals in Nos. 27-30 :

27. $\int x\, \text{sh}\, x\, dx$. 28. $\int x\, \text{ch}\, 2x\, dx$.

29. $\int x\, \text{sech}^2 x\, dx$. 30. $\int x^2\, \text{ch}\, x\, dx$.

The Rectangular Hyperbola. The trigonometrical functions are called *circular* functions because they are connected with the geometry of the circle. The coordinates of any point P on the circle $x^2 + y^2 = a^2$ (see Fig. 113) may be expressed in the form $(a \cos \theta,\ a \sin \theta)$ and, with this notation, the area of the sector bounded by the lines $\theta = 0$, $\theta = \theta_1$, is $\frac{1}{2}a^2 \theta_1$.

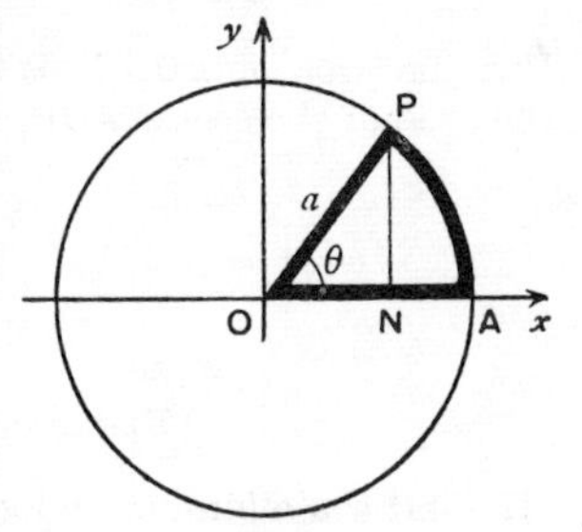

FIG. 113.

The functions ch u, sh u are connected in a similar way with the geometry of the rectangular hyperbola and this is the origin of the name "*hyperbolic* functions".

Since $\text{ch}^2 u - \text{sh}^2 u = 1$, the point $(a \text{ ch } u, a \text{ sh } u)$ lies on the rectangular hyperbola $x^2 - y^2 = a^2$; see Fig. 114.

As u increases from $-\infty$ to $+\infty$, the point $\mathsf{P}(a \text{ ch } u, a \text{ sh } u)$ traces out that part of the curve for which $x > 0$, assuming that $a > 0$; and the other part is given by $(-a \text{ ch } u, a \text{ sh } u)$.

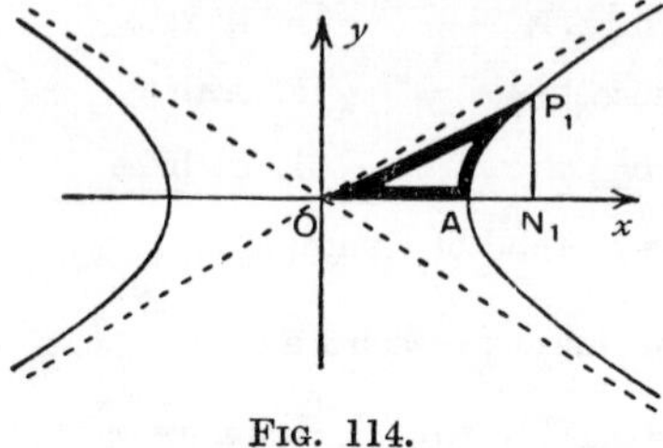

FIG. 114.

We shall prove as an example that if A, P_1 are the points corresponding to $u = 0$, $u = u_1$ respectively, the area of the sector AOP_1 is $\frac{1}{2}a^2 u_1$. An alternative method is given in Chapter XVIII, p. 363.

Example 6. If A, P_1 are the positions of the point $(a \text{ ch } u, a \text{ sh } u)$ on the rectangular hyperbola $x^2 - y^2 = a^2$, given by $u = 0$, $u = u_1$ respectively, prove that the area of the sector AOP_1 (see Fig. 114) is $\frac{1}{2}a^2 u_1$.

The area bounded by the arc AP_1, the ordinate $\mathsf{P}_1\mathsf{N}_1$, and AN_1 is $\int_a^{x_1} y\,dx$, where $x_1 = a \text{ ch } u_1$.

Also, $\quad y = a \text{ sh } u \quad$ and $\quad dx = d(a \text{ ch } u) = a \text{ sh } u\,du$;

$$\therefore \text{ area } \mathsf{AN}_1\mathsf{P}_1 = \int_0^{u_1} a \text{ sh } u \,.\, a \text{ sh } u\,du = \tfrac{1}{2}a^2 \int_0^{u_1} (\text{ch } 2u - 1)\,du$$

$$= \tfrac{1}{2}a^2 \Big[\tfrac{1}{2} \text{ sh } 2u - u \Big]_0^{u_1} = \tfrac{1}{2}a^2(\text{sh } u_1 \text{ ch } u_1 - u_1).$$

But the area of $\Delta\mathsf{ON}_1\mathsf{P}_1$ is $\frac{1}{2}a^2 \text{ sh } u_1 \text{ ch } u_1$; hence by subtraction the area of the sector AOP_1 is $\frac{1}{2}a^2 u_1$.

EXERCISE XIII d

(Miscellaneous Applications)

1. Find the minimum value of $5 \text{ ch } x + 3 \text{ sh } x$.

2. Find the area bounded by $\mathsf{O}x$, $\mathsf{O}y$, the catenary $y = c \text{ ch } (x/c)$, and the ordinate $x = c$.

3. Prove that the volume generated by the revolution about Ox of the area in No. 2 is $\frac{1}{8}\pi c^3(e^2 - e^{-2} + 4)$.

4. Find the maximum value of $x(\operatorname{ch} x - \operatorname{sh} x)$.

5. The space-time relation for a particle moving along Ox is $x = a \operatorname{sh} nt + b \operatorname{ch} nt$ where a, b, n are constants. Prove that the acceleration is proportional to x.

6. Prove that the area of the ellipse $x = a \cos t$, $y = b \sin t$ is πab.

7. Find the area bounded by Oy, the curves $y = \operatorname{ch} x$, $y = \operatorname{sh} x$, and the ordinate $x = 1$.

8. The space-time relation for a particle moving along Ox is $x = \tan^{-1}(\operatorname{ch} nt + \operatorname{sh} nt)$. Prove that its velocity at time t is $\frac{1}{2}n \operatorname{sech} nt$.
   What is its initial position and acceleration?

9. Sketch the graph of $y = \operatorname{sech} x$ and find the volume generated by the revolution about Ox of the area bounded by the curve and the ordinates $x = a$, $x = -a$. Does the expression for the volume tend to a limit when $a \to \infty$?

10. A particle falls in a resisting medium so that its velocity v at time t is given by $v = c \operatorname{th} ckt$ where c, k are constants. Prove that its acceleration is $k(c^2 - v^2)$ and find the distance it has fallen at time t.

Inverse Hyperbolic Functions

The graph of $y = \operatorname{sh} x$ in Fig. **111**, p. 266, illustrates the fact that when x increases steadily from $-\infty$ to $+\infty$, $\operatorname{sh} x$ also *increases steadily* from $-\infty$ to $+\infty$; this may be proved by noting that $\frac{d}{dx} \operatorname{sh} x = \operatorname{ch} x$ and that $\operatorname{ch} x$ is always positive. Hence we conclude that $\operatorname{sh} x$ assumes any assigned value y for one, and only one, value of x; this unique value of x is denoted by $\operatorname{sh}^{-1} y$. Thus,

$$\text{if } \operatorname{sh} x = y, \quad x = \operatorname{sh}^{-1} y.$$

The graph of $y = \operatorname{sh}^{-1} x$ can be obtained from that of $y = \operatorname{sh} x$ by interchanging the axes of x and y, or by taking the image of the curve in the line $y = x$; the reader should sketch it for himself in the margin.

$\operatorname{sh}^{-1} x$ is called the "*inverse* hyperbolic sine" of x.

Example 7. Find the value of $\text{sh}^{-1}\frac{3}{4}$.

$$\text{If } y=\text{sh}^{-1}\tfrac{3}{4}, \quad \text{sh}\, y=\tfrac{3}{4};$$

$$\therefore \text{ch}^2 y=1+\frac{3^2}{4^2}=\frac{5^2}{4^2};$$

$\therefore$ since $\text{ch}\, y$ is always positive, $\text{ch}\, y=+\frac{5}{4}$;

$$\therefore e^y \equiv \text{ch}\, y+\text{sh}\, y=2; \quad \therefore y=\log 2;$$

$$\therefore \text{sh}^{-1}\tfrac{3}{4}=\log 2.$$

The graph of $y=\text{ch}\, x$ in Fig. 111, p. 266, illustrates the fact that $\text{ch}\, x$ has no value less than 1, and shows that to any assigned value y of $\text{ch}\, x$, which is greater than 1, there correspond two values of x numerically equal but of opposite signs.

If $\text{ch}\, x=y \geqslant 1$, we may write $x=\text{Ch}^{-1} y$, and if $y>1$, $\text{Ch}^{-1} y$ has two (equal and opposite) values; the *positive* value is denoted by $\text{ch}^{-1} y$.

The graph of $y=\text{Ch}^{-1} x$ can be obtained from that of $y=\text{ch}\, x$ by interchanging the axes of x and y; the reader should sketch it in the margin and indicate in some way that the upper half of it is the graph of $y=\text{ch}^{-1} x$.

$\text{Ch}^{-1} x$ is called the "*inverse* hyperbolic cosine" of x.

Example 8. Evaluate $\text{Ch}^{-1}\frac{5}{4}$ and $\text{ch}^{-1}\frac{5}{4}$.

$$\text{If } y=\text{Ch}^{-1}\tfrac{5}{4}, \quad \text{ch}\, y=\tfrac{5}{4};$$

$$\therefore \text{sh}\, y= \pm\sqrt{(\text{ch}^2 y-1)}= \pm\tfrac{3}{4}.$$

Thus
$$e^y \equiv \text{ch}\, y+\text{sh}\, y=\tfrac{5}{4}\pm\tfrac{3}{4}=2 \text{ or } \tfrac{1}{2}.$$

$$\therefore y=\log 2 \text{ or } \log\tfrac{1}{2}, \text{ that is, } y= \pm\log 2;$$

$$\therefore \text{Ch}^{-1}\tfrac{5}{4}= \pm\log 2 \text{ and } \text{ch}^{-1}\tfrac{5}{4}=\log 2.$$

As shown on p. 266, $\text{th}\, x$ *increases steadily* from -1 to $+1$ as x increases from $-\infty$ to $+\infty$; this may also be proved by noting that $\frac{d}{dx}\text{th}\, x, =\text{sech}^2 x$, is always positive. Hence we conclude that if $y=\text{th}\, x$, then for any assigned value of y between -1 and $+1$, x has a unique value; we denote this by $\text{th}^{-1} y$ and call it the "*inverse* hyperbolic tangent" of y.

The reader should sketch the graph of $y=\text{th}^{-1} x$ in the margin.

The functions $\text{cosech}^{-1} x$, $\text{Sech}^{-1} x$, and $\text{coth}^{-1} x$ may be defined and discussed in a similar way, but they are not often used because $\text{cosech}^{-1} x = \text{sh}^{-1} \frac{1}{x}$, etc.

The values of $\text{sh}^{-1} x$ and $\text{ch}^{-1} x$, for given values of x, can be found from a table of natural logarithms by using the methods of Examples 7, 8, and this is also true of $\text{th}^{-1} x$. We proceed to obtain the general formulae, but the reader should first work Exercise XIII e, Nos. 1-3.

Inverse Hyperbolic Functions as Logarithms.

(i) $\mathbf{sh^{-1} x = \log \{x + \sqrt{(1 + x^2)}\}}$.

$$\text{If } y = \text{sh}^{-1} x, \text{ then } \text{sh}\, y = x;$$

$$\therefore \text{ch}^2 y = 1 + \text{sh}^2 y = 1 + x^2;$$

but $\text{ch}\, y$ is positive, $\therefore \text{ch}\, y = \sqrt{(1 + x^2)}$.

$$\therefore e^y = \text{sh}\, y + \text{ch}\, y = x + \sqrt{(1 + x^2)};$$

$$\therefore y = \log\{x + \sqrt{(1 + x^2)}\}.$$

(ii) $\mathbf{Ch^{-1} x = \pm \log\{x + \sqrt{(x^2 - 1)}\}, (x \geqslant 1)}$.

$$\text{If } y = \text{Ch}^{-1} x, \text{ then } \text{ch}\, y = x;$$

$$\therefore \text{sh}^2 y = \text{ch}^2 y - 1 = x^2 - 1; \quad \therefore \text{sh}\, y = \pm\sqrt{(x^2 - 1)} \text{ since } x \geqslant 1.$$

$$\therefore e^y = \text{ch}\, y + \text{sh}\, y = x \pm \sqrt{(x^2 - 1)}.$$

But $\{x + \sqrt{(x^2 - 1)}\}\{x - \sqrt{(x^2 - 1)}\} = x^2 - (x^2 - 1) = 1$,

$$\therefore e^y = x + \sqrt{(x^2 - 1)} \text{ or } 1/\{x + \sqrt{(x^2 - 1)}\};$$

$$\therefore y = \pm \log\{x + \sqrt{(x^2 - 1)}\}.$$

(iii) $\mathbf{th^{-1} x = \frac{1}{2} \log \frac{1 + x}{1 - x}, (-1 < x < 1)}$.

$$\text{If } y = \text{th}^{-1} x, \text{ then } x = \text{th}\, y = \frac{e^{2y} - 1}{e^{2y} + 1};$$

$$\therefore e^{2y} = \frac{1 + x}{1 - x}; \quad \therefore y = \tfrac{1}{2} \log \frac{1 + x}{1 - x} \text{ since } -1 < x < 1.$$

The results of (i)-(iii) emphasise the fact that the inverse hyperbolic functions are not many-valued functions like $\text{Sin}^{-1} x$, $\text{Cos}^{-1} x$, $\text{Tan}^{-1} x$; $\text{sh}^{-1} x$ and $\text{th}^{-1} x$ are one-valued and therefore $\text{Sh}^{-1} x$, $\text{Th}^{-1} x$ are not introduced. $\text{Ch}^{-1} x$ is two-valued.

Differentials of the Inverse Hyperbolic Functions.

If $y = \text{sh}^{-1} x$, $\text{sh}\, y = x$; $\therefore \text{ch}\, y\, dy = dx$.

But $\text{ch}^2 y = 1 + \text{sh}^2 y = 1 + x^2$ and $\text{ch}\, y$ is positive ;

$$\therefore \text{ch}\, y = \sqrt{(1 + x^2)} \quad \text{and} \quad \sqrt{(1 + x^2)}dy = dx.$$

$$\therefore \mathbf{d(sh^{-1}\, x) = \frac{dx}{\sqrt{(1+x^2)}}},$$

where, with the usual convention, $\sqrt{(1 + x^2)}$ denotes the positive square root of $1 + x^2$. If the reader looks at his sketch of the graph of $\text{sh}^{-1} x$ he should find that the gradient is everywhere positive.

It follows that

$$\mathbf{\int \frac{dx}{\sqrt{(1+x^2)}} = sh^{-1}\, x + C.}$$

This result may also be obtained by using the substitution $x = \text{sh}\, z$ as follows :

$$dx = \text{ch}\, z\, dz \quad \text{and} \quad \sqrt{(1 + x^2)} = \sqrt{(1 + \text{sh}^2 z)} = \text{ch}\, z\, ;$$

$$\therefore \int \frac{dx}{\sqrt{(1 + x^2)}} = \int \frac{\text{ch}\, z\, dz}{\text{ch}\, z} = \int 1\, dz$$

$$= z + c = \text{sh}^{-1} x + c.$$

If $y = \text{Ch}^{-1} x$, $(x \geqslant 1)$, $\text{ch}\, y = x$; $\therefore \text{sh}\, y\, dy = dx$.

But $\text{sh}\, y = \pm\sqrt{(\text{ch}^2 y - 1)} = \pm\sqrt{(x^2 - 1)}$,

$$\therefore \pm\sqrt{(x^2 - 1)}dy = dx\, ;$$

$$\therefore \mathbf{d(Ch^{-1}\, x) = \pm \frac{dx}{\sqrt{(x^2-1)}}}, \ (x > 1).$$

The ambiguity of sign is due to the fact that $\text{Ch}^{-1} x$ is a two-valued function. In the proof it is assumed that $\text{Ch}^{-1} x$ is confined either to positive or to negative values. If the reader looks at his sketch of $\text{Ch}^{-1} x$ he should find that the gradient is positive for the *positive* values of $\text{Ch}^{-1} x$, that is, for $\text{ch}^{-1} x$; hence

$$\mathbf{d(ch^{-1}\, x) = \frac{dx}{\sqrt{(x^2-1)}}}, \ (x > 1).$$

It follows that

$$\mathbf{\int \frac{dx}{\sqrt{(x^2-1)}} = ch^{-1}\, x + C}, \ (x > 1).$$

This result may also be obtained by using the substitution $x = \text{ch}\, z$.

Note. If $x < -1$ the corresponding substitution is $x = -\text{ch}\, z$ since $\text{ch}\, z$ is always positive; this gives

$$\int \frac{dx}{\sqrt{(x^2 - 1)}} = -\text{ch}^{-1}(-x) + c, \ (x < -1)$$
$$= -\log\{-x + \sqrt{(x^2 - 1)}\} + c, \ (x < -1).$$

The differential of the function, $\text{th}^{-1} x$, $(x^2 < 1)$, may be found in a similar way. See Exercise XIII e, Nos. 5, 6.

EXERCISE XIII e

Find, from first principles, the values in the form of logarithms of the expressions in Nos. 1-3:

1. $\text{sh}^{-1} 2\frac{2}{5}$.
2. $\text{sech}^{-1} \frac{1}{2}$.
3. $\text{th}^{-1} \frac{2}{7}$.
4. Evaluate $\displaystyle\int \frac{dx}{\sqrt{(x^2 - 1)}}$, $x > 1$, by the substitution $x = \text{ch}\, z$.
5. If $y = \text{th}^{-1} x$, $(x^2 < 1)$, prove that $dy = \dfrac{dx}{1 - x^2}$.

   What conclusions can be drawn about the gradient of the graph of $\text{th}^{-1} x$?
6. Use the formula $\text{th}^{-1} x = \frac{1}{2} \log\{(1 + x)/(1 - x)\}$, $(x^2 < 1)$, p. 275, to evaluate $\dfrac{d}{dx}(\text{th}^{-1} x)$.

Differentiate with respect to x the functions in Nos. 7-12 and state any necessary limitations on the values of x:

7. $\text{sh}^{-1}(2x/3)$.
8. $\text{ch}^{-1}(4x/5)$.
9. $\text{th}^{-1} 3x$.
10. $\text{cosech}^{-1} x$.
11. $\text{sech}^{-1} x$.
12. $\text{coth}^{-1} 2x$.

Express as logarithms the functions in Nos. 13-15 and state any necessary limitations on the values of x:

13. $\text{cosech}^{-1} x$.
14. $\text{sech}^{-1} x$.
15. $\text{coth}^{-1} x$.
16. Prove that $\text{th}^{-1}\{(x^2 - 1)/(x^2 + 1)\} = \log|x|$, $(x \neq 0)$.
17. Find the minimum value of $2 \log x - \text{ch}^{-1} x$.
18. Find the area bounded by the two branches of the curve $y^2(x^2 + 1) = 1$ and the lines $x = 1$, $x = -1$.

Applications to Integration.

The values of the integrals $\int \frac{dx}{\sqrt{(1+x^2)}}$ and $\int \frac{dx}{\sqrt{(x^2-1)}}$ were found on p. 276, and should be compared with that of $\int \frac{dx}{\sqrt{(1-x^2)}}$ on p. 125 (Vol. I).

The method of dealing with other integrals of the form $\int \frac{dx}{\sqrt{(a+bx^2)}}$ is illustrated in Examples 9, 10.

Example 9. Find the value of $\int \frac{dx}{\sqrt{(4+x^2)}}$.

$$\text{Put } x = 2 \operatorname{sh} z,$$

then $dx = 2 \operatorname{ch} z\, dz$ and $\sqrt{(4+x^2)} = \sqrt{\{4(1+\operatorname{sh}^2 z)\}} = 2 \operatorname{ch} z$;

$$\therefore \int \frac{dx}{\sqrt{(4+x^2)}} = \int \frac{2 \operatorname{ch} z\, dz}{2 \operatorname{ch} z} = \int 1\, dz$$

$$= z + c = \operatorname{sh}^{-1}(\tfrac{1}{2}x) + c.$$

Example 10. Find the value of $\int \frac{dx}{\sqrt{(9x^2-16)}}$, $(x > \frac{4}{3})$.

Put $\quad 3x = 4 \operatorname{ch} z$ where $z > 0$,

then $\quad 3\, dx = 4 \operatorname{sh} z\, dz$

and $\quad \sqrt{(9x^2-16)} = \sqrt{\{16(\operatorname{ch}^2 z - 1)\}} = 4 \operatorname{sh} z$, since $z > 0$;

$$\therefore \int \frac{dx}{\sqrt{(9x^2-16)}} = \int \frac{4 \operatorname{sh} z\, dz}{3 \,.\, 4 \operatorname{sh} z} = \int \tfrac{1}{3}\, dz$$

$$= \tfrac{1}{3} z + c = \tfrac{1}{3} \operatorname{ch}^{-1}(\tfrac{3}{4}x) + c.$$

An alternative method for Example 10 is to guess that the integral is $\operatorname{ch}^{-1} \frac{3}{4}x$ and differentiate this function with respect to x; this would lead to $3/\sqrt{(9x^2-16)}$, showing that the factor $\frac{1}{3}$ must be inserted.

Standard Substitutions. Every integral of the form $\int \frac{dx}{\sqrt{(a+bx^2)}}$ reduces after the removal of the factor $\sqrt{b}$ (b positive), or $\sqrt{(-b)}$ (b negative), to one of the forms

$$\int \frac{dx}{\sqrt{(k^2-x^2)}}, \quad \int \frac{dx}{\sqrt{(k^2+x^2)}}, \quad \int \frac{dx}{\sqrt{(x^2-k^2)}},$$

and these can be evaluated by the substitutions,

$$x=k\sin z,\ x=k\operatorname{sh} z,\ x=k\operatorname{ch} z \text{ respectively.}$$

The same substitutions are used for the integrals

$$\int\sqrt{(k^2-x^2)}\,dx,\ \int\sqrt{(k^2+x^2)}\,dx,\ \int\sqrt{(x^2-k^2)}\,dx.$$

Example 11. Find the value of $\int \frac{dx}{\sqrt{(x^2-a^2)}}$, $(0<a<x)$.

Put $x=a\operatorname{ch} z$ where $z>0$, then $\operatorname{sh} z>0$;

$\therefore\ \sqrt{(x^2-a^2)}=\sqrt{\{a^2(\operatorname{ch}^2 z-1)\}}=a\operatorname{sh} z$ since $a>0$, $\operatorname{sh} z>0$;

also, $$dx=a\operatorname{sh} z\,dz;$$

$$\therefore \int \frac{dx}{\sqrt{(x^2-a^2)}}=\int\frac{a\operatorname{sh} z\,dz}{a\operatorname{sh} z}=\int 1\,dz=z+c=\operatorname{ch}^{-1}(x/a)+c.$$

Note. The form of the integrand in Example 11 implies that x^2-a^2 is positive. If a is negative it can be changed into $-a$ without altering the integrand. If x is negative, as in the integral $\int_{-5}^{-4}\frac{dx}{\sqrt{(x^2-9)}}$, it is best to reduce the integral to $\int_4^5\frac{dy}{\sqrt{(y^2-9)}}$, by making the substitution $x=-y$. For this reason it is usual to regard x as positive in finding indefinite integrals.

$\int_{-5}^{+4}\frac{dx}{\sqrt{(x^2-9)}}$ is meaningless, because there are values of x between -5 and $+4$ for which x^2-9 is negative.

Example 12. Find the value of $\int\sqrt{(a^2+x^2)}\,dx$, $(a>0)$.

Put $$x=a\operatorname{sh} z, \text{ then } dx=a\operatorname{ch} z\,dz,$$

and $\sqrt{(a^2+x^2)}=\sqrt{\{a^2(1+\operatorname{sh}^2 z)\}}=a\operatorname{ch} z$ since $a>0$, $\operatorname{ch} z>0$;

$$\therefore \int\sqrt{(a^2+x^2)}\,dx=\int a\operatorname{ch} z\,.\,a\operatorname{ch} z\,dz=\tfrac{1}{2}a^2\int(\operatorname{ch} 2z+1)\,dz$$

$$=\tfrac{1}{2}a^2(\tfrac{1}{2}\operatorname{sh} 2z+z)+c=\tfrac{1}{2}a^2(\operatorname{sh} z\operatorname{ch} z+z)+c;$$

but $a\operatorname{ch} z=\sqrt{(a^2+x^2)}$ and $a\operatorname{sh} z=x$,

$$\therefore \int\sqrt{(a^2+x^2)}\,dx=\tfrac{1}{2}x\sqrt{(a^2+x^2)}+\tfrac{1}{2}a^2\operatorname{sh}^{-1}(x/a)+c.$$

EXERCISE XIII f

[Assume in this exercise that x takes only those *positive* values for which the integrand exists, except in No. 10.]

Integrate with respect to x the functions in Nos. 1-6 :

1. $\dfrac{1}{\sqrt{(x^2+9)}}$. 2. $\dfrac{1}{\sqrt{(x^2-16)}}$. 3. $\dfrac{1}{\sqrt{(4x^2-9)}}$.

4. $\dfrac{1}{\sqrt{(9x^2+1)}}$. 5. $\dfrac{1}{\sqrt{(16-9x^2)}}$. 6. $\dfrac{x}{\sqrt{(x^4+1)}}$.

Evaluate the integrals in Nos. 7-11 :

7. $\displaystyle\int_0^4 \frac{dx}{\sqrt{(x^2+9)}}$. 8. $\displaystyle\int_4^5 \frac{dx}{\sqrt{(x^2-9)}}$. 9. $\displaystyle\int_3^4 \frac{dx}{\sqrt{(25-x^2)}}$.

10. $\displaystyle\int_{-3}^{-2} \frac{dx}{\sqrt{(1+x^2)}}$. (Put $x=-y$.)

11. $\displaystyle\int_p^q \frac{dx}{\sqrt{(a^2+x^2)}}$, $(a<0)$. (Put $a=-b$.)

By hyperbolic substitutions, integrate with respect to x in 12-20:

12. $\sqrt{(x^2+9)}$. 13. $\sqrt{(x^2-9)}$. 14. $\sqrt{(9x^2+16)}$.

15. $\dfrac{x^2}{\sqrt{(1+x^2)}}$. 16. $\dfrac{1}{x^2\sqrt{(x^2-1)}}$. 17. $\dfrac{\sqrt{(x^2-1)}}{x^2}$.

18. $\dfrac{1}{\sqrt{\{(x-1)^2-1\}}}$. 19. $\sqrt{(x^2-2x)}$. 20. $\dfrac{1}{\sqrt{(x^2+2x+5)}}$.

Use integration by parts for Nos. 21-23 :

21. $\int \text{sh}^{-1} x\, dx$. 22. $\int x \,\text{ch}^{-1} x\, dx$. 23. $\int \text{th}^{-1} x\, dx$.

24. Evaluate $\displaystyle\int_0^3 \sqrt{(x^2+16)}\, dx$. 25. Evaluate $\displaystyle\int_0^1 \text{sh}^{-1}(\tfrac{1}{2}x)\, dx$.

26. Find the area of the segment cut off from the curve $y=\text{Ch}^{-1} x$ by the line $x=2$.

27. Prove that the area bounded by the curve $y=\sqrt{(x^2-4)}$, the x-axis, and the ordinate $x=2\frac{1}{2}$ is $\frac{15}{8}-2\log 2$.

28. A particle moves in the x-axis so that its velocity v at the point $(x, 0)$ is given by $v=k\sqrt{(a^2+x^2)}$ where k, a are positive constants. Find the time taken to move from (i) $x=0$ to $x=\frac{3}{4}a$, (ii) $x=0$ to $x=\frac{4}{3}a$.

CHAPTER XIV

SYSTEMATIC INTEGRATION

THERE is no general method of obtaining the integral (or primitive) of a given function, and in fact there are many comparatively simple algebraic functions whose integrals cannot be expressed in terms of the functions considered in this book.

In previous chapters we have evaluated a number of standard integrals to which many others can be reduced, and we have explained the method of substitution which is often used in the reduction.

The purpose of this chapter is to take stock of the results that have been obtained, and to show that certain types of functions can be integrated by a perfectly definite procedure.

We give here a list of standard forms; of these the student should be able to quote Nos. 1-12; and he must know how to obtain the remainder rapidly.

The constant of integration will be omitted in this chapter for the sake of brevity. For the same reason, if the logarithm of a function occurs, it is implied that only positive values of the function are considered. Thus in (2) we write $\int \frac{1}{x}\,dx = \log x$ instead of the more precise statement $\int \frac{1}{x}\,dx = \log|x| + c$.

Further, it is implied that x is confined to values for which the integrand exists.

Standard Integrals.

$$\int x^n\,dx = \frac{x^{n+1}}{n+1}, \quad (n \neq -1) \qquad (1)$$

$$\int \frac{1}{x}\,dx = \log x \qquad (2)$$

$$\int e^x\,dx = e^x \qquad (3)$$

$$\int \sin x\,dx = -\cos x; \quad \int \cos x\,dx = \sin x \qquad (4)$$

$$\int \sec^2 x\,dx = \tan x; \quad \int \operatorname{cosec}^2 x\,dx = -\cot x \qquad (5)$$

$$\int \operatorname{sh} x\,dx = \operatorname{ch} x; \quad \int \operatorname{ch} x\,dx = \operatorname{sh} x \qquad (6)$$

$$\int \operatorname{sech}^2 x\, dx = \operatorname{th} x\,; \qquad \int \operatorname{cosech}^2 x\, dx = -\coth x \quad . \quad (7)$$

$$\int \frac{1}{1+x^2}\, dx = \tan^{-1} x \quad . \quad . \quad . \quad (8)$$

$$\int \frac{1}{\sqrt{(1-x^2)}}\, dx = \sin^{-1} x \quad . \quad . \quad (9)$$

$$\int \frac{1}{\sqrt{(1+x^2)}}\, dx = \operatorname{sh}^{-1} x \quad . \quad . \quad . \quad (10)$$

$$\int \frac{1}{\sqrt{(x^2-1)}}\, dx = \operatorname{ch}^{-1} x, (x > 1). \quad . \quad (11)$$

$$\int \frac{f'(x)}{f(x)}\, dx = \log f(x) \quad . \quad . \quad . \quad (12)$$

(13) and (14) can be written down by inspection, using (12):

$$\int \tan x\, dx = -\log(\cos x)\,; \quad \int \cot x\, dx = \log(\sin x) \quad . \quad (13)$$

$$\int \operatorname{th} x\, dx = \log(\operatorname{ch} x)\,; \qquad \int \coth x\, dx = \log(\operatorname{sh} x) \quad . \quad (14)$$

For (15) and (16), see p. 297.

$$\int \frac{1}{\sin x}\, dx = \log(\tan \tfrac{1}{2}x) \quad . \quad . \quad . \quad . \quad (15)$$

$$\int \frac{1}{\cos x}\, dx = \log \tan(\tfrac{1}{4}\pi + \tfrac{1}{2}x) = \log(\sec x + \tan x) \quad . \quad (16)$$

(17) and (18) are the least important in this list; see p. 297 and p. 270.

$$\int \frac{1}{\operatorname{sh} x}\, dx = \log(\operatorname{th} \tfrac{1}{2}x) \quad . \quad . \quad . \quad . \quad . \quad (17)$$

$$\int \frac{1}{\operatorname{ch} x}\, dx = 2\tan^{-1} e^x \quad . \quad . \quad . \quad . \quad . \quad (18)$$

(19)-(22) can be deduced from (8)-(11) or evaluated by corresponding substitutions; see the note on (19). (20)-(22) can be treated in the same way as (19). After a little practice this can be done mentally.

$$\int \frac{1}{a^2+x^2}\, dx = \frac{1}{a} \tan^{-1} \frac{x}{a} \quad . \quad . \quad (19)$$

[Express it as $\int \frac{a d(x/a)}{a^2(1+x^2/a^2)}$ or put $x = a \tan \theta$.]

In (20)-(22) a is supposed *positive*.

$$\int \frac{1}{\sqrt{(a^2 - x^2)}}\,dx = \sin^{-1}\frac{x}{a} \quad . \quad . \quad . \quad . \quad (20)$$

$$\int \frac{1}{\sqrt{(a^2 + x^2)}}\,dx = \text{sh}^{-1}\frac{x}{a} \text{ or } \log\{x + \sqrt{(a^2 + x^2)}\} \quad . \quad (21)$$

$$\int \frac{1}{\sqrt{(x^2 - a^2)}}\,dx = \text{ch}^{-1}\frac{x}{a} \text{ or } \log\{x + \sqrt{(x^2 - a^2)}\}, \quad . \quad (22)$$

where, for (22), $0 < a < x$.
The logarithmic forms of (21), (22) may be obtained by using the formulae on p. 275 or by the methods suggested in Exercise XIV e, Nos. 22, 23.
To this list may be added the formulae

$$\int f(x)\,dx = \int f\{\phi(z)\}\phi'(z)\,dz \text{ where } x = \phi(z) \quad . \quad (23)$$

$$\int uv'\,dx = uv - \int vu'\,dx \quad . \quad . \quad . \quad . \quad (24)$$

Simple reductions to the standard integrals in this list can often be done mentally.

Thus applying mentally the substitution $x + a = z$ we have

$$\int \frac{1}{x + a}\,dx = \log(x + a).$$

Also, since

$$\frac{1}{x^2 - a^2} = \frac{1}{2a}\left(\frac{1}{x - a} - \frac{1}{x + a}\right),$$

$$\int \frac{1}{x^2 - a^2}\,dx = \frac{1}{2a}\log\frac{x - a}{x + a} \quad . \quad . \quad (25)$$

By using similar substitutions, which should be done mentally, we have such results as

$$\int \frac{dx}{(x + k)^2 - a^2} = \frac{1}{2a}\log\frac{x + k - a}{x + k + a}$$

$$\int \frac{dx}{(x + k)^2 + a^2} = \frac{1}{a}\tan^{-1}\frac{x + k}{a},$$

and so on.

No useful purpose is served by memorising such results, because they can be written down almost at sight, or at any rate with very little side-work.

EXERCISE XIV a

Write down the integrals with respect to x of the functions in Nos. 1-8:

1. (i) x^{2n}; (ii) $\frac{1}{x^n}$; (iii) $\frac{1}{x} - \frac{1}{x^2}$.

2. (i) e^{3x}; (ii) e^{-x}; (iii) $\sqrt{e^x}$.

3. (i) $\sin 2x$; (ii) $\cos \frac{1}{2}x$; (iii) $\tan 4x$.

4. (i) $\sec^2 3x$; (ii) $\operatorname{cosec}^2 2x$; (iii) $\cot \frac{1}{2}x$.

5. (i) $\operatorname{sh}(2x+1)$; (ii) $\operatorname{ch}(1-2x)$; (iii) $\operatorname{th}(2x+3)$.

6. (i) $\frac{1}{2x+3}$; (ii) $\frac{1}{5-4x}$; (iii) $\frac{x}{1+3x^2}$.

7. (i) $\frac{1}{\sqrt{(1-x^2)}}$; (ii) $\frac{x}{\sqrt{(1-x^2)}}$; (iii) $\frac{1}{\sqrt{(x^2-1)}}$.

8. (i) $\frac{x}{\sqrt{(x^2+1)}}$; (ii) $\frac{1}{\sqrt{(x^2+1)}}$; (iii) $\frac{x+1}{\sqrt{(x^2-1)}}$.

Write down or obtain the integrals with respect to x of the functions in Nos. 9-44:

9. $\frac{\sin x}{\cos^2 x}$. 10. $\frac{\cos x}{\sin^2 x}$. 11. $\frac{\sec^2 x}{\tan x}$.

12. $\tan^2 x$. 13. $\frac{1}{4+x^2}$. 14. $\frac{1}{\sqrt{(x^2-9)}}$.

15. $\frac{1}{4x^2+9}$. 16. $\frac{1}{\sqrt{(4x^2+9)}}$. 17. $\frac{\sin x}{1+\cos x}$.

18. $\frac{\operatorname{sh} x}{1+\operatorname{ch} x}$. 19. $\frac{1}{\sqrt{(1-4x^2)}}$. 20. $\frac{1}{\sqrt{(x^2-4)}}$.

21. $\frac{x}{4-x^2}$. 22. $\frac{e^x}{1+e^x}$. 23. $\frac{e^x}{1+e^{2x}}$.

24. $\frac{1}{1+9x^2}$. 25. $\operatorname{th} 2x$. 26. $\coth(1-x)$.

27. $\operatorname{sh} x \operatorname{sech}^2 x$. 28. $\operatorname{ch} x \operatorname{cosech}^2 x$. 29. $\frac{e^{2x}-1}{e^{2x}+1}$.

30. $\dfrac{1}{\sqrt{(9x^2-1)}}$. 31. $\dfrac{1}{\sqrt{(1-9x^2)}}$. 32. sech $2x$.

33. $\dfrac{1}{(x+2)^2+3}$. 34. $\dfrac{1}{(3-x)^2+25}$. 35. $\dfrac{1}{(2x+1)^2+9}$.

36. $\dfrac{1}{\sqrt{\{9-(x-2)^2\}}}$. 37. $\dfrac{1}{\sqrt{\{5-(2-3x)^2\}}}$. 38. $\dfrac{1}{\sqrt{\{4-(3x+1)^2\}}}$.

39. $\dfrac{1}{\sqrt{\{(x-2)^2-9\}}}$. 40. $\dfrac{1}{\sqrt{\{(x+2)^2+9\}}}$. 41. $\dfrac{1}{\sqrt{\{(3-x)^2+5\}}}$.

42. $\dfrac{1}{x^2-9}$. 43. $\dfrac{1}{25-x^2}$. 44. $\dfrac{1}{x^2+6x+7}$.

Rational Functions. A function of x of the form

$$\frac{a_0+a_1x+a_2x^2+\;.\;.\;.\;.\;.\;+a_mx^m}{b_0+b_1x+\;.\;.\;.\;.\;.\;+b_nx^n}$$

is called a rational function of x, and if the denominator can be resolved into linear or quadratic factors the function can be integrated by expressing it in partial fractions. It is beyond the scope of this volume to give an exhaustive account of the general theory, but the following examples will enable the reader to deal with such functions as occur in ordinary practice. A knowledge of the method of expressing a function in partial fractions will be assumed; see Durell, *Advanced Algebra*, vol. i. pp. 89-93.

When the rational function to be integrated is a fraction whose numerator is not of lower degree than its denominator, it is best to start by reducing the function (to a polynomial plus a proper fraction) by division or an equivalent process.

Example 1. Find $\displaystyle\int \frac{x^3}{x+2}\,dx$.

Since $x^3 \equiv x^3+8-8=(x+2)(x^2-2x+4)-8$,

$$\frac{x^3}{x+2}=x^2-2x+4-\frac{8}{x+2};$$

$$\therefore \int \frac{x^3}{x+2}\,dx=\tfrac{1}{3}x^3-x^2+4x-8\log(x+2).$$

In discussing the theory we shall assume that the degree of the numerator is less than that of the denominator.

If the denominator can be expressed as the product of any number of linear factors (including repeated factors), the function can be integrated by inspection as soon as it has been put into partial fractions.

We start with examples where the coefficients of each linear factor are rational, because in such cases the numerical work in obtaining the partial fractions is comparatively simple.

Example 2. Find $\int \frac{x-8}{x^2-x-2}\,dx$.

$$x^2-x-2 \equiv (x+1)(x-2)\text{ ; assume that}$$

$$\frac{x-8}{x^2-x-2} \equiv \frac{\mathsf{A}}{x+1} + \frac{\mathsf{B}}{x-2},$$

then $$x-8 \equiv \mathsf{A}(x-2) + \mathsf{B}(x+1).$$

Put $x=-1$, $\therefore -9=\mathsf{A}(-3)$, $\therefore \mathsf{A}=3$.

Put $x=2$, $\therefore -6=\mathsf{B}(3)$, $\therefore \mathsf{B}=-2$.

$$\therefore \int \frac{x-8}{x^2-x-2}\,dx = \int \left\{\frac{3}{x+1} - \frac{2}{x-2}\right\} dx.$$

$$= 3\log(x+1) - 2\log(x-2).$$

Example 3. Find $\int \frac{4x}{(x-1)(x+1)^2}\,dx$.

Assume that $$\frac{4x}{(x-1)(x+1)^2} \equiv \frac{\mathsf{A}}{x-1} + \frac{\mathsf{B}}{x+1} + \frac{\mathsf{C}}{(x+1)^2},$$

then $$4x \equiv \mathsf{A}(x+1)^2 + \mathsf{B}(x-1)(x+1) + \mathsf{C}(x-1).$$

Put $x=1$, $\therefore 4=\mathsf{A}\,.\,2^2$, $\therefore \mathsf{A}=1$.

Put $x=-1$, $\therefore -4=\mathsf{C}(-2)$, $\therefore \mathsf{C}=2$.

Equate coefficients of x^2, $\therefore 0=\mathsf{A}+\mathsf{B}$, $\therefore \mathsf{B}=-1$.

$$\therefore \int \frac{4x}{(x-1)(x+1)^2}\,dx = \int \left\{\frac{1}{x-1} - \frac{1}{x+1} + \frac{2}{(x+1)^2}\right\} dx$$

$$= \log(x-1) - \log(x+1) - \frac{2}{x+1}.$$

EXERCISE XIV b

Integrate with respect to x the functions in Nos. 1-21 :

1. $\dfrac{3}{4-5x}$.
2. $\dfrac{x}{1-x}$.
3. $\dfrac{x^2}{x-1}$.
4. $\dfrac{x^3}{x+1}$.
5. $\dfrac{1}{x^2-1}$.
6. $\dfrac{x}{x^2-4}$.
7. $\dfrac{6}{9-x^2}$.
8. $\dfrac{x+2}{x(x-1)}$.
9. $\dfrac{5(x+1)}{(x-1)(x+4)}$.
10. $\dfrac{3x+1}{(x+2)^2}$.
11. $\dfrac{x+10}{x^2-x-12}$.
12. $\dfrac{1}{x^2(x+3)}$.
13. $\dfrac{9x-8}{x(x-2)^2}$.
14. $\dfrac{6x^2}{(x^2-1)(x-2)}$.
15. $\dfrac{3x+1}{x^3+x^2}$.
16. $\dfrac{x^2-1}{(x-3)^3}$.
17. $\dfrac{2}{x(x+1)(x+2)}$.
18. $\dfrac{x^3}{(x-1)(x-2)}$.
19. $\dfrac{9x}{(x-1)^2(x+2)}$.
20. $\dfrac{x^2+2x-1}{x^2(x-1)^2}$.
21. $\dfrac{2x^2-x-1}{x^3(x+1)}$.
22. If $\int f(x)\,dx = \log\{1+f(x)\}$, find $f(x)$.
23. Use the substitution $e^x = y$ to find $\displaystyle\int \frac{dx}{1+3e^x+2e^{2x}}$.
24. If $(x^2-a^2)\dfrac{dy}{dx} = b^2$, express y in terms of x.

Quadratic Factors. If the denominator contains a quadratic factor which can be expressed as the product of two linear factors with irrational coefficients, the method of partial fractions can still be used, but it is shorter to proceed as in Examples 4, 5.

Example 4. Find $\int \frac{dx}{9x^2 - 12x + 2}$.

$$9x^2 - 12x + 2 \equiv 9(x^2 - \tfrac{4}{3}x + \tfrac{2}{9}) \equiv 9\{(x - \tfrac{2}{3})^2 - \tfrac{2}{9}\};$$

$$\therefore \int \frac{dx}{9x^2 - 12x + 2} = \tfrac{1}{9}\int \frac{dx}{(x - \tfrac{2}{3})^2 - (\tfrac{1}{3}\sqrt{2})^2}.$$

[*Side-work:* $\frac{1}{x^2 - a^2} = \frac{1}{2a}\left(\frac{1}{x-a} - \frac{1}{x+a}\right)$;

$$\therefore \int \frac{dx}{x^2 - a^2} = \frac{1}{2a}\log\frac{x-a}{x+a};$$

and $$\int \frac{dx}{(x+k)^2 - a^2} = \frac{1}{2a}\log\frac{x+k-a}{x+k+a}.]$$

$$\therefore \int \frac{dx}{9x^2 - 12x + 2} = \tfrac{1}{9}\,\frac{3}{2\sqrt{2}}\log\left\{\left(x - \tfrac{2}{3} - \frac{\sqrt{2}}{3}\right) \Big/ \left(x - \tfrac{2}{3} + \frac{\sqrt{2}}{3}\right)\right\}$$

$$= \frac{1}{6\sqrt{2}}\log\frac{3x - 2 - \sqrt{2}}{3x - 2 + \sqrt{2}}.$$

Note. The reader may at first find it necessary to write down in full the part of Example 4 labelled "side-work", but he will soon realise that he needs only to remind himself of the first step.

Example 5. Find $\int \frac{x+1}{x^2 + x - 1}\,dx$.

[Since $\frac{d}{dx}(x^2 + x - 1) = 2x + 1$, we express the numerator $x + 1$ in the form $\lambda(2x+1) + \mu$.]

By inspection, $x + 1 \equiv \tfrac{1}{2}(2x+1) + \tfrac{1}{2}$,

$$\therefore \int \frac{x+1}{x^2+x-1}\,dx = \tfrac{1}{2}\int \frac{2x+1}{x^2+x-1}\,dx + \tfrac{1}{2}\int \frac{dx}{x^2+x-1}$$

$$= \tfrac{1}{2}\log(x^2 + x - 1) + \tfrac{1}{2}\int \frac{dx}{x^2+x-1}.$$

$$x^2 + x - 1 \equiv (x + \tfrac{1}{2})^2 - \tfrac{5}{4} \equiv (x + \tfrac{1}{2})^2 - (\tfrac{1}{2}\sqrt{5})^2;$$

therefore, using the same side-work as in Example 4, we have

$$\int \frac{dx}{x^2+x-1} = \int \frac{dx}{(x+\tfrac{1}{2})^2 - (\tfrac{1}{2}\sqrt{5})^2}$$

$$= \frac{1}{\sqrt{5}}\log\{(x + \tfrac{1}{2} - \tfrac{1}{2}\sqrt{5})/(x + \tfrac{1}{2} + \tfrac{1}{2}\sqrt{5})\};$$

$$\therefore \int \frac{x+1}{x^2+x-1}\,dx = \tfrac{1}{2}\log(x^2+x-1) + \frac{1}{2\sqrt{5}}\log\frac{x + \tfrac{1}{2} - \tfrac{1}{2}\sqrt{5}}{x + \tfrac{1}{2} + \tfrac{1}{2}\sqrt{5}}.$$

Since $x^2+x-1 \equiv (x+\frac{1}{2}-\frac{1}{2}\sqrt{5})(x+\frac{1}{2}+\frac{1}{2}\sqrt{5})$, this can be written

$$\tfrac{1}{10}\{(5+\sqrt{5})\log(x+\tfrac{1}{2}-\tfrac{1}{2}\sqrt{5})+(5-\sqrt{5})\log(x+\tfrac{1}{2}+\tfrac{1}{2}\sqrt{5})\},$$

but the first form is more convenient for computation.

If the denominator contains a quadratic factor which cannot be expressed as the product of two linear factors (in real algebra), we proceed as in Examples 6, 7.

Example 6. Find $\displaystyle\int \frac{dx}{9x^2-12x+8}$.

$$9x^2-12x+8 \equiv 9\{x^2-\tfrac{4}{3}x+\tfrac{8}{9}\} \equiv 9\{(x-\tfrac{2}{3})^2+\tfrac{4}{9}\};$$

$$\therefore \int \frac{dx}{9x^2-12x+8} = \frac{1}{9}\int \frac{dx}{(x-\frac{2}{3})^2+(\frac{2}{3})^2}.$$

There are two ways of completing the solution :

(i) $\Bigg[$ *Side-work :* $\displaystyle\int \frac{dx}{x^2+a^2} = \int \frac{a\,d(x/a)}{a^2(x^2/a^2+1)}$

$$= \frac{1}{a}\tan^{-1}\frac{x}{a};$$

and $\displaystyle\int \frac{dx}{(x+k)^2+a^2} = \frac{1}{a}\tan^{-1}\frac{x+k}{a}.\Bigg]$

$$\therefore \int \frac{dx}{9x^2-12x+8} = \tfrac{1}{9}\cdot\tfrac{3}{2}\tan^{-1}\{(x-\tfrac{2}{3})/\tfrac{2}{3}\}$$

$$= \tfrac{1}{6}\tan^{-1}\{\tfrac{1}{2}(3x-2)\}.$$

(ii) Put $x-\frac{2}{3}=\frac{2}{3}\tan\theta$, then $dx=\frac{2}{3}\sec^2\theta\,d\theta$;

$$\therefore \int \frac{dx}{9x^2-12x+8} = \frac{1}{9}\int \frac{\frac{2}{3}\sec^2\theta\,d\theta}{(\frac{2}{3})^2\sec^2\theta}$$

$$= \tfrac{1}{9}\,\tfrac{3}{2}\theta = \tfrac{1}{6}\tan^{-1}\{\tfrac{1}{2}(3x-2)\}.$$

Note. At first sight method (i) appears to be the longer. It is however the shorter method, because with very little practice all the " side-work " will be done mentally.

Example 7. Find $\int \frac{5-4x}{x^2+6x+14}\,dx.$

[Since $\frac{d}{dx}(x^2+6x+14)=2x+6$, we express the numerator $5-4x$ in the form $\lambda(2x+6)+\mu$.]

By inspection, $5-4x=-2(2x+6)+17$,

$$\therefore \int \frac{5-4x}{x^2+6x+14}\,dx=-2\int \frac{2x+6}{x^2+6x+14}\,dx+17\int \frac{dx}{x^2+6x+14}$$

$$=-2\log(x^2+6x+14)+17\int \frac{dx}{(x+3)^2+5}.$$

Using the same side-work as in Example 6, we have

$$\int \frac{dx}{(x+3)^2+(\sqrt{5})^2}=\frac{1}{\sqrt{5}}\tan^{-1}\frac{x+3}{\sqrt{5}}.$$

(*Alternatively*, put $x+3=\sqrt{5}\tan\theta$.)

$$\therefore \int \frac{5-4x}{x^2+6x+14}\,dx=-2\log(x^2+6x+14)+\frac{17}{\sqrt{5}}\tan^{-1}\frac{x+3}{\sqrt{5}}.$$

When the denominator of the general rational function contains both linear and quadratic factors we begin by expressing it in partial fractions. The integration can then be carried out by the methods explained above unless there is a repeated quadratic factor. We give one example dealing with this exceptional case, but the general method involves the use of reduction formulae to which reference is made on p. 301.

Example 8. Find $\int \frac{4x-2}{x(x-1)(x^2+1)}\,dx.$

Assume that $\frac{4x-2}{x(x-1)(x^2+1)}\equiv\frac{A}{x}+\frac{B}{x-1}+\frac{Cx+D}{x^2+1}$,

then $4x-2\equiv A(x-1)(x^2+1)+Bx(x^2+1)+(Cx+D)x(x-1)$.

Put $x=0$, $\therefore -2=-A$; $\therefore A=2$.

Put $x=1$, $\therefore 2=2B$; $\therefore B=1$.

Equate coefficients of x^3, $\therefore 0=A+B+C$; $\therefore C=-3$.

Equate coefficients of x^2, $\therefore 0=-A-C+D$; $\therefore D=-1$.

$$\therefore \int \frac{4x-2}{x(x-1)(x^2+1)}\,dx=\int\left(\frac{2}{x}+\frac{1}{x-1}-\frac{3x+1}{x^2+1}\right)dx$$

$$=2\log x+\log(x-1)-\tfrac{3}{2}\int\frac{2x\,dx}{x^2+1}-\int\frac{dx}{x^2+1}$$

$$=2\log x+\log(x-1)-\tfrac{3}{2}\log(x^2+1)-\tan^{-1}x.$$

Example 9. Find $\int \frac{6x+5}{(x^2+1)^2}\,dx.$

$$\int \frac{6x+5}{(x^2+1)^2}\,dx = \int \frac{6x\,dx}{(x^2+1)^2} + \int \frac{5\,dx}{(x^2+1)^2}$$
$$= -\frac{3}{x^2+1} + 5\int \frac{dx}{(x^2+1)^2}.$$

Putting $x = \tan\theta$, we have

$$\int \frac{dx}{(x^2+1)^2} = \int \frac{\sec^2\theta\,d\theta}{\sec^4\theta} = \int \cos^2\theta\,d\theta$$
$$= \tfrac{1}{2}\int (1+\cos 2\theta)\,d\theta = \tfrac{1}{2}(\theta + \tfrac{1}{2}\sin 2\theta);$$

but $\sin 2\theta = 2\sin\theta\cos\theta = \frac{2\tan\theta}{\tan^2\theta+1} = \frac{2x}{x^2+1},$

$$\therefore \int \frac{dx}{(x^2+1)^2} = \tfrac{1}{2}\left(\tan^{-1}x + \frac{x}{x^2+1}\right);$$

$$\therefore \int \frac{6x+5}{(x^2+1)^2}\,dx = -\frac{3}{x^2+1} + \tfrac{5}{2}\left(\tan^{-1}x + \frac{x}{x^2+1}\right)$$
$$= \frac{5x-6}{2(x^2+1)} + \tfrac{5}{2}\tan^{-1}x.$$

EXERCISE XIV c

Integrate with respect to x the following functions :

1. $\frac{1}{x^2+2}$.
2. $\frac{x}{x^2+2}$.
3. $\frac{1}{x^2-3}$.
4. $\frac{1}{2x^2-1}$.
5. $\frac{1}{2x^2+5}$.
6. $\frac{1}{3x^2-2}$.
7. $\frac{x^2}{x^2+9}$.
8. $\frac{x+3}{x^2+3}$.
9. $\frac{x+2}{2x^2+1}$.
10. $\frac{2x-9}{4x^2-9}$.
11. $\frac{(x+1)^2}{x^2+1}$.
12. $\frac{x^3}{x^2+2}$.
13. $\frac{x^3}{x^4+1}$.
14. $\frac{1}{x^2+2x+5}$.
15. $\frac{1}{x^2+x+1}$.
16. $\frac{1}{(2x-3)^2+25}$.
17. $\frac{1}{(2x-3)^2-25}$.
18. $\frac{1}{3-(x-2)^2}$.

19. $\dfrac{3x+2}{x^2+2x+10}$. 20. $\dfrac{x^2-1}{x^2-2x-2}$. 21. $\dfrac{1}{1+x-x^2}$.

22. $\dfrac{1}{x(x^2+1)}$. 23. $\dfrac{x+3}{(x-1)(x^2+1)}$. 24. $\dfrac{(1+x)^3}{1+x^2}$.

25. $\dfrac{x^2}{x^4-1}$. 26. $\dfrac{1}{(x+1)^2(x^2+1)}$. 27. $\dfrac{2+x}{(4+x^2)(1-x)}$.

28. $\dfrac{1}{(x^2+9)^2}$. 29. $\dfrac{x-1}{(x^2+4)^2}$. 30. $\dfrac{1}{(x^2+2x+5)^2}$.

Functions involving $\sqrt{(ax+b)}$ or $\sqrt{(ax^2+2bx+c)}$.

If an algebraic function contains only a single irrational expression and if that is of the form $\sqrt{(ax+b)}$, its integral can be transformed into that of a rational function by the substitution $\mathbf{z=\sqrt{(ax+b)}}$; see Example 10.

When there is a single irrational expression $\sqrt{(ax^2+2bx+c)}$ the substitution $z=\sqrt{(ax^2+2bx+c)}$ is not usually successful. By using the process of " completing the square " the irrationality can be reduced to one of the types $\sqrt{(k^2-z^2)}$, $\sqrt{(k^2+z^2)}$, $\sqrt{(z^2-k^2)}$ where $z=x+\dfrac{b}{a}$. If the integration cannot then be effected by reference to one of the standard forms a trigonometric or hyperbolic substitution should be tried; see Example 16, p. 295

Example 10. Find $\displaystyle\int\frac{x\,dx}{(x+3)\sqrt{(x-1)}}$.

Put $z=\sqrt{(x-1)}$, then $x-1=z^2$, $\therefore\ dx=2z\,dz$;

$$\therefore \int\frac{x\,dx}{(x+3)\sqrt{(x-1)}}=\int\frac{(1+z^2)\,.\,2z\,dz}{(4+z^2)z}=2\int\frac{z^2+1}{z^2+4}dz$$

$$=2\int\left(1-\frac{3}{z^2+4}\right)dz$$

$$=2z-6\int\frac{1}{z^2+4}dz,$$

$$\left[\textit{Side-work}:\ \int\frac{dz}{z^2+a^2}=\int\frac{a\,d(z/a)}{a^2(z^2/a^2+1)}=\frac{1}{a}\tan^{-1}\frac{z}{a},\right]$$

$$\therefore \int\frac{x\,dx}{(x+3)\sqrt{(x-1)}}=2z-3\tan^{-1}\frac{z}{2}$$

$$=2\sqrt{(x-1)}-3\tan^{-1}\{\tfrac{1}{2}\sqrt{(x-1)}\}.$$

Example 11. Find $\int \frac{dx}{\sqrt{(x^2+4x+9)}}$.

$$\int \frac{dx}{\sqrt{(x^2+4x+9)}} = \int \frac{dx}{\sqrt{\{(x+2)^2+5\}}},$$

[*Side-work:* $\int \frac{dx}{\sqrt{(x^2+a^2)}} = \int \frac{a\,d(x/a)}{a\sqrt{(x^2/a^2+1)}} = \text{sh}^{-1}\frac{x}{a}$.]

$$\therefore \int \frac{dx}{\sqrt{(x^2+4x+9)}} = \text{sh}^{-1}\frac{x+2}{\sqrt{5}}.$$

Example 12. Find $\int \frac{dx}{\sqrt{(10x^2+8x+1)}}$.

$$10x^2+8x+1 \equiv 10(x^2+\tfrac{8}{10}x+\tfrac{1}{10}) \equiv 10\{(x+\tfrac{2}{5})^2-\tfrac{6}{100}\};$$

$$\therefore \int \frac{dx}{\sqrt{(10x^2+8x+1)}} = \frac{1}{\sqrt{10}}\int \frac{dx}{\sqrt{\{(x+\frac{2}{5})^2-\frac{6}{100}\}}},$$

[*Side-work:* $\int \frac{dx}{\sqrt{(x^2-a^2)}} = \int \frac{a\,d(x/a)}{a\sqrt{(x^2/a^2-1)}} = \text{ch}^{-1}\frac{x}{a}$.]

$$\therefore \int \frac{dx}{\sqrt{(10x^2+8x+1)}} = \frac{1}{\sqrt{10}}\,\text{ch}^{-1}\left\{(x+\tfrac{2}{5})\Big/\frac{\sqrt{6}}{10}\right\}$$

$= \frac{1}{\sqrt{10}}\,\text{ch}^{-1}\frac{10x+4}{\sqrt{6}}$. It is assumed in the working that $x+\frac{2}{5} > \frac{1}{10}\sqrt{6}$; if $x+\frac{2}{5} < -\frac{1}{10}\sqrt{6}$, the result is

$-(1/\sqrt{10})\,\text{ch}^{-1}\{(-10x-4)/\sqrt{6}\}$; see p. 277.

Example 13. Find $\int \frac{dx}{\sqrt{(1-x-x^2)}}$.

$$1-x-x^2 \equiv 1-(x+\tfrac{1}{2})^2+\tfrac{1}{4} \equiv \tfrac{5}{4}-(x+\tfrac{1}{2})^2;$$

$$\therefore \int \frac{dx}{\sqrt{(1-x-x^2)}} = \int \frac{dx}{\sqrt{\{\frac{5}{4}-(x+\frac{1}{2})^2\}}},$$

[*Side-work:* $\int \frac{dx}{\sqrt{(a^2-x^2)}} = \int \frac{a\,d(x/a)}{a\sqrt{(1-x^2/a^2)}} = \sin^{-1}\frac{x}{a}$.]

$$\therefore \int \frac{dx}{\sqrt{(1-x-x^2)}} = \sin^{-1}\{(x+\tfrac{1}{2})/(\tfrac{1}{2}\sqrt{5})\}$$

$$= \sin^{-1}\frac{2x+1}{\sqrt{5}}.$$

Any integral of the form $\int \frac{dx}{\sqrt{(ax^2+2bx+c)}}$ belongs to one of the three types illustrated by Examples 11-13. The more general integral $\int \frac{(px+q)\,dx}{\sqrt{(ax^2+2bx+c)}}$ is found, as in Example 7, p. 290, by using the identity $px+q \equiv \frac{p}{2a}(2ax+2b)+q-\frac{pb}{a}$ to split up the integral into two parts.

Example 14. Find $\int \frac{x-5}{\sqrt{(4x-x^2)}}\,dx.$

Since $\frac{d}{dx}(4x-x^2)=4-2x$, we write $x-5 \equiv -\frac{1}{2}(4-2x)-3.$

Also, $$4x-x^2 \equiv 4-(x-2)^2\,;$$

$$\therefore \int \frac{x-5}{\sqrt{(4x-x^2)}}\,dx = -\tfrac{1}{2}\int \frac{4-2x}{\sqrt{(4x-x^2)}}\,dx - 3\int \frac{dx}{\sqrt{\{4-(x-2)^2\}}}\,;$$

therefore, using the same side-work as in Example 13, we have

$$\int \frac{x-5}{\sqrt{(4x-x^2)}}\,dx = -\sqrt{(4x-x^2)} - 3\sin^{-1}\{\tfrac{1}{2}(x-2)\}.$$

Alternatively, as follows:

$$\int \frac{x-5}{\sqrt{(4x-x^2)}}\,dx = \int \frac{x-5}{\sqrt{\{4-(x-2)^2\}}}\,dx.$$

Put $$x-2=2\sin\theta, \text{ then } dx = 2\cos\theta\,d\theta\,;$$

also, $$4-(x-2)^2 = 4(1-\sin^2\theta) = 4\cos^2\theta\,;$$

$$\therefore \int \frac{x-5}{\sqrt{(4x-x^2)}}\,dx = \int \frac{(2\sin\theta-3)\,.\,2\cos\theta\,d\theta}{2\cos\theta}$$

$$= \int (2\sin\theta-3)\,d\theta = -2\cos\theta - 3\theta$$

$$= -\sqrt{(4x-x^2)} - 3\sin^{-1}\{\tfrac{1}{2}(x-2)\}.$$

The integral $\int \frac{dx}{(x-k)\sqrt{(ax^2+2bx+c)}}$ can be reduced to one of the forms already discussed by the substitution $x-k=\frac{1}{z}$.

Example 15. Find $\int \frac{dx}{x\sqrt{(9x^2+4x+1)}}$, $(x>0)$.

Put $x=\frac{1}{z}$, then $\log x = -\log z$, $\therefore \frac{dx}{x} = -\frac{dz}{z}$;

also, $9x^2+4x+1=\frac{1}{z^2}(9+4z+z^2)$;

$$\therefore \int \frac{dx}{x\sqrt{(9x^2+4x+1)}} = -\int \frac{dz}{\sqrt{(z^2+4z+9)}}$$

$$= -\operatorname{sh}^{-1}\frac{z+2}{\sqrt{5}} \text{ (see Example 11)}$$

$$= -\operatorname{sh}^{-1}\frac{1+2x}{x\sqrt{5}}.$$

Example 16. Find $\int \frac{x^2\,dx}{(x^2-1)^{3/2}}$, $(x>1)$,

Put $x=\operatorname{ch}\theta$, $(\operatorname{sh}\theta>0)$, then $dx=\operatorname{sh}\theta\,d\theta$ and $x^2-1=\operatorname{sh}^2\theta$;

$\therefore$ for the positive value of the integrand,

$$\int \frac{x^2\,dx}{(x^2-1)^{3/2}} = \int \frac{\operatorname{ch}^2\theta\operatorname{sh}\theta\,d\theta}{\operatorname{sh}^3\theta} = \int \frac{\operatorname{ch}^2\theta}{\operatorname{sh}^2\theta}\,d\theta$$

$$= \int (1+\operatorname{cosech}^2\theta)\,d\theta = \theta - \coth\theta$$

$$= \operatorname{ch}^{-1}x - x/\sqrt{(x^2-1)}.$$

EXERCISE XIV d

Integrate with respect to x the functions in Nos. 1-33:

1. $\frac{1+x}{\sqrt{x}}$.
2. $\frac{x}{\sqrt{(1+x)}}$.
3. $\frac{1}{1-\sqrt{x}}$.
4. $\frac{2x-1}{\sqrt{(1-x)}}$.
5. $\frac{1}{(x+2)\sqrt{(1-x)}}$.
6. $\frac{x^2}{\sqrt{(x-2)}}$.
7. $\frac{1}{\sqrt{(x^2+2x+10)}}$.
8. $\frac{1}{\sqrt{(5+4x-x^2)}}$.
9. $\frac{1}{\sqrt{(x^2+6x+5)}}$.
10. $\frac{1}{\sqrt{(x^2+x)}}$.
11. $\frac{2x+1}{\sqrt{(1-x^2)}}$.
12. $\frac{x+1}{\sqrt{(x^2+4)}}$.
13. $\frac{2x-1}{\sqrt{(x^2+3x+2)}}$.
14. $\frac{1-x}{\sqrt{(1-x+x^2)}}$.
15. $\frac{1+x}{\sqrt{(1-x-x^2)}}$.

16. $\sqrt{\dfrac{1-x}{1+x}}$. 17. $\sqrt{\dfrac{x}{2+x}}$. 18. $\dfrac{x^3+1}{\sqrt{(x^2+4)}}$.

19. $\dfrac{1}{x\sqrt{(x^2+9)}}$. 20. $\dfrac{1}{x\sqrt{(9-x^2)}}$. 21. $\dfrac{1}{(x-2)\sqrt{(x^2-1)}}$.

22. $\dfrac{\sqrt{(1+x^2)}}{x}$. 23. $\dfrac{x}{\sqrt{(1+x^4)}}$. 24. $\dfrac{1}{(x^2+1)^{3/2}}$.

25. $\dfrac{x^2}{(x^2+1)^{3/2}}$. 26. $\dfrac{x-1}{x\sqrt{(x^2-1)}}$. 27. $\dfrac{1}{x^2\sqrt{(1+x^2)}}$.

28. $\sqrt{(x^2+1)}$. 29. $\sqrt{(4-x^2)}$. 30. $\sqrt{(x^2-9)}$.

31. $\dfrac{x^2}{\sqrt{(1+x^2)}}$. 32. $\dfrac{x^2+1}{x^2\sqrt{(x^2-1)}}$. 33. $\dfrac{x^2}{\sqrt{(x^2+2x+2)}}$.

34. Use the substitution $x=2\cos^2\theta+3\sin^2\theta$ to evaluate $\displaystyle\int\frac{dx}{\sqrt{\{(x-2)(3-x)\}}}$.

35. Use the substitution $x=3\,\text{ch}^2\,\theta-2\,\text{sh}^2\,\theta$ to evaluate $\int\sqrt{\{(x-2)(x-3)\}}\,dx$, where $x>3$.

Trigonometrical Functions. A substitution which is of great theoretical importance for the integration of trigonometrical functions is

$$\tan\tfrac{1}{2}\mathbf{x}=\mathbf{t}.$$

This gives $\frac{1}{2}\sec^2\frac{1}{2}x\,dx=dt$, $\quad\therefore\ dx=\dfrac{2dt}{1+t^2}$.

Also, $\quad\sin x=\dfrac{2t}{1+t^2},\quad\cos x=\dfrac{1-t^2}{1+t^2}$.

Hence this substitution will transform the integral with respect to x of any rational function of $\sin x$, $\cos x$ into the integral with respect to t of a rational function of t, which, we have seen, can be evaluated by the method of partial fractions.

It must, however, be remembered that this substitution is often not the simplest one to employ in practice, but it can be used with advantage for $\int\text{cosec}\,x\,dx$ and for the more general integral $\displaystyle\int\frac{dx}{a+b\cos x+c\sin x}$.

One case in which the substitution $\tan \frac{1}{2}x = t$ is unnecessary is when the integrand can be expressed as a homogeneous function of even degree in $\sin x$ and $\cos x$; the substitution $\tan x = z$ is then better, as in Example 17.

Example 17. Find $\displaystyle\int \frac{dx}{4 - 5\sin^2 x}$.

$$4 - 5\sin^2 x \equiv 4(\cos^2 x + \sin^2 x) - 5\sin^2 x \equiv 4\cos^2 x - \sin^2 x.$$

Put $\tan x = z$, then $\sec^2 x\, dx = dz$,

$$\therefore \int \frac{dx}{4 - 5\sin^2 x} = \int \frac{dx}{4\cos^2 x - \sin^2 x} = \int \frac{\sec^2 x\, dx}{4 - \tan^2 x}.$$

$$= \int \frac{dz}{4 - z^2} = \int \tfrac{1}{4}\left(\frac{1}{2 + z} + \frac{1}{2 - z}\right) dz$$

$$= \tfrac{1}{4} \log \frac{2 + \tan x}{2 - \tan x}.$$

$\int$ cosec x dx and $\int$ sec x dx.

Put $\tan \frac{1}{2}x = t$, then from p. 296

$$\int \frac{dx}{\sin x} = \int \frac{1 + t^2}{2t} \cdot \frac{2dt}{1 + t^2} = \int \frac{1}{t}\, dt = \log t;$$

$$\therefore \int \operatorname{cosec} x\, dx = \log(\tan \tfrac{1}{2}x).$$

Hence, since $\sec x = \operatorname{cosec}(\frac{1}{2}\pi + x)$,

$$\int \sec x\, dx = \log\{\tan(\tfrac{1}{4}\pi + \tfrac{1}{2}x)\},$$

which may be written $\log \dfrac{1 + \sin x}{\cos x}$, that is, $\log(\sec x + \tan x)$.

It should also be noted that $\int \operatorname{cosech} x\, dx$ may be evaluated by a similar method:

$$\int \frac{dx}{\operatorname{sh} x} = \int \frac{dx}{2 \operatorname{sh} \frac{1}{2}x \operatorname{ch} \frac{1}{2}x} = \int \frac{\operatorname{sech}^2 \frac{1}{2}x\, dx}{2 \operatorname{th} \frac{1}{2}x}$$

$$= \int \frac{d(\operatorname{th} \frac{1}{2}x)}{\operatorname{th} \frac{1}{2}x} = \log(\operatorname{th} \tfrac{1}{2}x).$$

$\int \operatorname{sech} x\, dx$ was evaluated on p. 270.

Example 18. Find (i) $\int \frac{dx}{5+4\cos x}$; (ii) $\int \frac{dx}{4+5\cos x}$.

Put $\tan \frac{1}{2}x = t$, then $dx = \frac{2dt}{1+t^2}$ and $\cos x = \frac{1-t^2}{1+t^2}$.

$$\text{(i)} \quad \int \frac{dx}{5+4\cos x} = \int \frac{2dt}{5(1+t^2)+4(1-t^2)} = 2\int \frac{dt}{9+t^2},$$

$$\left[\textit{Side-work}: \int \frac{dx}{a^2+x^2} = \int \frac{ad(x/a)}{a^2(1+x^2/a^2)} = \frac{1}{a}\tan^{-1}\frac{x}{a}.\right]$$

$$\therefore \int \frac{dx}{5+4\cos x} = \tfrac{2}{3}\tan^{-1}\frac{t}{3} = \tfrac{2}{3}\tan^{-1}\left(\tfrac{1}{3}\tan \tfrac{1}{2}x\right).$$

$$\text{(ii)} \quad \int \frac{dx}{4+5\cos x} = \int \frac{2dt}{4(1+t^2)+5(1-t^2)} = 2\int \frac{dt}{9-t^2}$$

$$= 2\int \tfrac{1}{6}\left(\frac{1}{3-t} + \frac{1}{3+t}\right) dt$$

$$= \tfrac{1}{3}\log \frac{3+\tan\frac{1}{2}x}{3-\tan\frac{1}{2}x}.$$

The integral $\int \frac{\alpha\cos x + \beta\sin x + \gamma}{a\cos x + b\sin x + c}\,dx$ can be simplified by writing the numerator in the form

$$\lambda(-a\sin x + b\cos x) + \mu(a\cos x + b\sin x + c) + \nu,$$

where the coefficient of λ is the derivative of the denominator; compare pp. 288, 290, 294 and see Example 19.

Example 19. Find $\int \frac{11\cos x - 16\sin x}{2\cos x + 5\sin x}\,dx$.

Suppose that

$$11\cos x - 16\sin x \equiv \lambda(-2\sin x + 5\cos x) + \mu(2\cos x + 5\sin x);$$

then $5\lambda + 2\mu = 11$ and $-2\lambda + 5\mu = -16$; hence $\lambda = 3$, $\mu = -2$.

$$\therefore \int \frac{11\cos x - 16\sin x}{2\cos x + 5\sin x}\,dx = 3\int \frac{-2\sin x + 5\cos x}{2\cos x + 5\sin x}\,dx - \int 2\,dx$$

$$= 3\log(2\cos x + 5\sin x) - 2x.$$

EXERCISE XIV e

Integrate with respect to x the functions in Nos. 1-21 :

1. $\sec 2x$.
2. $\operatorname{cosec} \frac{1}{2}x$.
3. $\operatorname{cosec} x \sec x$.
4. $\cos x \cot x$.
5. $\cos 2x \sec x$.
6. $\sin 3x \operatorname{cosec}^2 x$.
7. $\dfrac{1}{1+\cos x}$.
8. $\dfrac{\cos x}{(1+\sin x)^2}$.
9. $\dfrac{1}{1+\sin x}$.
10. $\dfrac{1}{5+3\cos x}$.
11. $\dfrac{1}{3+5\cos x}$.
12. $\dfrac{1}{5-3\cos x}$.
13. $\dfrac{\tan x}{1+\tan x}$.
14. $\dfrac{\sin^2 x}{1+\cos^2 x}$.
15. $\dfrac{1}{\sin x+\cos x}$.
16. $\dfrac{1}{1+\sin^2 x}$.
17. $\dfrac{5}{1+2\cot x}$.
18. $\dfrac{1}{13+5\sin x}$.
19. $\dfrac{1}{1-\cos x+\sin x}$.
20. $\dfrac{\sin x+2\cos x}{\cos x+2\sin x}$.
21. $\dfrac{1}{5-13\sin x}$.

22. Express $\displaystyle\int\frac{dx}{\sqrt{(1+x^2)}}$, $(x>0)$, in the form of a logarithm by the substitution $x=\cot\theta$.

23. Express $\displaystyle\int\frac{dx}{\sqrt{(x^2-4)}}$, $(x>2)$, in the form of a logarithm by the substitution $x=2\operatorname{cosec}\theta$.

24. Find $\int\sqrt{(\sec x-1)}\,dx$ by the substitution $\cos\frac{1}{2}x=z$.

25. Find $\displaystyle\int\frac{(1+x)\,dx}{x^2\sqrt{(1-x^2)}}$ by the substitution $x=\sin\theta$.

$\int \sin^m x\,dx$, $\int \cos^n x\,dx$, $\int \sin^m x \cos^n x\,dx$.

If m and n are positive integers, these integrals can be found by expressing the integrands as a sum of sines or cosines of multiples of x. This can be done by elementary trigonometry if m and n are small. For the general method see Durell and Robson, *Advanced Trigonometry*, p. 169.

If n or m is odd it is easier to use a substitution $\sin x=s$ or $\cos x=c$, as in Examples 21, 22 respectively.

Example 20. Find $\int \sin^2 x \cos^2 x \, dx$.

$$\sin^2 x \cos^2 x = \tfrac{1}{4} \sin^2 2x = \tfrac{1}{8}(1 - \cos 4x);$$

$$\therefore \int \sin^2 x \cos^2 x \, dx = \tfrac{1}{8}\int(1 - \cos 4x)\, dx = \tfrac{1}{8}(x - \tfrac{1}{4}\sin 4x).$$

Example 21. Find $\int \sin^2 x \cos^5 x \, dx$.

Put $\sin x = s$, then $\cos x \, dx = ds$,

$$\begin{aligned}\therefore \int \sin^2 x \cos^5 x \, dx &= \int s^2(1 - s^2)^2 \, ds \\ &= \int (s^2 - 2s^4 + s^6)\, ds \\ &= \tfrac{1}{3}\sin^3 x - \tfrac{2}{5}\sin^5 x + \tfrac{1}{7}\sin^7 x.\end{aligned}$$

Example 22. Find $\int \sin^7 x \sec^8 x \, dx$.

Put $\cos x = c$, then $-\sin x \, dx = dc$,

$$\begin{aligned}\therefore \int \sin^7 x \sec^8 x \, dx &= -\int (1 - c^2)^3 \left(\frac{1}{c}\right)^8 dc \\ &= -\int \left(\frac{1}{c^8} - \frac{3}{c^6} + \frac{3}{c^4} - \frac{1}{c^2}\right) dc \\ &= \frac{1}{7c^7} - \frac{3}{5c^5} + \frac{1}{c^3} - \frac{1}{c} \\ &= \tfrac{1}{7}\sec^7 x - \tfrac{3}{5}\sec^5 x + \sec^3 x - \sec x.\end{aligned}$$

EXERCISE XIV f

Integrate with respect to x the functions in Nos. 1-12:

1. $\sin^2 2x$.
2. $\sin^3 x$.
3. $\cos^4 x$.
4. $\sin^6 x \cos^3 x$.
5. $\sin^2 x \cos^4 x$.
6. $\cos^2 x \sin 3x$.
7. $\sin x \sin^2 2x$.
8. $\sin^5 x \sec^6 x$.
9. $\cos^3 x \operatorname{cosec}^4 x$.
10. $\tan^3 x$.
11. $\sin x \cos^2 x \cos 2x$.
12. $\sin^2 x \sec 2x$.

Obtain the following results:

13. $\displaystyle\int \frac{dx}{\sin x \cos^2 x} = \sec x + \log \tan \tfrac{1}{2}x.$

14. $\displaystyle\int \frac{dx}{\sin^3 x} = -\frac{1}{2}\frac{\cos x}{\sin^2 x} + \frac{1}{2}\log \tan \tfrac{1}{2}x.$

15. $\int \tan^4 x \, dx = \tfrac{1}{3}\tan^3 x - \tan x + x.$

16. $\int \cot^3 x \, dx = -\tfrac{1}{2}\cot^2 x - \log \sin x.$

Reduction Formulae. When the integrand involves an integer n (which is not a function of the variable) it may be possible to express the integral in terms of a similar integral with $n-1$ (or $n-2$, $n-3$, etc.) instead of n. Such a relation is called a *reduction formula.*

Example 23. If $u_n = \int x^n e^x \, dx$, find a relation between u_n and u_{n-1}. Hence find $\int x^3 e^x \, dx$.

$$u_n = \int x^n \, d(e^x) = x^n e^x - \int e^x \, d(x^n)$$
$$= x^n e^x - n\int e^x x^{n-1} \, dx \, ;$$
$$\therefore \; u_n = x^n e^x - n u_{n-1}.$$

Hence
$$u_3 = x^3 e^x - 3u_2, \quad u_2 = x^2 e^x - 2u_1,$$
and
$$u_1 = xe^x - u_0 = xe^x - e^x \, ;$$
$$\therefore \; u_3 = x^3 e^x - 3x^2 e^x + 6xe^x - 6e^x \, ;$$
$$\therefore \; \int x^3 e^x \, dx = e^x(x^3 - 3x^2 + 6x - 6).$$

Example 24. Find a reduction formula for $\displaystyle\int \frac{dx}{(x^2+1)^n}$.

$$u_n \equiv \int (x^2+1)^{-n} \, dx = x(x^2+1)^{-n} - \int x \, d(x^2+1)^{-n}$$
$$= x(x^2+1)^{-n} + 2n\int x^2(x^2+1)^{-n-1} \, dx$$
$$= x(x^2+1)^{-n} + 2n\int \{(x^2+1) - 1\}(x^2+1)^{-n-1} \, dx$$
$$= x(x^2+1)^{-n} + 2n\int \{(x^2+1)^{-n} - (x^2+1)^{-n-1}\} \, dx \, ;$$
$$\therefore \; u_n = x(x^2+1)^{-n} + 2nu_n - 2nu_{n+1} \, ;$$
$$\therefore \; 2nu_{n+1} = x(x^2+1)^{-n} + (2n-1)u_n.$$

By writing $n-1$ for n a relation between u_n, u_{n-1} can be obtained. By writing in succession $n-1$, $n-2$, $n-3$, . . ., 2, 1 for n the value of u_n can be found in terms of u_1, and $u_1 = \tan^{-1} x$.

$\int \sin^n \theta\, d\theta$ and $\int \cos^n \theta\, d\theta$.

For brevity we put $\sin\theta = s$ and $\cos\theta = c$.

Then if $u_n = \int \sin^n \theta\, d\theta$, we have

$$u_n = \int s^{n-1}\, d(-c) = -cs^{n-1} + \int c\, d(s^{n-1})$$

$$= -cs^{n-1} + (n-1)\int c^2 s^{n-2}\, d\theta$$

$$= -cs^{n-1} + (n-1)\int (1-s^2) s^{n-2}\, d\theta;$$

$$\therefore u_n = -cs^{n-1} + (n-1)u_{n-2} - (n-1)u_n,$$

$$\therefore nu_n = -cs^{n-1} + (n-1)u_{n-2}.$$

By successive applications of this result we can express u_n in terms of u_1 if n is odd and in terms of u_0 if n is even.

A reduction formula for $\int \cos^n \theta\, d\theta$ can be found in the same way. Alternatively, it may be deduced by changing θ into $\frac{1}{2}\pi + \phi$ in the previous result:

$$\text{if } v_n = \int \cos^n \theta\, d\theta,\ nv_n = sc^{n-1} + (n-1)v_{n-2}.$$

$\int \sin^m \theta \cos^n \theta\, d\theta$. A similar method could be applied to this integral. *Alternatively*, instead of integrating by parts, it is sometimes convenient to start by differentiating a suitable function:

$$\frac{d}{d\theta}(s^p c^q) = ps^{p-1}c^{q+1} - qs^{p+1}c^{q-1}$$

$$= ps^{p-1}c^{q-1}(1-s^2) - qs^{p+1}c^{q-1} = ps^{p-1}c^{q-1} - (p+q)s^{p+1}c^{q-1}.$$

Integrating and denoting $\int s^m c^n\, d\theta$ by $u_{m,n}$

$$s^p c^q = pu_{p-1,q-1} - (p+q)u_{p+1,q-1}.$$

Writing m for $p+1$ and n for $q-1$, we have

$$(m+n)u_{m,n} = -s^{m-1}c^{n+1} + (m-1)u_{m-2,n}.$$

In a similar way, or by writing $\frac{1}{2}\pi - \theta$ for θ and interchanging m and n, $u_{m,n}$ can be connected with $u_{m,n-2}$; see Exercise XIV g, No. 8.

By successive applications of these formulae, $u_{m,n}$ can be reduced to $u_{1,n}$ if m is odd, and to $u_{m,1}$ if n is odd, and to $u_{m,0}$ if m and n are even.

$$u_{1,n} = \int \sin\theta \cos^n\theta \, d\theta = -\frac{1}{n+1}\cos^{n+1}\theta\,;$$

$$u_{m,1} = \int \sin^m\theta \cos\theta \, d\theta = \frac{1}{m+1}\sin^{m+1}\theta.$$

$u_{m,0}$ is $\int \sin^m\theta \, d\theta$ and is found by the reduction formula on p. 302.

EXERCISE XIV g

Use reduction formulae to find the integrals in Nos. 1-5:

1. $\int x^4 e^{2x}\, dx$.
2. $\int \sin^6\theta \, d\theta$.
3. $\int \cos^7\theta \, d\theta$.
4. $\int \cos^5\theta \sin^6\theta \, d\theta$.
5. $\int \cos^4\theta \sin^6\theta \, d\theta$.

6. If $u_n = \int (\log x)^n \, dx$, prove that $u_n + nu_{n-1} = x(\log x)^n$. Hence find $\int (\log x)^3 \, dx$.

7. If $u_n = \int \tan^n\theta \, d\theta$, prove that $u_n + u_{n-2} = \dfrac{\tan^{n-1}\theta}{n-1}$. Hence find $\int \tan^6\theta \, d\theta$ and $\int \tan^5\theta \, d\theta$.

8. If $u_{m,n} = \int \sin^m\theta \cos^n\theta \, d\theta$, prove that
$$(m+n)u_{m,n} = \sin^{m+1}\theta \cos^{n-1}\theta + (n-1)u_{m,n-2}.$$

9. Obtain a reduction formula for $\int \operatorname{sh}^n x \, dx$.

10. Find $\int \sec^6\theta \, d\theta$ and $\int \sec^5\theta \, d\theta$.

11. If $u_n = \int (x^2+a^2)^n \, dx$, prove that
$$(2n+1)u_n = x(x^2+a^2)^n + 2na^2 u_{n-1}.$$

12. If $u_n = \int x^n \operatorname{ch} x \, dx$, prove that
$$u_n = x^n \operatorname{sh} x - nx^{n-1}\operatorname{ch} x + n(n-1)u_{n-2}.$$
Hence find $\int x^6 \operatorname{ch} x \, dx$.

13. Find $\displaystyle\int \frac{dx}{(x^2+2)^4}$.

14. Find $\displaystyle\int \frac{6x-1}{(x^2-1)^3}\, dx$.

15. Find a reduction formula for $\displaystyle\int \frac{dx}{(x^2+2x+2)^n}$ and evaluate $\displaystyle\int \frac{dx}{(x^2+2x+2)^3}$.

CHAPTER XV

DEFINITE INTEGRALS

THE ordinary method of evaluating the definite integral $\int_a^b f(x)\,dx$ is to find a function $\mathsf{F}(x)$ such that $\mathsf{F}'(x)=f(x)$ for all values of x from a to b, and then use the relation

$$\int_a^b f(x)\,dx=\mathsf{F}(b)-\mathsf{F}(a).$$

It may happen, however, that although $\mathsf{F}(x)$ is complicated, as in Example 2 (ii), or even unknown, as in Example 9, p. 321, the definite integral $\mathsf{F}(b)-\mathsf{F}(a)$ can be found for special values of the limits a, b by some other method not involving an explicit statement, or even knowledge, of the general value of $\mathsf{F}(x)$. This is of practical importance because in most applications of integration we are concerned with definite rather than indefinite integrals.

$\int_0^{\pi/2}$ **$\sin^m\theta\cos^n\theta\,d\theta$, where m and n are positive integers or zero.**

Put $\theta=\frac{1}{2}\pi-\phi$, then

$$\int_0^{\pi/2}\sin^m\theta\cos^n\theta\,d\theta=\int_{\pi/2}^0\cos^m\phi\sin^n\phi(-1)\,d\phi$$

$$=\int_0^{\pi/2}\sin^n\phi\cos^m\phi\,d\phi\,;$$

$$\therefore \int_0^{\pi/2}\sin^m\theta\cos^n\theta\,d\theta=\int_0^{\pi/2}\sin^n\theta\cos^m\theta\,d\theta\,;$$

hence the indices in the integral are interchangeable, and in particular, putting $n=0$, we have

$$\int_0^{\pi/2}\sin^m\theta\,d\theta=\int_0^{\pi/2}\cos^m\theta\,d\theta.$$

By the reduction formula on p. 302, if $u_{m,n}=\int_0^{\pi/2}\sin^m\theta\cos^n\theta\,d\theta$,

$$(m+n)u_{m,n}=-\Big[\sin^{m-1}\theta\cos^{n+1}\theta\Big]_0^{\pi/2}+(m-1)u_{m-2,n};$$

$$\therefore \text{ for } m>1, \quad (m+n)u_{m,n}=(m-1)u_{m-2,n};$$

$$\therefore \int_0^{\pi/2}\sin^m\theta\cos^n\theta\,d\theta=\frac{m-1}{m+n}\int_0^{\pi/2}\sin^{m-2}\theta\cos^n\theta\,d\theta.$$

(In the same way, or by interchanging the indices, it may be shown that,

$$\text{for } n>1, \quad \int_0^{\pi/2}\sin^m\theta\cos^n\theta\,d\theta=\frac{n-1}{m+n}\int_0^{\pi/2}\sin^m\theta\cos^{n-2}\theta\,d\theta.)$$

By the special case of the general formula when $n=0$, we have

$$\int_0^{\pi/2}\sin^m\theta\,d\theta=\frac{m-1}{m}\int_0^{\pi/2}\sin^{m-2}\theta\,d\theta;$$

$$\therefore \text{ by p. 304, } \int_0^{\pi/2}\cos^m\theta\,d\theta=\frac{m-1}{m}\int_0^{\pi/2}\cos^{m-2}\theta\,d\theta.$$

These special results may be obtained from the reduction formulae on p. 302 for $\int\sin^m\theta\,d\theta$ and $\int\cos^m\theta\,d\theta$ or directly by integration by parts.

By repeated application of these relations we can evaluate $\int_0^{\pi/2}\sin^m\theta\cos^n\theta\,d\theta$ where m and n are any positive integers (or zero). If this method is used, it is sufficient for the reader to remember the single relation $u_{m,n}=\dfrac{m-1}{m+n}u_{m-2,n}$ as it comprises the others, or else to know how to obtain it. Integrals of this type may alternatively be found as in Example 2.

Example 1. Find (i) $\int_0^{\pi/2}\cos^7\theta\,d\theta$; (ii) $\int_0^{\pi/2}\cos^6\theta\,d\theta$.

If $u_n=\int_0^{\pi/2}\cos^n\theta\,d\theta$,

(i) $u_7=\frac{6}{7}u_5=\frac{6}{7}\,\frac{4}{5}u_3=\frac{6}{7}\,\frac{4}{5}\,\frac{2}{3}u_1$;

and $u_1=\int_0^{\pi/2}\cos\theta\,d\theta=1$; $\therefore u_7=\dfrac{6\,.\,4\,.\,2}{7\,.\,5\,.\,3}=\frac{16}{35}$.

(ii) $u_6=\frac{5}{6}u_4=\frac{5}{6}\,\frac{3}{4}u_2=\frac{5}{6}\,\frac{3}{4}\,\frac{1}{2}u_0$;

and $u_0=\int_0^{\pi/2}1\,d\theta=\frac{1}{2}\pi$; $\therefore u_6=\dfrac{5\,.\,3\,.\,1}{6\,.\,4\,.\,2}\cdot\frac{1}{2}\pi=\frac{5}{32}\pi$.

Example 2. Find (i) $\int_0^{\pi/2} \sin^4\theta\cos^5\theta\, d\theta$;

(ii) $\int_0^{\pi/2} \sin^6\theta\cos^4\theta\, d\theta$.

Denote $\int_0^{\pi/2} \sin^m\theta\cos^n\theta\, d\theta$ by $u_{m,n}$.

(i) If there is an *odd* index, select it for reduction.

$$u_{4,5} = \tfrac{4}{9}u_{4,3} = \tfrac{4}{9}\,\tfrac{2}{7}u_{4,1};$$

and $u_{4,1} = \int_0^{\pi/2} \sin^4\theta\cos\theta\, d\theta = \left[\tfrac{1}{5}\sin^5\theta\right]_0^{\pi/2} = \tfrac{1}{5}$;

$$\therefore u_{4,5} = \tfrac{4}{9}\,\tfrac{2}{7}\,\tfrac{1}{5} = \tfrac{8}{315}.$$

(ii) $u_{6,4} = \tfrac{5}{10}u_{4,4} = \tfrac{5}{10}\,\tfrac{3}{8}u_{2,4} = \tfrac{5}{10}\,\tfrac{3}{8}\,\tfrac{1}{6}u_{0,4}$;

and $u_{0,4} = \int_0^{\pi/2} \cos^4\theta\, d\theta = \tfrac{3}{4}\,\tfrac{1}{2}\int_0^{\pi/2} 1\, d\theta = \tfrac{3}{4}\,\tfrac{1}{2}\,\tfrac{1}{2}\pi$;

$$\therefore u_{6,4} = \tfrac{5}{10}\,\tfrac{3}{8}\,\tfrac{1}{6}\,\tfrac{3}{4}\,\tfrac{1}{2}\,\tfrac{1}{2}\pi = \tfrac{3}{512}\pi.$$

Note. The student will find it easy to abbreviate the working as soon as he has become accustomed to the process.

Alternatively, the integrals may be evaluated as follows:

(i) Put $\sin\theta = s$, $\therefore \cos\theta\, d\theta = ds$ and $\cos^4\theta = (1-s^2)^2$;

$$\therefore \int_0^{\pi/2} \sin^4\theta\cos^5\theta\, d\theta = \int_0^1 s^4(1-s^2)^2\, ds$$

$$= \int_0^1 (s^4 - 2s^6 + s^8)\, ds$$

$$= \tfrac{1}{5} - \tfrac{2}{7} + \tfrac{1}{9} = \tfrac{8}{315}.$$

(ii) $\sin^6\theta\cos^4\theta = \tfrac{1}{2}(1-\cos 2\theta)(\tfrac{1}{2}\sin 2\theta)^4$

$$= \tfrac{1}{32}(\sin^4 2\theta - \sin^4 2\theta\cos 2\theta);$$

and $\sin^4 2\theta = \tfrac{1}{4}(1-\cos 4\theta)^2 = \tfrac{1}{4} - \tfrac{1}{2}\cos 4\theta + \tfrac{1}{8}(1+\cos 8\theta)$;

$$\therefore \int_0^{\pi/2} \sin^6\theta\cos^4\theta\, d\theta$$

$$= \tfrac{1}{32}\int_0^{\pi/2} (\tfrac{3}{8} - \tfrac{1}{2}\cos 4\theta + \tfrac{1}{8}\cos 8\theta - \sin^4 2\theta\cos 2\theta)\, d\theta$$

$$= \tfrac{1}{32}\left[\tfrac{3}{8}\theta - \tfrac{1}{8}\sin 4\theta + \tfrac{1}{64}\sin 8\theta - \tfrac{1}{10}\sin^5 2\theta\right]_0^{\pi/2}$$

$$= \tfrac{1}{32}\,\tfrac{3}{8}\,\tfrac{1}{2}\pi = \tfrac{3}{512}\pi.$$

Example 3. Find $\int_0^1 x^4\sqrt{(1-x^2)^3}\,dx$.

Put $x = \sin\theta$, $\therefore dx = \cos\theta\,d\theta$ and $1 - x^2 = \cos^2\theta$;

$$\therefore \int_0^1 x^4\sqrt{(1-x^2)^3}\,dx = \int_0^{\pi/2} \sin^4\theta\cos^3\theta\,.\cos\theta\,d\theta$$

$$= \int_0^{\pi/2} \sin^4\theta\cos^4\theta\,d\theta$$

$$= \tfrac{3}{8}\,\tfrac{1}{6}\,\tfrac{3}{4}\,\tfrac{1}{2}\,\tfrac{1}{2}\pi = \tfrac{3}{256}\pi.$$

Example 4. Find $\int_0^{\pi} \cos^n\theta\,d\theta$ where n is any positive integer.

$$\int_0^{\pi} \cos^n\theta\,d\theta = \int_0^{\pi/2} \cos^n\theta\,d\theta + \int_{\pi/2}^{\pi} \cos^n\theta\,d\theta.$$

In the second integral put $\theta = \pi - \phi$, $\therefore d\theta = -d\phi$, $\cos\theta = -\cos\phi$

$$\therefore \int_{\pi/2}^{\pi} \cos^n\theta\,d\theta = \int_{\pi/2}^{0} (-1)^n\cos^n\phi(-1)\,d\phi = (-1)^n\int_0^{\pi/2} \cos^n\phi\,d\phi.$$

Hence if n is odd, $\int_0^{\pi} \cos^n\theta\,d\theta = 0$;

and if n is even, $\int_0^{\pi} \cos^n\theta\,d\theta = 2\int_0^{\pi/2} \cos^n\theta\,d\theta$

$$= 2\frac{n-1}{n}\,\frac{n-3}{n-2}\,\ldots\ldots\,\tfrac{3}{4}\,\tfrac{1}{2}\,\tfrac{1}{2}\pi = \frac{(n-1)(n-3)\ldots\ldots 3.1}{n(n-2)\ldots\ldots 4.2}\pi.$$

This is sometimes denoted by $\dfrac{(n-1)!!}{n!!}\pi$.

EXERCISE XV a

Find the values of the integrals in Nos. 1-16:

1. $\int_0^{\pi/2} \sin^5\theta\,d\theta$.
2. $\int_0^{\pi/2} \cos^4\theta\,d\theta$.
3. $\int_0^{\pi/2} \cos^9\theta\,d\theta$.
4. $\int_0^{\pi/2} \sin^8\theta\,d\theta$.
5. $\int_0^{\pi/4} \sin^7 2\theta\,d\theta$.
6. $\int_0^{\pi} \cos^6 \tfrac{1}{2}\theta\,d\theta$.
7. $\int_0^{\pi/2} \sin^3\theta\cos^4\theta\,d\theta$.
8. $\int_0^{\pi/2} \sin^2\theta\cos^6\theta\,d\theta$.

9. $\int_0^{\pi/2} \sin^4\theta \cos^8\theta \, d\theta.$ 10. $\int_0^{\pi/2} \sin^4\theta\cos^7\theta\,d\theta.$

11. $\int_0^{\pi} (1-\cos\theta)^3\,d\theta.$ 12. $\int_0^{\pi} \sin^2\theta\,(1+\cos\theta)^4\,d\theta.$

13. $\int_0^1 x^6\sqrt{(1-x^2)}\,dx.$ 14. $\int_0^1 \sqrt{(x-x^2)^3}\,dx.$

15. $\int_0^2 x^3\sqrt{(2-x)}\,dx.$ 16. $\int_0^{1/2} x^3\sqrt{(1-4x^2)}\,dx.$

17. Find the value of $\int_0^1 x(1-x)^{3/2}\,dx$ by the substitutions (i) $x=\sin^2\theta$, (ii) $1-x=y^2$.

18. If $v_n=\int_0^{\pi/4}\tan^n\theta\,d\theta$, find the relation between v_n and v_{n-2}, $(n \geqslant 2)$. Hence find the value of v_6.

Find the values of the integrals in Nos. 19-25 :

19. $\int_0^{\pi/2} x^2\sin x\,dx.$ 20. $\int_0^1 x\tan^{-1}x\,dx.$

21. $\int_0^{\pi/2} x(1+\cos x)^{-1}\,dx.$ 22. $\int_0^{\pi} x\sin x\cos^4 x\,dx.$

23. $\int_{-1}^{+1}(1-x^2)\cos\tfrac{1}{2}\pi x\,dx.$ 24. $\int_0^{\pi/4} x\tan^2 x\,dx.$

25. $\int_1^2 \sqrt{\{(x-1)(2-x)\}}\,dx$; put $x=\cos^2\theta+2\sin^2\theta.$

26. Prove that $\int_0^{\pi/2}\frac{dx}{\text{ch}^2 a-\cos^2 x}=\pi/|\text{sh}\,2a|.$

27. Find the whole area enclosed by $a^4y^2=x^5(2a-x)$.

28. Prove that the area of the loop of $ay^2=(x-b)(x-c)^2$, $c>b>0$, is $\frac{8}{15}\sqrt{\{(c-b)^5/a\}}$.

29. AB is a diameter of a given circle, centre O ; PQ is a variable chord parallel to AB. Prove that the volume generated by the revolution of the quadrilateral APQB about AB is a maximum if $\cos \text{AOP}=\frac{1}{6}(\sqrt{13}-1)$.

30. A particle of mass m moves in Ox under a variable force so that its position $(x, 0)$ at time t is given by $x = 2a(2t - \sin 2t)$ where a is constant. Prove that (i) the force is $8ma \sin 2t$, (ii) the work done up to time t is $32ma^2 \sin^4 t$, (iii) the power at time t is $128ma^2 \sin^3 t \cos t$.

31. The density of a semicircular lamina, radius a, varies as the distance from the bounding diameter. Prove that the centre of mass is at distance $3\pi a/16$ from the centre.

32. Evaluate $\int_0^1 \frac{dx}{(x^2 - 2x + 2)^3}$ by the substitution $x = 1 + \tan\theta$.

33. Prove that the area bounded by the x-axis, the ordinates $x = 0$, $x = \pi$, and the curve $y(5 + 3\cos x) = 1$ is $\frac{1}{4}\pi$.

34. Prove that $\int_{-1}^{+1} \frac{\sin\alpha \, dx}{1 - 2x\cos\alpha + x^2} = \frac{1}{2}\pi$, $0 < \alpha < \pi$.
Find its value for $\pi < \alpha < 2\pi$.

Infinite Integrals.

We have supposed hitherto, in discussing definite integrals, that the range of integration is finite and that the integrand is continuous. It is sometimes convenient to assign a meaning to an integral in which one or both of these conditions does not hold; such integrals are called " improper " or " infinite ".

Infinite Integrals : infinite Limits.

Suppose that $t > 1$; then $\int_1^t \frac{1}{x^2} dx = \left[-\frac{1}{x}\right]_1^t = 1 - \frac{1}{t}$.

When $t \to \infty$, $1 - \frac{1}{t} \to 1$,

$$\therefore \lim_{t\to\infty} \int_1^t \frac{1}{x^2} dx = 1.$$

This is expressed more shortly by

$$\int_1^\infty \frac{1}{x^2} dx = 1,$$

and this is an example of an infinite integral which has a meaning.

On the other hand, $\int_1^t \frac{1}{x} dx = \log t$, and this increases indefinitely when $t \to \infty$; hence the infinite integral $\int_1^\infty \frac{1}{x} dx$ *does not exist.*

In general, if $\lim_{t\to\infty}\int_a^t f(x)\,dx = l,$

we write $\int_a^\infty f(x)\,dx = l\,;$

it is understood that the integral $\int_a^t f(x)\,dx$ exists; otherwise there could be no question of the existence of its limit.

Also, if $\int_{t_1}^b f(x)\,dx$ tends to a limit when $t_1 \to -\infty$, we denote this limit by $\int_{-\infty}^b f(x)\,dx.$

And if $\int_{t_1}^t f(x)\,dx$ tends to a limit when $t \to \infty$ and $t_1 \to -\infty$ independently, this limit is denoted by $\int_{-\infty}^{+\infty} f(x)\,dx.$

Example 5. Discuss the existence of the following:

$$\text{(i)} \int_1^\infty \frac{1}{\sqrt{x}}\,dx\,; \quad \text{(ii)} \int_0^\infty \cos x\,dx\,; \quad \text{(iii)} \int_{-\infty}^{+\infty} \frac{dx}{1+x^2}.$$

(i) If $t > 1$, $\int_1^t \frac{1}{\sqrt{x}}\,dx = \Big[\,2\sqrt{x}\,\Big]_1^t = 2\sqrt{t} - 2.$

But when $t \to \infty$, $2\sqrt{t} \to \infty$; therefore the integral does not tend to a limit when $t \to \infty$;

$$\therefore \int_1^\infty \frac{1}{\sqrt{x}}\,dx \text{ has no meaning.}$$

(ii) $$\int_0^t \cos x\,dx = \Big[\,\sin x\,\Big]_0^t = \sin t.$$

But there are indefinitely large values of t for which $\sin t$ takes any assigned value between -1 and $+1$; therefore $\sin t$ does not tend to a limit when $t \to \infty$; it is said to oscillate finitely.

$$\therefore \int_0^\infty \cos x\,dx \text{ has no meaning.}$$

(iii) $$\int_{t_1}^t \frac{dx}{1+x^2} = \Big[\,\tan^{-1} x\,\Big]_{t_1}^t = \tan^{-1} t - \tan^{-1} t_1.$$

When $t \to \infty$, $\tan^{-1} t \to \frac{1}{2}\pi$ and when $t_1 \to -\infty$, $\tan^{-1} t_1 \to -\frac{1}{2}\pi$; therefore $\int_{t_1}^t \frac{dx}{1+x^2}$ tends to $\frac{1}{2}\pi - (-\frac{1}{2}\pi)$, that is to π.

$$\therefore \int_{-\infty}^{+\infty} \frac{dx}{1+x^2} \text{ exists and equals } \pi.$$

EXERCISE XV b

Discuss the following infinite integrals and, when possible, find their values:

1. $\int_1^\infty \frac{dx}{x^3}$.
2. $\int_1^\infty \frac{dx}{x^{2/3}}$.
3. $\int_0^\infty \frac{dx}{x^2+9}$.
4. $\int_{-\infty}^{+\infty} \frac{x\,dx}{x^4+1}$.
5. $\int_0^\infty \sin x\,dx$.
6. $\int_{-\infty}^0 e^x\,dx$.
7. $\int_2^\infty \frac{dx}{x^2-1}$.
8. $\int_{-\infty}^{+\infty} \frac{dx}{x^2-2x+2}$.
9. $\int_1^\infty \frac{dx}{x^2(x^2+1)}$.
10. $\int_0^\infty xe^{-x}\,dx$.
11. $\int_0^\infty \frac{dx}{\operatorname{ch} x}$.
12. $\int_1^\infty \frac{\log x}{x}\,dx$.
13. $\int_1^\infty \frac{dx}{x(x^2+1)}$.
14. $\int_0^\infty e^{-x}(\cos x - \sin x)\,dx$.
15. Prove that $\int_0^\infty \frac{dx}{x^2+2x\cos\alpha+1} = 2\int_0^1 \frac{dx}{x^2+2x\cos\alpha+1}$.

Infinite Integrals : discontinuous Integrand.

Suppose that $t<1$, then

$$\int_0^t \frac{dx}{\sqrt{(1-x)}} = \Big[-2\sqrt{(1-x)}\Big]_0^t = 2-2\sqrt{(1-t)}.$$

When $t\to 1$ *through values less than* 1, $\sqrt{(1-t)}\to 0$, and this is sometimes written, $\lim_{t\to 1-}\sqrt{(1-t)}=0$;

hence
$$\lim_{t\to 1-}\int_0^t \frac{dx}{\sqrt{(1-x)}}=2.$$

This is expressed more shortly by

$$\int_0^1 \frac{dx}{\sqrt{(1-x)}}=2$$

and is an example of a second kind of infinite integral which has a meaning.

Note. It is necessary to restrict t to values less than 1 because $\sqrt{(1-t)}$ has no meaning when $t>1$, and the integrand is undefined for $x\geqslant 1$. It is just because the integrand is undefined for $x=1$ that the integral is not an ordinary one.

In general if $f(x)$ is defined from a inclusive to b exclusive, that is for $a \leqslant x < b$, and if

$$\lim_{t \to b-} \int_a^t f(x)\,dx = l, \text{ we write } \int_a^b f(x)\,dx = l.$$

Also, if $f(x)$ is defined from a exclusive to b inclusive, that is for $a < x \leqslant b$, and if $\lim\limits_{t_1 \to a+} \int_{t_1}^b f(x)\,dx$ exists, we denote its value by $\int_a^b f(x)\,dx$.

And if $f(x)$ is defined from a exclusive to b exclusive, that is for $a < x < b$, and if $\int_{t_1}^t f(x)\,dx$ tends to a limit when $t \to b-$ and $t_1 \to a+$ independently, we denote the limit by $\int_a^b f(x)\,dx$.

More generally, if $f(x)$ is undefined for a value $x = c$ which lies between a and b,

$$\int_a^b f(x)\,dx \text{ is defined to mean } \int_a^c f(x)\,dx + \int_c^b f(x)\,dx,$$

provided that each of these integrals has a meaning.

Example 6. Discuss the existence of the following:

$$\text{(i) } \int_0^1 \frac{dx}{\sqrt[3]{x}}; \quad \text{(ii) } \int_{-1}^{+1} \frac{dx}{x^2}.$$

(i) If $t > 0$, $\displaystyle\int_t^1 \frac{dx}{\sqrt[3]{x}} = \left[\tfrac{3}{2}x^{2/3}\right]_t^1 = \tfrac{3}{2}(1 - t^{2/3})$.

But when $t \to 0$, $t^{2/3} \to 0$, $\therefore \displaystyle\lim_{t \to 0} \int_t^1 \frac{dx}{\sqrt[3]{x}} = \tfrac{3}{2}$;

$\therefore \displaystyle\int_0^1 \frac{dx}{\sqrt[3]{x}}$ exists and equals $\tfrac{3}{2}$.

(ii) The integrand is undefined for $x = 0$, and for the existence of the integral it is necessary that each of the infinite integrals $\displaystyle\int_{-1}^0 \frac{dx}{x^2}$ and $\displaystyle\int_0^1 \frac{dx}{x^2}$ should exist.

But for $t > 0$, $\displaystyle\int_t^1 \frac{dx}{x^2} = \left[-\frac{1}{x}\right]_t^1 = \frac{1}{t} - 1$;

and when $t \to 0+$, $\dfrac{1}{t} \to +\infty$, $\therefore \displaystyle\int_t^1 \frac{dx}{x^2}$ has no limit.

Hence $\displaystyle\int_{-1}^{+1} \frac{dx}{x^2}$ has no meaning.

These examples emphasise the fact that the relation $\int_a^b f(x)\,dx = F(b) - F(a)$ cannot be used without special examination unless $F'(x) = f(x)$ for *all* values of x from a to b, both inclusive.

Thus $\frac{d}{dx}\left(-\frac{1}{x}\right) = \frac{1}{x^2}$ usually, but this equation has no meaning if $x = 0$, and, since this is a value between -1 and $+1$, we cannot say that $\int_{-1}^{+1} \frac{1}{x^2}\,dx$ equals $\left[-\frac{1}{x}\right]_{-1}^{+1}$, i.e. -2; and in fact we have proved in Example 6 (ii) that this integral has no meaning.

Transformation of Infinite Integrals. If a substitution is used to evaluate an infinite integral it is necessary to show that the transformation is legitimate. This is best done by reasoning from first principles.

Example 7. Find $\int_0^1 \sqrt{\frac{x}{1-x}}\,dx.$

Put $\sqrt{x} = \sin\theta$ where θ is acute; then θ increases from 0 to $\frac{1}{2}\pi$ when x increases from 0 to 1;

$$x = \sin^2\theta, \quad \therefore\ dx = 2\sin\theta\cos\theta\,d\theta.$$

$$\therefore \int_0^1 \sqrt{\frac{x}{1-x}}\,dx = \int_0^{\pi/2} \frac{\sin\theta}{\cos\theta}\,2\sin\theta\cos\theta\,d\theta$$

$$= 2\int_0^{\pi/2} \sin^2\theta\,d\theta = \tfrac{1}{2}\pi.$$

To justify the process we may argue that, for $0 < t < 1$,

$$\int_0^t \sqrt{\frac{x}{1-x}}\,dx = 2\int_0^{\alpha} \sin^2\theta\,d\theta \text{ where } \alpha = \sin^{-1}\sqrt{t},$$

because these are ordinary integrals to which the substitution method can properly be applied. But

$$2\int_0^{\alpha} \sin^2\theta\,d\theta = \int_0^{\alpha} (1 - \cos 2\theta)\,d\theta = \alpha - \sin\alpha\cos\alpha,$$

and $\alpha - \sin\alpha\cos\alpha \to \frac{1}{2}\pi$ when $\alpha \to \frac{1}{2}\pi$, that is when $t \to 1-$.

Hence $$\lim_{t\to 1-} \int_0^t \sqrt{\frac{x}{1-x}}\,dx = \tfrac{1}{2}\pi.$$

In ordinary cases it is easy to see whether the process is valid without writing down the detailed proof.

Example 8. Evaluate the integral $\int_0^{\pi} \frac{d\theta}{1+e\cos\theta}$ where $0<e<1$. What happens when $e \geqslant 1$?

Put $\tan\frac{1}{2}\theta = t$, then $\frac{1}{2}\sec^2\frac{1}{2}\theta\, d\theta = dt$, $\therefore d\theta = \frac{2dt}{1+t^2}$;

and $\cos\theta = (\cos^2\frac{1}{2}\theta - \sin^2\frac{1}{2}\theta)/(\cos^2\frac{1}{2}\theta + \sin^2\frac{1}{2}\theta) = \frac{1-t^2}{1+t^2}$; see p. 296.

Also, as θ increases from 0 to π, t increases from 0 to ∞.

$$\therefore \int_0^{\pi} \frac{d\theta}{1+e\cos\theta} = \int_0^{\infty} \frac{2dt}{(1+t^2)+e(1-t^2)}$$

$$= \frac{2}{1-e}\int_0^{\infty} \frac{dt}{c^2+t^2}, \text{ where } c = \sqrt{\frac{1+e}{1-e}}.$$

$$\left[\textit{Side work: } \int \frac{dt}{c^2+t^2} = \int \frac{c\,d(t/c)}{c^2(1+t^2/c^2)} = \frac{1}{c}\tan^{-1}\frac{t}{c}.\right]$$

$$\therefore \int_0^{\pi} \frac{d\theta}{1+e\cos\theta} = \frac{2}{1-e}\left[\frac{1}{c}\tan^{-1}\frac{t}{c}\right]_0^{\infty}$$

$$= \frac{2}{1-e}\sqrt{\frac{1-e}{1+e}}\,\frac{\pi}{2} = \frac{\pi}{\sqrt{(1-e^2)}}.$$

If $e \geqslant 1$ there exists a value α of θ in the range of integration for which $1+e\cos\theta = 0$, and the integrand is meaningless for that value of θ. It can be shown that the integrals $\int_0^{\alpha} \frac{d\theta}{1+e\cos\theta}$ and $\int_{\alpha}^{\pi} \frac{d\theta}{1+e\cos\theta}$ do not exist, and so $\int_0^{\pi} \frac{d\theta}{1+e\cos\theta}$ has no meaning.

EXERCISE XV c

Discuss the infinite integrals in Nos. 1-18, and, when possible, find their values:

1. $\int_0^2 \frac{dx}{\sqrt{(2-x)}}$.
2. $\int_0^2 \frac{dx}{2-x}$.
3. $\int_{-1}^{+1} \frac{dx}{x^3}$.
4. $\int_{-1}^{+1} x^{-2/3}\,dx$.
5. $\int_{-1}^{+1} \frac{dx}{\sqrt{(1-x^2)}}$.
6. $\int_0^{\pi/2} \frac{\cos x\,dx}{\sqrt{(\sin x)}}$.

7. $\int_1^3 \frac{dx}{\sqrt[3]{(x-2)}}$.

8. $\int_1^\infty \frac{dx}{x^2-1}$.

9. $\int_{-\pi/2}^{+\pi/2} \operatorname{cosec} x \, dx$.

10. $\int_0^\infty \frac{dx}{(x+1)(x+2)}$.

11. $\int_0^a \frac{\cos x \, dx}{\sqrt{(\sin^2 a - \sin^2 x)}}$.

12. $\int_0^1 \frac{x^3 \, dx}{\sqrt{(1-x^2)}}$.

13. $\int_0^\infty \frac{x^2 \, dx}{(1+x^2)^2}$.

14. $\int_0^1 x \log x \, dx$.

15. $\int_1^\infty \frac{dx}{x(x+1)^2}$.

16. $\int_{-1}^{+1} \sqrt{\frac{1+x}{1-x}} \, dx$.

17. $\int_0^\infty x^2 e^{-x} \, dx$.

18. $\int_1^3 \frac{dx}{\sqrt{\{(3-x)(x-1)\}}}$.

19. If $a > 0$, prove that $\int_0^\infty \frac{dx}{(a+x)(a^2+x^2)} = \frac{\pi}{4a^2}$.

What happens if $a < 0$?

20. If $0 < b < a$, prove that $\int_0^\pi \frac{d\theta}{a + b\cos\theta} = \frac{\pi}{\sqrt{(a^2-b^2)}}$.

21. If $u_n = \int_0^\infty x^n e^{-x} \, dx$, prove that if $n \geqslant 1$, $u_n = n u_{n-1}$.

Find the value of u_n when n is a positive integer.

22. If $u_n = \int_0^\infty \frac{dx}{(1+x^2)^n}$, prove that if $n > 1$, $u_n = \frac{2n-3}{2n-2} u_{n-1}$.

Find the value of $\int_0^\infty \frac{dx}{(1+x^2)^6}$.

Generalised Measure of Area.

Fig. 115 represents part of the graph of $y = 1/(1 + x^2)$. The area

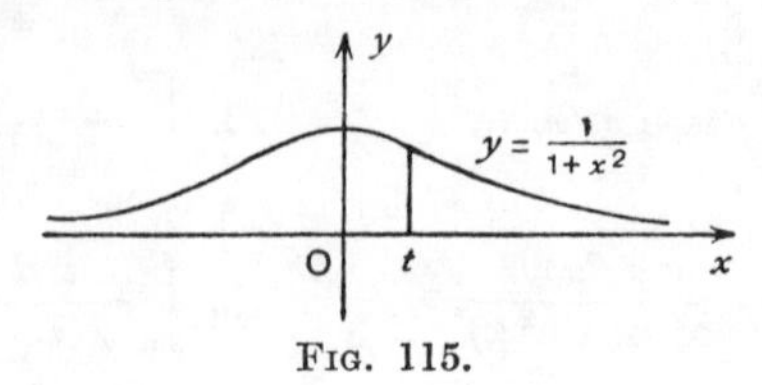

FIG. 115.

bounded by Ox, Oy, the ordinate $x = t$, $(t > 0)$, and the curve is measured by $\int_0^t \frac{dx}{1+x^2}$, that is by $\tan^{-1} t$.

When $t \to \infty$, $\tan^{-1} t \to \frac{1}{2}\pi$; therefore the infinite integral $\int_0^\infty \frac{dx}{1+x^2}$ exists and equals $\frac{1}{2}\pi$. We say that this infinite integral measures the "area" between Ox, Oy, and the part of the curve for which x is positive, and therefore that the "area" is $\frac{1}{2}\pi$.

Here the word "area" is used, in a new sense, of a portion of the plane which is not in fact bounded and is defined as the limit of an ordinary area.

More generally, if the infinite integral $\int_a^\infty f(x)\,dx$ exists we say that it measures the "area" between the x-axis, the ordinate $x = a$, and the part of the curve $y = f(x)$ for which $x > a$. Other types of infinite integrals can be associated with "areas" in the same way.

Also, a generalised measure of "volume" can be defined as a limit of an ordinary volume.

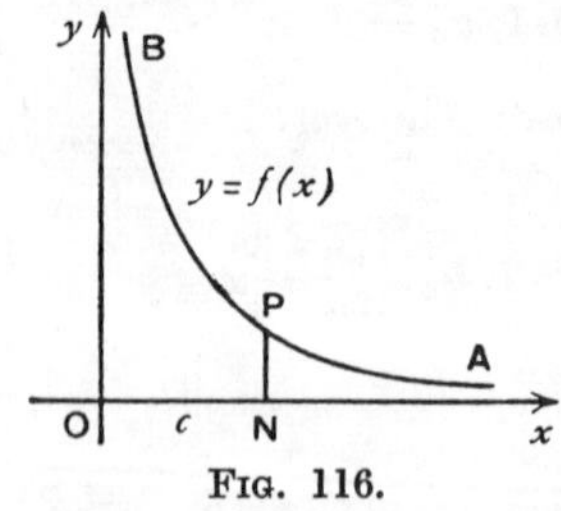

FIG. 116.

The two kinds of infinite integrals which have been discussed on p. 309 and p. 311 are not essentially distinct.

Suppose Fig. 116 represents the graph of $y = f(x)$, where $f(x) \to \infty$ when $x \to 0+$, and $f(x) \to 0$ when $x \to \infty$. Then $\int_c^\infty f(x)\,dx$ is an infinite integral *of the first kind*, and, if it exists, measures the "area" $APNx$; also, $\int_0^c f(x)\,dx$ is an infinite integral *of the second kind* and, if it exists, measures the "area" $yONPB$.

The first of these could be transformed into an integral of the second kind by the substitution $x=\frac{1}{z}$ provided the substitution is valid.

If $f(x)\equiv 1/\sqrt{x}$ it is easy to see that the area yONPB exists and that the area APNx does not:

$$\text{If } t_1>0,\ \int_{t_1}^{c}\frac{dx}{\sqrt{x}}=\Big[2\sqrt{x}\Big]_{t_1}^{c}=2\sqrt{c}-2\sqrt{t_1};$$

$$\therefore \text{ area } y\text{ONPB}=\lim_{t_1\to 0+}(2\sqrt{c}-2\sqrt{t_1})=2\sqrt{c}.$$

But $\int_c^t\frac{dx}{\sqrt{x}}=2\sqrt{t}-2\sqrt{c}$, $\quad\therefore \lim\limits_{t\to\infty}\int_c^t\frac{dx}{\sqrt{x}}$ does not exist.

EXERCISE XV d

1. Find the area between the x-axis, the line $x=2$, and the part of the curve $y=1/x^2$ for which $x>2$.
2. Find the area between the x-axis and the curve $y(x^2+a^2)=a^3$.
3. Sketch the curve $xy^2=4(2-x)$ and find the total area between the curve and the y-axis.
4. Sketch the curve $y^2(1-x)=x$ and prove that the area between the curve and the line $x=1$ is π.
5. Prove that the area between the curve $y(1+x^2)^4=x^4$ and the positive part of the x-axis is $\frac{1}{32}\pi$.
6. Sketch the curve $y^2=1/\{(2-x)(x-1)\}$ and prove that the area between the curve and the lines $x=1$, $x=2$ is 2π.
7. Sketch the curve $y^4=x^4/(1-x^2)$ and find the area between the line $x=1$ and the part of the curve for which x is positive.
8. Sketch the curve $y^2=(x-1)/(4-x)$ and prove that the area between the curve and the line $x=4$ is 3π.
9. Prove that $\int_{-1}^{\infty}e^{-x}(x^2-1)\,dx$ is zero and interpret the result geometrically.

10. Find the total area between the curve $y = \operatorname{sech} x$ and the x-axis.

11. Sketch the graph of $y = 1/(1 + 2\cos x)$ from $x = 0$ to $x = \pi$. Prove that the area between the axes, the ordinate $x = \frac{1}{2}\pi$, and the curve is $\frac{1}{3}\sqrt{3}\log(2+\sqrt{3})$. Is there a measure for the area between the x-axis, the ordinates $x = \frac{1}{2}\pi$, $x = \pi$, and the curve?

12. Prove that the volume of the solid formed by the revolution about Ox of the area between Ox, the ordinate $x = 1$ and the part of the curve $xy = 1$ for which $x > 1$ is π.

13. Find the volume of the solid formed by revolving about Ox the area between the axes and the part of the curve $y = e^{-x}$ for which x is positive.

14. P is any point on a circle, diameter AB; AP meets the tangent at B in T, and AQ is taken on AT so that AQ = PT. Prove that the area contained between the locus of Q and the tangent at B is 3 times the area of the circle.

15. Find the centre of gravity of the area bounded by the x-axis, the ordinates $x = 1$, $x = n$, $(n > 1)$, and the part of the curve $y = x^{-3}$ between those ordinates. Has this point a definite limiting position when $n \to +\infty$? (If so, the limiting position can be called the "centre of gravity" of the generalised area.)

    Can any meaning be assigned to the centre of gravity of the whole area between the axes and the part of the curve $y = x^{-3}$ for which x is positive?

16. Find the "centre of gravity" of the volume of the solid formed by revolving about Ox the area between the x-axis, the ordinate $x = 1$, and the part of the curve $yx\sqrt{x} = 1$ for which $x > 1$.

General Properties of Definite Integrals.

It was shown on p. 102 (Vol. I) that

$$\int_a^b f(x)\,dx = \int_a^c f(x)\,dx + \int_c^b f(x)\,dx$$

for all values of a, b, c for which the integrals exist.

We shall now prove some other general properties. All of these should be illustrated graphically by the reader.

(i) $\int_{-a}^{+a} \mathbf{f(x)\,dx} = \int_0^a \{\mathbf{f(x) + f(-x)}\}\,\mathbf{dx}.$

$$\int_{-a}^{+a} f(x)\,dx = \int_{-a}^{0} f(x)\,dx + \int_0^{+a} f(x)\,dx\,;$$

and, putting $x = -z$

$$\int_{-a}^{0} f(x)\,dx = \int_a^0 f(-z)(-1)\,dz = \int_0^a f(-z)\,dz = \int_0^a f(-x)\,dx\,;$$

hence the result follows.

Two special cases should be noticed:

If $f(x)$ is an *odd function* of x, that is if $f(-x) \equiv -f(x)$, then $\int_{-a}^{+a} f(x)\,dx = 0.$

If $f(x)$ is an *even function* of x, that is if $f(-x) \equiv f(x)$, then $\int_{-a}^{+a} f(x)\,dx = 2\int_0^a f(x)\,dx.$

Thus, for example,

$$\begin{aligned}\int_{-\pi/2}^{+\pi/2} (1+\sin x)^2\,dx &= \int_{-\pi/2}^{+\pi/2} (1+\sin^2 x)\,dx + \int_{-\pi/2}^{+\pi/2} 2\sin x\,dx \\ &= 2\int_0^{\pi/2} (1+\sin^2 x)\,dx + 0 \\ &= 2(\tfrac{1}{2}\pi + \tfrac{1}{4}\pi) = 3\pi/2.\end{aligned}$$

(ii) $\int_0^a \mathbf{f(x)\,dx} = \int_0^a \mathbf{f(a-x)\,dx}.$

Put $x = a - z$, then $dx = -dz$;

$$\therefore \int_0^a f(x)\,dx = -\int_a^0 f(a-z)\,dz = \int_0^a f(a-z)\,dz.$$

Thus, for example,

$$\int_0^{\pi/2} f(\sin x, \cos x)\,dx = \int_0^{\pi/2} f(\cos x, \sin x)\,dx$$

and
$$\int_0^1 x^m(1-x)^n\,dx = \int_0^1 x^n(1-x)^m\,dx.$$

(iii) If $m \leqslant f(x) \leqslant \mathsf{M}$ for all values of x in the interval $a \leqslant x \leqslant b$, we shall prove that

$$m(b-a) \leqslant \int_a^b f(x)\,dx \leqslant \mathsf{M}(b-a).$$

As on p. 131 (Vol. I), let the interval from a to b be divided into n parts at $x_1, x_2, \ldots, x_{n-1}$, so that $a < x_1 < x_2 < \ldots < x_{n-1} < b$.

Then $\int_a^b f(x)\,dx$ is the limit, I, of

$$(x_1 - a)f(a) + (x_2 - x_1)f(x_1) + \ldots\ldots + (b - x_{n-1})f(x_{n-1}),$$

when $n \to \infty$ and the length of even the longest sub-interval tends to zero. But this sum always lies between $m(b-a)$ and $\mathsf{M}(b-a)$;

$$\therefore\ \mathsf{I} \geqslant m(b-a) \quad \text{and} \quad \mathsf{I} \leqslant \mathsf{M}(b-a).$$

If $f(x)$ is continuous in the interval $a \leqslant x \leqslant b$, and $m \neq \mathsf{M}$, it can actually be proved that $\mathsf{I} > m(b-a)$ and $\mathsf{I} < \mathsf{M}(b-a)$; hence $\mathsf{I} = \kappa(b-a)$ where $m < \kappa < \mathsf{M}$ and not merely $m \leqslant \kappa \leqslant \mathsf{M}$.

Also, it can be proved that a continuous function $f(x)$ assumes any value κ between m and M for a value ξ of x inside the interval (a, b); that is a value ξ of x exists such that

$$f(\xi) = \kappa,\ a < \xi < b, \text{ if } m < \kappa < \mathsf{M}.$$

These results, which we shall not prove in this book, may be stated:

$$\mathbf{m(b-a)} < \int_a^b \mathbf{f(x)\,dx} < \mathbf{M(b-a)}$$

and
$$\int_a^b \mathbf{f(x)\,dx = (b-a)f(\xi)}, \quad \mathbf{(a < \xi < b)}.$$

This second result is called a *mean-value theorem* for integrals and it may also be expressed in the form

$$\int_a^b f(x)\,dx = (b-a)f\{a + \lambda(b-a)\}, \quad (0 < \lambda < 1).$$

The mean-value theorem also holds when $a > b$, since the exchange of a and b merely alters the sign on both sides of the equation.

(iv) If $f(x)$ is a continuous function for the interval $a \leqslant x \leqslant b$, and

$$\textbf{if } \varphi(x) = \int_a^x f(x)\,dx, \textbf{ then } \frac{d}{dx}\varphi(x) = f(x), \quad a < x < b.$$

If x and $x + \delta x$ lie in the interval (a, b),

$$\phi(x+\delta x) - \phi(x) = \int_a^{x+\delta x} f(x)\,dx - \int_a^x f(x)\,dx$$

$$= \int_x^{x+\delta x} f(x)\,dx$$

$$= \delta x f(x + \lambda\delta x), \quad 0 < \lambda < 1;$$

$$\therefore \frac{\phi(x+\delta x) - \phi(x)}{\delta x} = f(x + \lambda\delta x), \quad 0<\lambda<1.$$

But, by hypothesis, $f(x)$ is continuous in (a, b);

$$\therefore \text{ when } \delta x \to 0,\ f(x + \lambda\delta x) \to f(x);$$

$$\therefore \frac{\phi(x+\delta x) - \phi(x)}{\delta x} \to f(x); \quad \therefore \frac{d}{dx}\phi(x) = f(x).$$

Example 9. Find $\int_0^\pi \frac{x \sin x}{1 + \cos^2 x}\,dx$.

From (ii), p. 319, or by writing $\pi - x$ for x,

$$\int_0^\pi \frac{x \sin x}{1+\cos^2 x}\,dx = \int_0^\pi \frac{(\pi - x)\sin x}{1+\cos^2 x}\,dx$$

$$= \int_0^\pi \frac{\pi \sin x}{1+\cos^2 x}\,dx - \int_0^\pi \frac{x \sin x}{1+\cos^2 x}\,dx;$$

$$\therefore 2\int_0^\pi \frac{x \sin x}{1+\cos^2 x}\,dx = \int^\pi \frac{\pi \sin x}{1+\cos^2 x}\,dx$$

$$= \int_{+1}^{-1} \frac{\pi(-1)\,dz}{1+z^2}, \text{ where } z = \cos x$$

$$= -\pi\Big[\tan^{-1} z\Big]_{+1}^{-1} = -\pi(-\tfrac{1}{4}\pi - \tfrac{1}{4}\pi);$$

$$\therefore \int_0^\pi \frac{x \sin x}{1+\cos^2 x}\,dx = \tfrac{1}{4}\pi^2.$$

EXERCISE XVe

Evaluate the integrals in Nos. 1-8 :

1. $\int_0^{\pi} \cos^n x\,dx$ if (i) $n=5$, (ii) $n=6$.

2. $\int_{-\pi/2}^{+\pi/2} \sin^n x\,dx$ if (i) $n=7$, (ii) $n=8$.

3. $\int_0^{2\pi} \cos^n x\,dx$ if (i) $n=4$, (ii) $n=5$.

4. $\int_0^{\pi} \sin^3 x \cos^n x\,dx$ if (i) $n=3$, (ii) $n=4$.

5. $\int_{-a}^{+a} \frac{x^5\,dx}{a^2+x^2}$.

6. $\int_{-\pi}^{+\pi} x \sin x\,dx$.

7. $\int_0^{\pi} \frac{\sin 4x}{\sin x}\,dx$.

8. $\int_0^{\pi} \sin 4x \cot x\,dx$.

9. Prove that $\int_a^{2a} f(x)\,dx = \int_0^a f(2a-x)\,dx$, and deduce that

$$\int_0^{2a} f(x)\,dx = \int_0^a \{f(x)+f(2a-x)\}\,dx.$$

10. Prove that $\int_0^{\pi} f(\sin x)\,dx = 2\int_0^{\pi/2} f(\sin x)\,dx$.

11. If $f(\cos x)$ is an odd function of $\cos x$, prove that

$$\int_0^{\pi} f(\cos x)\,dx = 0.$$

12. Prove that $\int_a^b f(nx)\,dx = \frac{1}{n}\int_{na}^{nb} f(x)\,dx$.

13. Give alternative forms for

(i) $\int_a^b f(a+b-x)\,dx$; (ii) $\int_{a-c}^{b-c} f(x+c)\,dx$.

14. If $\phi(c-x)=\phi(x)$, prove that

$$\int_c^c x^3\phi(x)\,dx = \tfrac{3}{2}c\int_0^c x^2\phi(x)\,dx - \tfrac{1}{4}c^3\int_0^c \phi(x)\,dx.$$

In Nos. 15-17 find values of ξ within the limits of integration for which the results are true :

15. $\int_0^1 x^2\,dx = \xi^2$. 16. $\int_0^{\pi/2} \sin^2 x\,dx = \frac{1}{2}\pi \sin^2 \xi$. 17. $\int_1^t \frac{dx}{x} = \frac{t-1}{\xi}$.

18. Prove that $\int_0^{1/2} \frac{dx}{\sqrt{(1-x^4)}}$ lies between 0·5 and 0·53.

19. Use the inequality, $\sin x < \sqrt{(\sin x)} < \sqrt{x}$, $(0 < x < \frac{1}{2}\pi)$, to prove that $\int_0^{\pi/4} \sqrt{(\sin x)}\,dx$ lies between 0·29 and 0·47.

20. Prove that $\int_0^{\pi} x \sin^2 x\,dx = \frac{1}{4}\pi^2$, by dividing the range into two parts, 0 to $\frac{1}{2}\pi$ and $\frac{1}{2}\pi$ to π.

21. Prove that, if $t > 0$, $\int_0^t \frac{dx}{1+x^2} + \int_0^{1/t} \frac{dx}{1+x^2}$ is independent of t without evaluating the integrals separately.

22. Point out the mistake in the following :

$$\int_0^{\pi} \frac{dx}{\cos^2 x + \sin^2 x} = \int_0^{\pi} \frac{\sec^2 x\,dx}{1+\tan^2 x} = \Big[\tan^{-1}(\tan x)\Big]_0^{\pi} = 0.$$

What is the true value of the integral ?

23. Prove that $\int_{\pi/2}^{\pi} \frac{x\,dx}{1+\cos\alpha \sin x} = \int_0^{\pi/2} \frac{(\pi - x)\,dx}{1+\cos\alpha\sin x}$, and deduce that

$$\int_0^{\pi} \frac{x\,dx}{1+\cos\alpha\sin x} = \frac{\pi\alpha}{\sin\alpha}, \text{ where } 0 < \alpha < \tfrac{1}{2}\pi.$$

CHAPTER XVI

GEOMETRICAL APPLICATIONS

Length of an Arc. It is impossible to identify any portion of a *curve*, however small, with a *straight* line, and when the word " length " is associated with the arc of a curve it is used in a new sense which needs to be defined.

Suppose that any open n-sided polygon $AP_1P_2 \ldots P_{n-1}B$ is inscribed in the arc AB of a curve, and denote its perimeter, which is the sum of the lengths of the n chords, AP_1, P_1P_2, . . ., $P_{n-1}B$, by s_n.

FIG. 117.

Now, keeping the end-points A, B fixed, let the inscribed polygon vary in any manner such that $n \to \infty$ and the length of every side tends to zero. If, then, s_n tends to a definite limit s, the arc AB is said to be of *length* s.

If the length of the arc measured from a fixed point A to a point P on the curve is denoted by s, the direction in which s increases can be chosen arbitrarily. If a curve is given in the form $y=f(x)$, it is convenient to measure s so that it increases with x, and if given by the parametric equations $x=f(t)$, $y=g(t)$, to measure s so that it increases with t.

A line PQ drawn along the tangent at P in the direction of s increasing is called the *positive direction* of the tangent at P, and one of the angles from Ox to this positive direction, measured in accordance with the usual convention of trigonometry, is denoted by ψ. The actual values of ψ are chosen if possible so that ψ changes continuously. See Fig. 118.

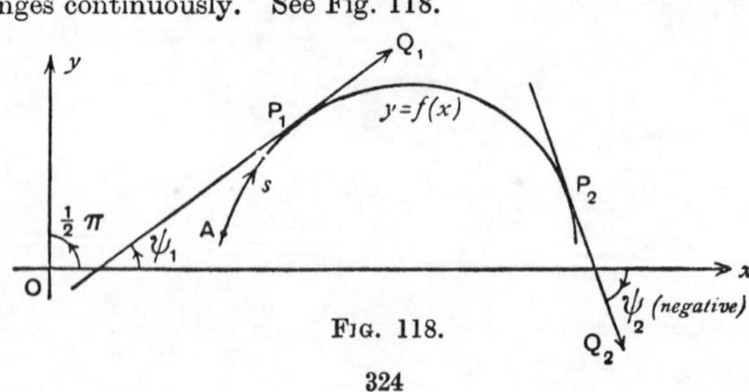

FIG. 118.

Consider now the arc AB of a curve $y=f(x)$, and first suppose that the angle ψ between the tangent at (x, y) and the x-axis is not only a continuous function of x but is also acute and increases steadily as (x, y) moves from A to B.

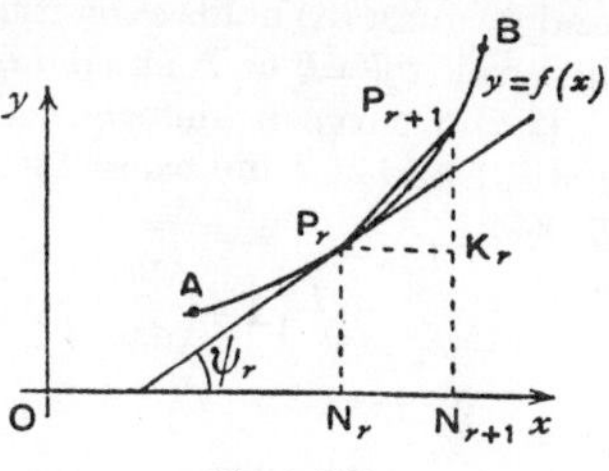

FIG. 119.

Let P_rP_{r+1} be any side of the open polygon inscribed in the curve, let the coordinates of P_r be (x_r, y_r) and let ψ_r be the value of ψ corresponding to P_r. Draw P_rK_r parallel to the x-axis.

Since ψ increases steadily and remains acute, the angle $K_rP_rP_{r+1}$, which equals the angle that P_rP_{r+1} makes with Ox, lies between ψ_r and ψ_{r+1}. But

$$P_rP_{r+1}=(x_{r+1}-x_r)\sec K_rP_rP_{r+1}\,;$$

$\therefore$ P_rP_{r+1} lies between $(x_{r+1}-x_r)\sec\psi_r$ and $(x_{r+1}-x_r)\sec\psi_{r+1}$,

and therefore the perimeter of the open polygon $AP_1P_2\ldots P_{n-1}B$ lies between

$$\sum_0^{n-1}(x_{r+1}-x_r)\sec\psi_r \quad\text{and}\quad \sum_0^{n-1}(x_{r+1}-x_r)\sec\psi_{r+1}$$

where x_0, x_n stand for the abscissae a, b of A, B.

But by the theory of definite integration, see p. 132 (Vol. I), the limit of each of these sums is $\int_a^b \sec\psi\,dx$, and it is known that this integral exists, in virtue of the assumed continuity of $\sec\psi$.

$$\therefore \text{ the length of the arc } AB\equiv s=\int_a^b \sec\psi\,dx.$$

But $\sec^2\psi=1+\tan^2\psi=1+\left(\dfrac{dy}{dx}\right)^2$, and $\sec\psi$ is positive because s increases with x,

$$\therefore s=\int_a^b\sqrt{\left\{1+\left(\frac{dy}{dx}\right)^2\right\}}\,dx \quad . \quad . \quad . \quad \text{(i)}$$

Similarly if α, β are the ordinates of A, B, it can be proved that the length of the arc AB is given by

$$s=\int_\alpha^\beta \operatorname{cosec}\psi\,dy=\int_\alpha^\beta\sqrt{\left\{1+\left(\frac{dx}{dy}\right)^2\right\}}dy \quad . \quad \text{(ii)}$$

Formula (i) holds even if $\psi=0$ at A, but involves an improper integral if $\psi=\frac{1}{2}\pi$ at B, since dy/dx would not then exist at $x=b$; and formula (ii) holds even if $\psi=\frac{1}{2}\pi$ at B, but involves an improper integral if $\psi=0$ at A since dx/dy would not then exist at $y=\alpha$.

If the curve is determined by parametric equations, $x=f(t)$, $y=g(t)$ and if t increases steadily from t_1 at A to t_2 at B, from p. 244,

$$\int_a^b \sqrt{\left\{1+\left(\frac{dy}{dx}\right)^2\right\}}\,dx=\int_{t_1}^{t_2}\sqrt{\left\{1+\left(\frac{dy}{dx}\right)^2\right\}}\,\frac{dx}{dt}\,dt.$$

But, since x and t both increase with s, $\dfrac{dx}{dt}$ is positive and therefore

$$\frac{dx}{dt}\sqrt{\left\{1+\left(\frac{dy}{dx}\right)^2\right\}}=\sqrt{\left\{\left(\frac{dx}{dt}\right)^2+\left(\frac{dy}{dt}\right)^2\right\}}.$$

$$\therefore \int_a^b \sqrt{\left\{1+\left(\frac{dy}{dx}\right)^2\right\}}\,dx=\int_{t_1}^{t_2}\sqrt{\left\{\left(\frac{dx}{dt}\right)^2+\left(\frac{dy}{dt}\right)^2\right\}}\,dt;$$

and similarly for $\displaystyle\int_\alpha^\beta \sqrt{\left\{1+\left(\frac{dx}{dy}\right)^2\right\}}\,dy$.

Hence both (i) and (ii) become

$$s=\int_{t_1}^{t_2}\sqrt{\left\{\left(\frac{dx}{dt}\right)^2+\left(\frac{dy}{dt}\right)^2\right\}}\,dt \qquad \text{(iii)}$$

Therefore (iii) holds for the arc AB even if $\psi=0$ at A and $\psi=\frac{1}{2}\pi$ at B.

By a similar argument it can be proved that (i) and (ii) hold for any arc along which $\dfrac{dy}{dx}$ and $\dfrac{dx}{dy}$ respectively are continuous and (iii) for any arc along which $\dfrac{dx}{dt}$ and $\dfrac{dy}{dt}$ are continuous and t increases steadily, s being measured so that it increases with t.

If s is the length of the arc of a curve measured from some fixed point on it to an arbitrary point P, and if ψ is the angle which the tangent at P makes with a fixed line, the relation between s and ψ is called the **intrinsic equation** of the curve.

Arbitrary constants may be added to the s and ψ of an intrinsic equation since the fixed point and fixed line are arbitrary.

Example 1. P is a marked point on the circumference of a circle of radius a which rolls, without sliding, on the axis of x, starting with P at the origin. The curve traced out by P is called a *cycloid*. Write down parametric equations of the cycloid and hence find its intrinsic equation.

Fig. 120 shows the position of the tracing point P when the rolling circle touches Ox at B. Denote the angle BAP through

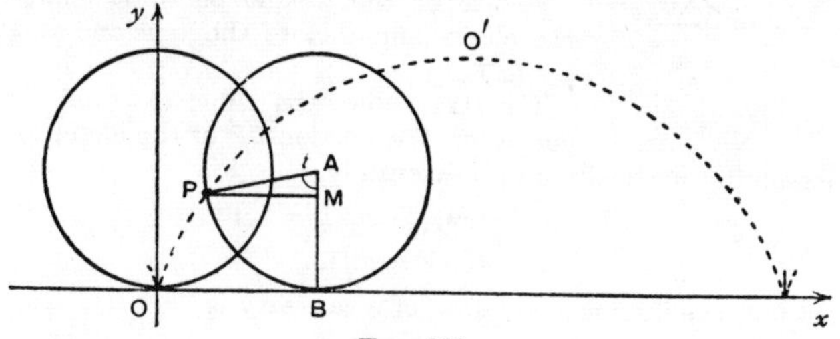

FIG. 120.

which the radius AP has turned by t, and draw PM perpendicular to AB.

Since there is no sliding, OB = arc PB $=at$; hence if P is the point (x, y),

$$x=\text{OB}-\text{PM}=at-a\sin t \quad \text{and} \quad y=\text{BA}-\text{MA}=a-a\cos t;$$

these are parametric equations of the cycloid.

Hence $$dx=(a-a\cos t)dt \text{ and } dy=a\sin t\, dt,$$

$$\therefore \left(\frac{dx}{dt}\right)^2+\left(\frac{dy}{dt}\right)^2=a^2\{(1-\cos t)^2+\sin^2 t\}=a^2(2-2\cos t).$$

$\therefore$ the length of the arc of the cycloid from O to the point t is

$$s=\int_0^t \sqrt{\{a^2(2-2\cos t)\}}\,dt=\int_0^t 2a\sin\tfrac{1}{2}t\,dt$$

$$=\Big[-4a\cos\tfrac{1}{2}t\Big]_0^t=4a(1-\cos\tfrac{1}{2}t).$$

Also $$\tan\psi=dy/dx=\sin t/(1-\cos t)=\cot\tfrac{1}{2}t;$$

$$\therefore \psi=\tfrac{1}{2}\pi-\tfrac{1}{2}t; \quad \therefore s=4a-4a\sin\psi.$$

If O′ is the point on the cycloid where $\psi=0$, OO′ $=4a$. Hence if s is measured from O′ and in the reverse direction, the intrinsic equation may be written

$$\mathbf{s=4a\sin\psi.}$$

Example 2. Find the intrinsic equation of a *catenary*, that is, the curve assumed by a uniform thin flexible chain, whose ends are fixed.

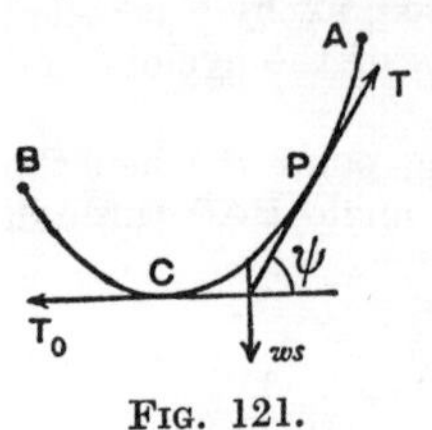

FIG. 121.

Let C be the point at which the tangent is horizontal and P any point (s, ψ) on the curve, where s is measured from C, and ψ from the horizontal.

Let w be the weight per unit length of the chain, and denote the tensions at P, C by T, T_0.

Fig. 121 shows the three external forces acting on the portion CP of the chain.

Resolving vertically and horizontally,

$$T \sin \psi = ws, \quad T \cos \psi = T_0;$$

$$\therefore \tan \psi = ws/T_0.$$

Therefore the intrinsic equation of a catenary is

$$\mathbf{s = c \tan \psi},$$

where c is the constant T_0/w.

EXERCISE XVI a

1. Find the length of the arc of the curve $y = c \operatorname{ch}(x/c)$ measured from $(0, c)$ to (x, y) and prove that the intrinsic equation is $s = c \tan \psi$ (a catenary).

2. Sketch the *astroid* given by $x = a \cos^3 t$, $y = a \sin^3 t$ and find the total length of the curve.

3. Prove that the length of the arc of the parabola given by $x = a \operatorname{sh}^2 u$, $y = 2a \operatorname{sh} u$, measured from the vertex $u = 0$, is $a(u + \frac{1}{2} \operatorname{sh} 2u)$.

4. Sketch the curve $y = \log \sec x$ for $-\frac{1}{2}\pi < x < \frac{1}{2}\pi$ and prove that $s = \log \tan(\frac{1}{4}\pi + \frac{1}{2}x)$ if s is measured from the origin.

5. Find the length of the arc of the curve $y = \log \coth \frac{1}{2}x$, from $x = a$ to $x = b$, where a and b are positive.

6. Find the intrinsic equation of the curve given by

$$x : y : a = (2 \cos \theta + \cos 2\theta) : (2 \sin \theta + \sin 2\theta) : 1.$$

7. Find the length of the arc of the curve given by $x = \cos t \sin^2 t$, $y = \sin t(1 + \cos^2 t)$ from $t = 0$ to $t = \frac{1}{2}\pi$.

8. If s is the length of the arc of the curve $3ay^2 = x(x-a)^2$, measured from the origin to the point (x, y), prove that $3s^2 = 4x^2 + 3y^2$.

Differential Relations. In this section we shall suppose for the sake of simplicity that the arcs considered contain no points where the tangent is parallel to Ox or Oy. This limitation involves no practical inconvenience because the length of arc s may be measured from any convenient fixed point on the curve.

If s denotes the length of the arc of a curve $y = f(x)$ from the fixed point A up to any point P (x, y), s is a function of x given by

$$s = \int_a^x \sec \psi \, dx.$$

Hence from p. 321, $\frac{ds}{dx} = \sec \psi$; but $\frac{dy}{dx} = \tan \psi$,

$$\therefore \mathbf{dx : dy : ds = \cos \psi : \sin \psi : 1.}$$

and $$\mathbf{(ds)^2 = (dx)^2 + (dy)^2.}$$

In Fig. 122, which is essentially the same as Fig. 119, p. 325,

$$PR = \delta x = dx,$$
$$RQ = \delta y,$$
$$RK = dy;$$

also arc $AP = s$, arc $PQ = \delta s$.

$$\therefore PK = \sqrt{(PR^2 + RK^2)}$$
$$= \sqrt{\{(dx)^2 + (dy)^2\}} = ds.$$

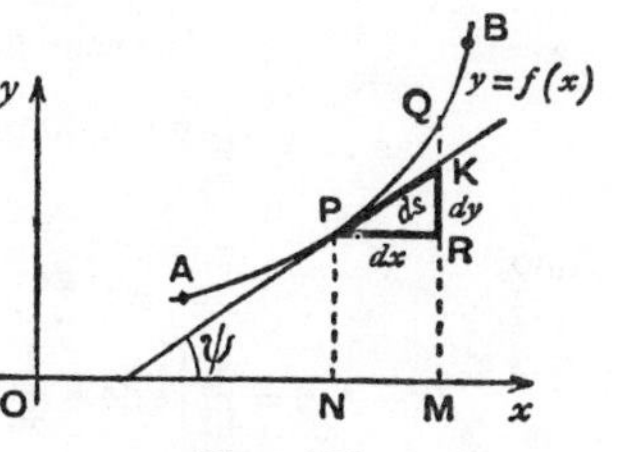

FIG. 122.

The results proved above are best remembered by thinking of the triangle PRK, and since the formula for $(ds)^2$ may be expressed in any of the forms,

$$\left(\frac{ds}{dx}\right)^2 = 1 + \left(\frac{dy}{dx}\right)^2, \quad \left(\frac{ds}{dy}\right)^2 = 1 + \left(\frac{dx}{dy}\right)^2, \quad \left(\frac{ds}{dt}\right)^2 = \left(\frac{dx}{dt}\right)^2 + \left(\frac{dy}{dt}\right)^2,$$

the figure also serves to recall the three formulae for s on pp. 325, 326.

In Fig. 122, chord $\mathsf{PQ} = \sqrt{(\mathsf{PR}^2 + \mathsf{RQ}^2)} = \sqrt{\{(\delta x)^2 + (\delta y)^2\}}$

and arc $\mathsf{PQ} = \delta s$.

$$\therefore \frac{\text{chord PQ}}{\text{arc PQ}} = \sqrt{\left\{\left(\frac{\delta x}{\delta s}\right)^2 + \left(\frac{\delta y}{\delta s}\right)^2\right\}}.$$

$$\lim_{\delta s \to 0} \frac{\text{chord PQ}}{\text{arc PQ}} = \sqrt{\left\{\left(\frac{dx}{ds}\right)^2 + \left(\frac{dy}{ds}\right)^2\right\}} = 1.$$

Fig. 122.

Hence, if δs is small,

$$(\delta s)^2 \simeq (\delta x)^2 + (\delta y)^2.$$

This approximate expression for $(\delta s)^2$ is a deduction from the *exact* relation $(ds)^2 = (dx)^2 + (dy)^2$, which is itself a deduction from the integral formula for s. It is important to note that we cannot argue in the reverse direction.

Example 3. Deduce from the intrinsic equation, $s = 4a \sin \psi$, of a cycloid (p. 327), parametric equations for x and y, taking the origin at the point $\psi = 0$, $s = 0$.

$$\text{Since } s = 4a \sin \psi, \quad ds = 4a \cos \psi \, d\psi,$$

$$\therefore \sec \psi \, dx = ds = 4a \cos \psi \, d\psi$$

$$\therefore x = \int 4a \cos^2 \psi \, d\psi = \int 2a(1 + \cos 2\psi) \, d\psi;$$

$$\therefore x = a(2\psi + \sin 2\psi) \text{ since } x = 0 \text{ when } \psi = 0,$$

Also,
$$\frac{dy}{ds} = \sin \psi = \frac{s}{4a},$$

$$\therefore y = \frac{1}{4a} \int s \, ds = \frac{1}{8a} s^2, \text{ since } y = 0 \text{ when } s = 0;$$

$$\therefore y = \frac{1}{8a} 16a^2 \sin^2 \psi = a(1 - \cos 2\psi).$$

$\therefore$ writing t for 2ψ,

$$x = a(t + \sin t), \quad y = a(1 - \cos t).$$

Note. This example illustrates the two methods of finding parametric cartesian equations of the curve $s = f(\psi)$: sometimes it is simplest to work in terms of ψ and sometimes in terms of s.

EXERCISE XVI b

1. For the parabola $4ay = x^2$, prove that $\dfrac{ds}{dx} = \sqrt{\left(1 + \dfrac{y}{a}\right)}$.

2. For the ellipse given by $x = a\cos t$, $y = b\sin t$, prove that $ds/dt = a\sqrt{(1 - e^2\cos^2 t)}$ where $a^2e^2 = a^2 - b^2$.

3. For the curve $y = \log\sec x$, prove that $ds/dx = \sec x$.

4. If the intrinsic equation of a curve is $s = \log\tan\frac{1}{2}\psi$, prove that with a suitable origin the cartesian equation is $x = \log\sin y$.

5. Express dx and dy in terms of s, ds for the catenary $s = c\tan\psi$. Also if $x = 0$ and $y = c$ when $s = 0$, express x and y in terms of s and show that $y = c\,\mathrm{ch}\,(x/c)$.

6. For the curve $y = \log(\mathrm{ch}\, s)$, prove that $\sin\psi = \mathrm{th}\, s$ and $\tan\psi = \mathrm{sh}\, s$. Deduce that the cartesian equation can be expressed in the form $y = \log\sec x$.

7. For the curve $s = ae^{x/a}$, prove that $s\cos\psi$ is constant and express dy/dx in terms of x.

8. The tangent at any point P on the curve $y = ae^{s/c}$ cuts the x-axis at T. Prove that the length of PT is constant.

9. If the intrinsic equation of a curve is $s = a\psi^2$, express x and y in terms of ψ.

10. If $x = a\log\sin(y/a)$, prove that $s = a\log\tan(y/2a)$ if $s = 0$ when $x = 0$.

11. For the curve $y = a\tan\psi$, prove that $ye^{-x/a}$ is constant and express x and s in terms of ψ.

Polar Coordinates. We assume that the reader is familiar with the use of polar coordinates. The relations by which we can transform from cartesian to polar coordinates, and *vice versa*, may be written down from Fig. 123; they are

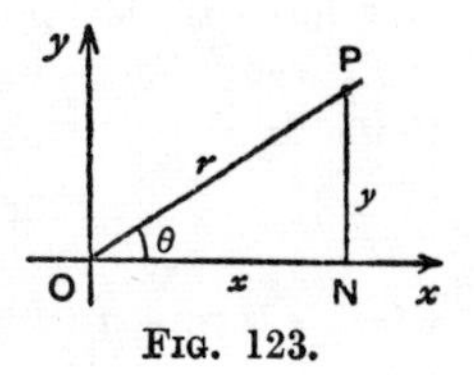

Fig. 123.

$$x = r\cos\theta, \quad y = r\sin\theta$$

and

$$r = \sqrt{(x^2 + y^2)},$$

$$\cos\theta : \sin\theta : 1 = x : y : \sqrt{(x^2 + y^2)}.$$

According to this definition, r is intrinsically positive, and for many purposes this restriction is desirable. But the polar equations of some curves can be expressed more concisely by allowing r to take both positive and negative values ; for example, the curve $r^2 = a^2 \cos^2 2\theta$, which consists of 4 loops, can be written in the simpler form $r = a \cos 2\theta$ if r is allowed to assume both positive and negative values.

Length of an Arc in Polar Coordinates.

If (r, θ) are the polar coordinates of any point on a curve and (x, y) its cartesian coordinates,

$$dx = d(r\cos\theta) = \cos\theta\, dr - r\sin\theta\, d\theta,$$

and

$$dy = d(r\sin\theta) = \sin\theta\, dr + r\cos\theta\, d\theta\ ;$$

squaring and adding

$$(dx)^2 + (dy)^2 = (dr)^2 + r^2(d\theta)^2.$$

Hence if r and θ are given functions of a parameter t,

$$\left(\frac{dx}{dt}\right)^2 + \left(\frac{dy}{dt}\right)^2 = \left(\frac{dr}{dt}\right)^2 + \left(r\frac{d\theta}{dt}\right)^2.$$

Therefore from (iii) on p. 326, if t increases steadily from t_1 at A to t_2 at B and if $\frac{dr}{dt}$ and $\frac{d\theta}{dt}$ are continuous, the length of the arc AB is given by

$$s = \int_{t_1}^{t_2} \sqrt{\left\{\left(\frac{dr}{dt}\right)^2 + \left(r\frac{d\theta}{dt}\right)^2\right\}}\, dt$$

and, in particular, putting $\theta = t$ for the curve $r = f(\theta)$ and $r = t$ for the curve $\theta = g(r)$, it follows that the length of arc from (r_1, θ_1) to (r_2, θ_2) is

$$s = \int_{\theta_1}^{\theta_2} \sqrt{\left\{r^2 + \left(\frac{dr}{d\theta}\right)^2\right\}}\, d\theta$$

where s is measured so that it increases with θ;

and

$$s = \int_{r_1}^{r_2} \sqrt{\left\{1 + \left(r\frac{d\theta}{dr}\right)^2\right\}}\, dr$$

where s is measured so that it increases with r.

Example 4. Find the length of the arc of the parabola $r = a \sec^2 \frac{1}{2}\theta$ from $\theta = 0$ to $\theta = \frac{1}{2}\pi$.

$$\log r = \log a - 2 \log \cos \tfrac{1}{2}\theta, \quad \therefore \frac{1}{r}\frac{dr}{d\theta} = \tan \tfrac{1}{2}\theta;$$

$$\therefore \left(\frac{ds}{d\theta}\right)^2 \equiv r^2 + \left(\frac{dr}{d\theta}\right)^2 = r^2(1 + \tan^2 \tfrac{1}{2}\theta);$$

$$\therefore \frac{ds}{d\theta} = r \sec \tfrac{1}{2}\theta = a \sec^3 \tfrac{1}{2}\theta;$$

$$\therefore s = a\int_0^{\pi/2} \sec^3 \tfrac{1}{2}\theta \, d\theta = 2a\int_0^{\pi/4} \sec^3 u \, du, \text{ where } \theta = 2u.$$

But $\dfrac{d}{du}\dfrac{\sin u}{\cos^2 u} = \dfrac{1}{\cos u} + \dfrac{2\sin^2 u}{\cos^3 u} = \dfrac{2}{\cos^3 u} - \dfrac{1}{\cos u}$,

$$\therefore 2\int \sec^3 u \, du = \int \sec u \, du + \sin u \sec^2 u$$
$$= \log(\sec u + \tan u) + \sin u \sec^2 u + c.$$

$$\therefore s = a\Big[\log(\sec u + \tan u) + \sin u \sec^2 u\Big]_0^{\pi/4}$$
$$= a\{\log(\sqrt{2} + 1) + \sqrt{2}\}.$$

Note. Alternatively, $\int \sec^3 u \, du = \int \sec u \, d(\tan u)$
$= \int \sqrt{(1 + t^2)}\, dt$, where $t = \tan u$,

and this integral is evaluated by putting $t = \text{sh}\, z$.

EXERCISE XVI c

Express in polar coordinates the cartesian equations in Nos. 1-3:

1. $xy = c^2$.
2. $x^2 + y^2 = 2ax$.
3. $x^2y^2 = a^2(x^2 + y^2)$.

Express in cartesian coordinates the polar equations in Nos. 4-6:

4. $r \cos(\theta - \alpha) = p$.
5. $r^2 \cos 2\theta = a^2$.
6. $r^2 = a^2 \sin 2\theta$.

7. Draw the circle $r = a \cos \theta$ and use it to sketch the cardioid $r = a(1 + \cos\theta)$. Prove that for this curve $ds = 2a \cos \frac{1}{2}\theta \, d\theta$, and find its total length.

8. Find the length of the arc of the curve $r = a \sin \theta$ from $\theta = 0$ to $\theta = \alpha$. Explain the result geometrically.

9. Prove that the length of the arc of the equiangular spiral $r = ae^{\theta \cot \alpha}$ from (r_1, θ_1) to (r_2, θ_2) is $(r_1 \sim r_2) \sec \alpha$.

10. For the curve $r^m = a^m \sin m\theta$, prove that $\dfrac{dr}{ds} = \pm \cos m\theta$.

11. Find the length of the arc of the curve $r = a\cos^4 \frac{1}{4}\theta$ from $\theta = 0$ to $\theta = 2\pi$.

12. Sketch the limaçon $r = 1 + 2\cos\theta$, and show that its total length is $2\int_0^\pi \sqrt{(5 + 4\cos\theta)}\, d\theta$.

13. Sketch the rectangular hyperbola $r^2 \cos 2\theta = a^2$ and the lemniscate $r^2 = a^2 \cos 2\theta$, and obtain integral formulae for the lengths of their arcs from $\theta = 0$ to $\theta = \theta_1$.

14. The polar coordinates of any point on a curve are given by $r = 2a \sec t$, $\theta = \tan t - t$, $-\frac{1}{2}\pi < t < \frac{1}{2}\pi$. Prove that $s = a\tan^2 t$, where s is measured from the point $t = 0$.

15. Sketch the curve $r = a\theta$ and find the length of the arc from $\theta = 0$ to $\theta = \frac{3}{4}$.

Differential Relations in Polar Coordinates

In Fig. 124, P is a point (r, θ) on a curve whose polar equation is $r = f(\theta)$, and s denotes the length of arc from a fixed point A to P, measured so that s increases with θ. (It is possible to arrange that s increases with θ for all ordinary curves, though it may be necessary to consider various portions of a curve separately each with its own origin for s.) A line TP drawn along the tangent at P in the direction of s increasing has been called (see Fig. 124) the *positive* direction of the tangent at P, and one of the angles from the initial line $\theta = 0$ to this positive direction is denoted by ψ. Thus ψ has been defined as on p. 324 with $\theta = 0$ and $\theta = \frac{1}{2}\pi$ as cartesian axes. But t and θ need not increase together and so the positive tangent may be different for the equations $\{x = f(t),\ y = g(t)\}$ and $r = h(\theta)$ of the same curve.

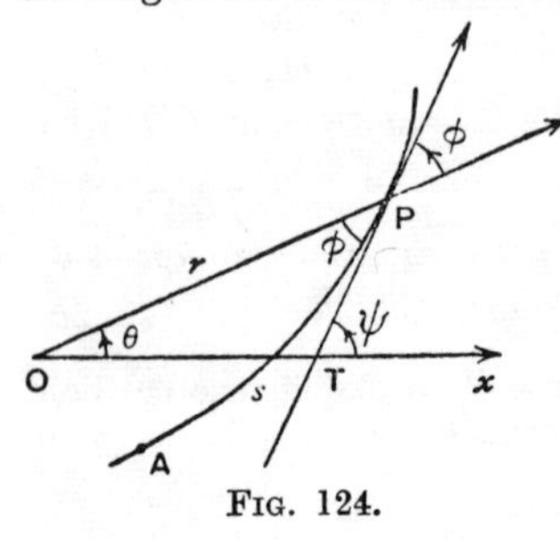

FIG. 124.

The angle from the radius vector (half-line) which makes an angle θ with Ox to the positive direction of the tangent at P (r, θ) is denoted by ϕ; it is measured positively (i.e. counter-clockwise); see Fig. 124. It is convenient to limit ϕ to the range 0 (inclusive) to 2π (exclusive). Thus

$$0 \leqslant \varphi < 2\pi \quad \text{and} \quad \theta + \varphi = \psi \ (\text{mod } 2\pi).$$

Hence $$\cos\phi = \cos\psi\cos\theta + \sin\psi\sin\theta$$

$$= \frac{dx}{ds}\frac{x}{r} + \frac{dy}{ds}\frac{y}{r} = \frac{xdx + ydy}{rds};$$

but $x^2 + y^2 = r^2$, $\therefore\ xdx + ydy = rdr$;

$$\therefore\ \cos\phi = \frac{dr}{ds}.$$

Also, $$\sin\phi = \sin\psi\cos\theta - \cos\psi\sin\theta$$

$$= \frac{dy}{ds}\frac{x}{r} - \frac{dx}{ds}\frac{y}{r} = \frac{xdy - ydx}{rds};$$

but $\frac{y}{x} = \tan\theta$, $\therefore\ \frac{xdy - ydx}{x^2} = \sec^2\theta\, d\theta = \frac{r^2 d\theta}{x^2}$;

$$\therefore\ \sin\phi = \frac{rd\theta}{ds}.$$

Hence $$\tan\phi = \frac{rd\theta}{ds} \div \frac{dr}{ds} = \frac{rd\theta}{dr}.$$

These important results,

$$\tan\phi = \frac{rd\theta}{dr}, \quad \sin\phi = \frac{rd\theta}{ds}, \quad \cos\phi = \frac{dr}{ds},$$

are best remembered by means of a right-angled triangle PQR, which is suggested by the approximate triangle $P_1Q_1R_1$ in Fig. 125.

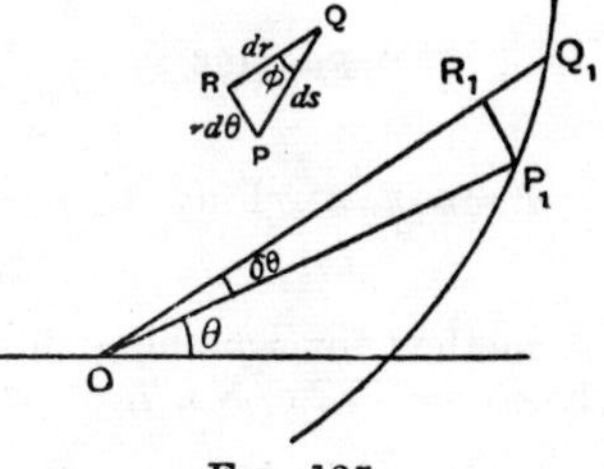

FIG. 125.

If P_1 is (r, θ) and Q_1 is $(r + \delta r,\ \theta + \delta\theta)$ and P_1R_1 is a circular arc, centre O, we have

$$P_1R_1 = r\delta\theta,$$
$$R_1Q_1 = \delta r,$$
$$\text{arc } P_1Q_1 = \delta s;$$

also $\phi = \lim \angle P_1Q_1R_1$; and these facts suggest that

$$\tan\phi \simeq \frac{r\delta\theta}{\delta r}, \sin\phi \simeq \frac{r\delta\theta}{\delta s}, \cos\phi \simeq \frac{\delta r}{\delta s}$$

and so serve to recall the accurate results proved above.

Since s is measured so as to increase with θ, $\sin\phi$ has the same sign as r; hence $0 \leqslant \phi \leqslant \pi$ if $r > 0$ and $\pi \leqslant \phi \leqslant 2\pi$ if $r < 0$.

p, r Equation of a Curve.

The length of the perpendicular OY from the pole O to the tangent at P is denoted by p, and the projection YP of OP on the positive tangent at P is denoted by t.

Hence $$p = r \sin \varphi, \quad t = r \cos \varphi.$$

[p may be defined formally as the projection of $\overrightarrow{OP}$ on the directed line which makes an angle $-\frac{1}{2}\pi$ with the positive direction of the tangent at P.]

Since $p = r \sin \phi = r^2 d\theta/ds$ and since s is measured so as to increase with θ, p is never negative; $(p, \psi - \frac{1}{2}\pi)$ are polar coordinates of Y.
The relation between p and r which holds for a given curve is sometimes called the *pedal equation* of the curve.

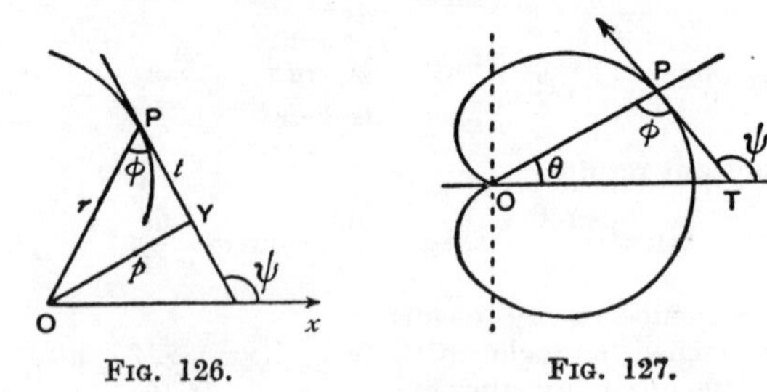

FIG. 126. FIG. 127.

Example 5. Find the p, r equation of the cardioid
$$r = a(1 + \cos \theta).$$

A method for sketching this curve, see Fig. 127, is indicated in Exercise XVI c, No. 7.

By logarithmic differentiation, since $r = 2a \cos^2 \frac{1}{2}\theta$,

$$\frac{1}{r}\frac{dr}{d\theta} = -\tan \tfrac{1}{2}\theta;$$

$$\therefore \cot \phi = -\tan \tfrac{1}{2}\theta = \cot(\tfrac{1}{2}\pi + \tfrac{1}{2}\theta);$$

$$\therefore \text{in Fig. 127, } \angle \text{OPT} \equiv \phi = \tfrac{1}{2}\pi + \tfrac{1}{2}\theta.$$

Hence $p = r \sin \phi = r \cos \frac{1}{2}\theta$; $\therefore p^2 = r^2 \cos^2 \frac{1}{2}\theta = \frac{1}{2}r^2(1 + \cos \theta)$

$$\therefore 2ap^2 = r^3.$$

The method of Example 5 may be used to express p in terms of r, θ for any given curve.

Since $p = r\sin\phi$, $\dfrac{1}{p^2} = \dfrac{1}{r^2}\operatorname{cosec}^2\phi = \dfrac{1}{r^2}(1 + \cot^2\phi)$;

but $\tan\phi = \dfrac{rd\theta}{dr}$, $\quad \therefore \dfrac{1}{p^2} = \dfrac{1}{r^2} + \dfrac{1}{r^4}\left(\dfrac{dr}{d\theta}\right)^2$.

It is often convenient to put $\dfrac{1}{r} = u$;

then
$$\frac{dr}{d\theta} = \frac{dr}{du}\,\frac{du}{d\theta} = -\frac{1}{u^2}\,\frac{du}{d\theta};$$
$$\therefore \frac{1}{p^2} = u^2 + \left(\frac{du}{d\theta}\right)^2.$$

EXERCISE XVI d

Find ϕ and ψ in terms of θ for the curves in Nos. 1-3; also sketch the curves and find their p, r equations:

1. $r = a(1 - \cos\theta)$. **2.** $\log r = \theta$. **3.** $r^2 = a^2 \sin 2\theta$.

4. Prove that the p, r equation of the curve $r\theta = c$ is $p^{-2} - r^{-2} = c^{-2}$.

5. Prove that the p, r equation of the curve $x^2 - y^2 = a^2$ is $pr = a^2$.

6. Prove that the curves $r = a\cos\theta$, $r = a(1 - \cos\theta)$ intersect at the point $(\frac{1}{2}a, \frac{1}{3}\pi)$ and find their angle of intersection.

Find the p, r equations of the curves in Nos. 7-9:

7. $r = a\theta$. **8.** $r = a\operatorname{sech}\theta$. **9.** $r^n \sin n\theta = a^n$.

10. Use the relation, $\dfrac{1}{p^2} = u^2 + \left(\dfrac{du}{d\theta}\right)^2$, to find the p, r equation of the conic, $lu = 1 + e\cos\theta$, where $u = \dfrac{1}{r}$.

11. If the tangents at P, Q to $r = a(1 - \cos\theta)$ are parallel, and if O is the pole, prove that $\angle POQ = \frac{2}{3}\pi$.

12. Prove that the curves $r^2\cos(2\theta - \alpha) = c^2 \sin 2\alpha$ and $r^2 = 2c^2 \sin(2\theta + \alpha)$ cut orthogonally.

p, ψ Equations.

If Q is any point (r, θ) on the tangent at P to a given curve, $\angle \text{OQP} = \psi - \theta$, see Fig. 128,

and $$\text{OY} = \text{OQ} \sin(\psi - \theta).$$

Therefore the equation of the tangent at P is

$$r \sin(\psi - \theta) = p$$

or, in cartesian coordinates,

$$x \sin\psi - y \cos\psi = p,$$

where p is positive, retaining the polar conventions for p, ψ.

These forms are often convenient for curves for which there is a simple relation between p and ψ, and this relation is sometimes called the *tangential-polar equation* of the curve. Methods of finding it are illustrated in Examples 6, 7.

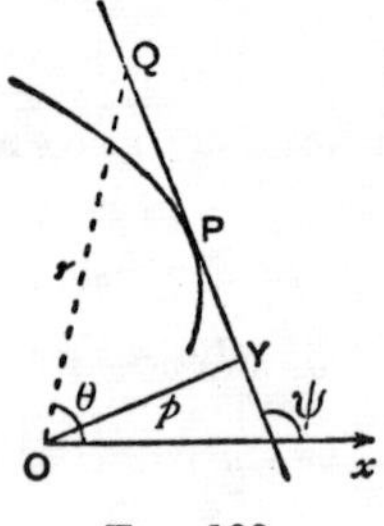

FIG. 128.

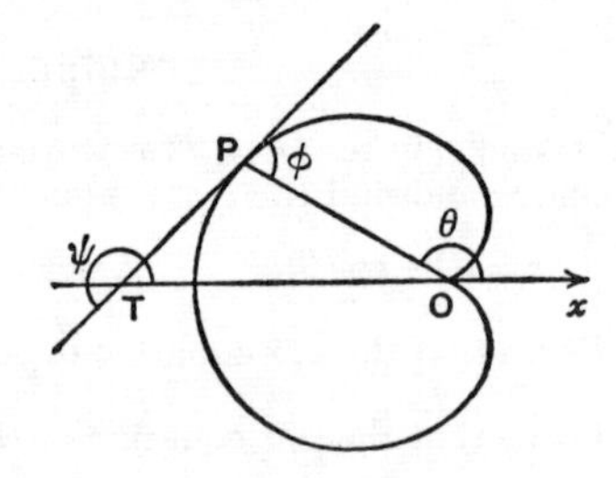

FIG. 129.

Example 6. Find the p, ψ equation of the cardioid,

$$r = a(1 - \cos\theta), \ (a > 0).$$

The curve can be described completely by allowing θ to vary continuously from 0 to 2π.
By logarithmic differentiation, since $r = 2a \sin^2 \frac{1}{2}\theta$

$$\cot\phi \equiv \frac{1}{r}\frac{dr}{d\theta} = \cot\tfrac{1}{2}\theta;$$

and r is positive,

$$\therefore 0 \leqslant \phi \leqslant \pi; \quad \therefore \phi = \tfrac{1}{2}\theta, \ (0 \leqslant \theta \leqslant 2\pi); \quad \therefore \psi = \theta + \phi = \tfrac{3}{2}\theta,$$

since we suppose ψ to vary continuously from 0 to 3π as θ varies from 0 to 2π.

Also, $p = r\sin\phi = 2a\sin^2\tfrac{1}{2}\theta\sin\tfrac{1}{2}\theta = 2a\sin^3\tfrac{1}{2}\theta$;

$$\therefore p = 2a\sin^3\tfrac{1}{3}\psi \ (0 \leqslant \psi \leqslant 3\pi).$$

Note. If the equation of a curve is given in the form $p = f(\psi)$, it is implied that ψ is confined to values for which $f(\psi) \geqslant 0$. Therefore, in the equation $p = 2a\sin^3\tfrac{1}{3}\psi$, $(a > 0)$, values between 3π and 6π are not assigned to ψ; these values would be used for $p = -2a\sin^3\tfrac{1}{3}\psi$ which represents the cardioid $r = -a(1-\cos\theta)$, $(a > 0)$.

Example 7. Find the p, ψ equation of the ellipse, $\dfrac{x^2}{a^2} + \dfrac{y^2}{b^2} = 1$.

Let (x_1, y_1) be the point of contact of the tangent,

$$x\sin\psi - y\cos\psi = p.$$

But the tangent at (x_1, y_1) is $\dfrac{xx_1}{a^2} + \dfrac{yy_1}{b^2} = 1$; therefore comparing coefficients

$$\frac{x_1}{a^2\sin\psi} = \frac{y_1}{-b^2\cos\psi} = \frac{1}{p}.$$

But $\left(\dfrac{x_1}{a}\right)^2 + \left(\dfrac{y_1}{b}\right)^2 = 1$, $\quad\therefore p^2 = a^2\sin^2\psi + b^2\cos^2\psi$.

Geometrical Meaning of $\dfrac{dp}{d\psi}$.

Since $p = r\sin\phi$, $\quad dp = \sin\phi\, dr + r\cos\phi\, d\phi$;

but $\sin\phi\, dr = r\cos\phi\, d\theta$, (p. 335),

$$\therefore dp = r\cos\phi\, d\theta + r\cos\phi\, d\phi = r\cos\phi\, d(\theta+\phi)$$
$$= r\cos\phi\, d\psi;$$

$$\therefore \frac{dp}{d\psi} = r\cos\phi = t = \mathsf{YP}, \text{ see Fig. 126, p. 336.}$$

Pedal Curves. Given a curve and a fixed point O, a new curve can be derived by taking the locus of the foot of the perpendicular from O to a variable tangent of the given curve. This locus is called the *pedal* of the given curve with respect to O.

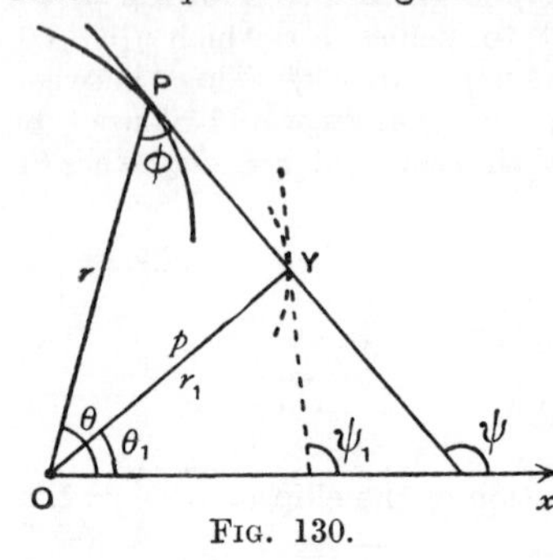

FIG. 130.

Thus in Fig. 130 the locus of Y is the pedal of the locus of P with respect to O.

An important property of pedals is that, if $r > 0$,

the φ of the pedal equals the φ of the given curve.

Denote the polar coordinates of Y by r_1, θ_1;

then $r_1 = p, \quad \therefore dr_1 = dp$;

and $\theta_1 = \psi - \frac{1}{2}\pi$,

$\therefore d\theta_1 = d\psi$.

$$\therefore r_1 \frac{d\theta_1}{dr_1} = p\frac{d\psi}{dp} = p/(r \cos \phi) = \tan \phi.$$

$\therefore$ the ϕ of the pedal equals the ϕ of the original curve; $(\phi < \pi)$.

A geometrical proof of this result is suggested in Exercise XVI e, No. 11.

By means of this property it is easy to deduce the p, r equation of the pedal locus from the p, r equation of the given curve. The method is illustrated in the next example.

Example 8. Find the p, r equation of the pedal with respect to the pole of the cardioid $r = a(1 + \cos \theta)$, $(a > 0)$.

We start by finding the p, r equation of the cardioid; this has been done in Example 5, p. 336; the equation is $2ap^2 = r^3$.
If suffixes refer to the pedal locus,

$$\frac{p_1}{r_1} = \sin \phi_1 = \sin \phi = \frac{p}{r};$$

$$\text{also } p = r_1, \quad \therefore r = \frac{r_1^2}{p_1};$$

$$\therefore 2ar_1^2 = \left(\frac{r_1^2}{p_1}\right)^3, \text{ that is, } 2ap_1^3 = r_1^4.$$

Hence, dropping the suffixes, the required equation is

$$2ap^3 = r^4.$$

The polar equation of the pedal locus can be deduced immediately from the p, ψ equation of the given curve by writing r for p and $\theta + \frac{1}{2}\pi$ for ψ.

Thus, from Example 6, it follows that the polar equation of the pedal locus of the cardioid $r = a(1 - \cos\theta)$ with respect to the pole is $r = 2a\sin^3(\frac{1}{3}\theta + \frac{1}{6}\pi)$, $(-\frac{1}{2}\pi < \theta < \frac{5}{2}\pi)$, where $a > 0$.

Miscellaneous Transformations.

(i) Given the p, ψ equation $p = f(\psi)$, to find parametric cartesian equations. (It is implied that $p > 0$).

From pp. 336, 339, $r\sin\phi = p = f(\psi)$ and $r\cos\phi = \dfrac{dp}{d\psi} = f'(\psi)$;

$$\therefore x \equiv r\cos\theta = r\cos(\psi - \phi) = r\cos\psi\cos\phi + r\sin\psi\sin\phi;$$

$$\therefore x = \cos\psi f'(\psi) + \sin\psi f(\psi)$$

Similarly $$y = \sin\psi f'(\psi) - \cos\psi f(\psi).$$

(ii) Given the p, r equation $p = f(r)$, to find the polar equation.

Since $p = r\sin\phi$ and $\tan\phi = r\,d\theta/dr$, or from the equivalent relation (p. 337), $\dfrac{1}{p^2} = \dfrac{1}{r^2} + \dfrac{1}{r^4}\left(\dfrac{dr}{d\theta}\right)^2$, we obtain $\dfrac{d\theta}{dr}$ in terms of r. The relation between r and θ is then found by integration.

Example 9. Find the polar equation of the curve $p^2 = ar$.

$$\sin^2\phi = \frac{p^2}{r^2} = \frac{a}{r}, \quad \therefore \cos^2\phi = \frac{r-a}{r};$$

$$\therefore r\frac{d\theta}{dr} \equiv \tan\phi = \pm\sqrt{\frac{a}{r-a}}; \quad \therefore \theta = \pm\sqrt{a}\int\frac{dr}{r\sqrt{(r-a)}}.$$

Put $r - a = az^2$, then $dr = 2az\,dz$,

$$\therefore \theta = \pm\sqrt{a}\int\frac{2az\,dz}{a(1+z^2)z\sqrt{a}} = \pm 2\int\frac{dz}{1+z^2};$$

$$\therefore \theta = \pm 2\tan^{-1}z = \pm 2\tan^{-1}\sqrt{\frac{r-a}{a}}, \text{ if } \theta = 0 \text{ when } r = a.$$

$$\therefore r - a = a\tan^2\tfrac{1}{2}\theta;$$

$$\therefore r = a\sec^2\tfrac{1}{2}\theta.$$

EXERCISE XVI e

1. For the cardioid $r = a(1 + \cos\theta)$, prove that $p = 2a\cos^3(\frac{1}{3}\psi - \frac{1}{6}\pi)$, and deduce the polar equation of its pedal with respect to the pole.

2. For the curve whose polar equation is $r^2 = a^2 \sin 2\theta$, prove that $p^2 = a^2 \sin^3 \frac{2}{3}\psi$.

3. Find the relation between p and ψ for the astroid $x = a\cos^3 t$, $y = a\sin^3 t$. What is the polar equation of its pedal with respect to the origin? Prove that perpendicular tangents to the astroid intersect on the curve $r^2 = \frac{1}{2}a^2 \cos^2 2\theta$.

4. Prove that $ds/d\psi = r\,dr/dp$.

Find the p, r equations of the pedals with respect to the pole of the curves in Nos. 5-7:

5. $p^2 = ar$. 6. $p = r\sin\alpha$. 7. $r^2 = a^2\cos 2\theta$.

Find the polar equations of the pedals with respect to the pole of the curves in Nos. 8-10:

8. $r = ae^\theta$. 9. $r = 2a\cos\theta$. 10. $r^2 \sin 2\theta = a^2$.

11. Draw a figure showing two neighbouring points Y, Y′ on the pedal corresponding to neighbouring points P, P′ on the original curve. Let the tangents at P, P′ meet at K, and produce Y′Y to T′. Prove that OYY′K is a cyclic quadrilateral and that ∠OYT′ = ∠OKY′. What result is obtained by making P′ → P along the curve?

12. Find a cartesian equation of the curve $p = a\sin 2\psi$.

13. Obtain for the curve $p = a\cos 3\psi$ the parametric equations $x = a(\sin 2t + 2\sin t)$, $y = a(\cos 2t - 2\cos t)$, and show that the p, r equation is $r^2 + 8p^2 = 9a^2$.

Find polar equations of the curves in Nos. 14-16:

14. $2ap = r^2$. 15. $p = r\sin\alpha$. 16. $2ap^2 = r^3$.

17. Prove that $p^2 + \left(\dfrac{dp}{d\psi}\right)^2 = r^2$ and deduce that $p + \dfrac{d^2p}{d\psi^2} = r\dfrac{dr}{dp}$.

18. The p, ψ equation of a curve is $p^2 = a^2\cos 2\psi$; prove that the p, ψ equation of the pedal with respect to the origin is $p^2 = a^2\cos^3\frac{2}{3}\psi$.

CHAPTER XVII

CURVATURE AND ENVELOPES

FIG. 131 shows a tangent to a curve at P. Using the sign conventions adopted in Chapter XVI, p. 324, denote the length of the arc AP, measured from some fixed point A, by s, and the angle which the tangent makes with the x-axis by ψ.

As P moves along the curve, s and ψ vary and the rate at which ψ increases relative to s is called the **curvature** of the curve at the corresponding point P. It is measured by $\frac{d\psi}{ds}$ and is denoted by κ, thus

$$\kappa = \frac{d\psi}{ds}.$$

If ψ increases in the direction in which s is measured, the curvature is positive, and if ψ decreases as s increases, the curvature

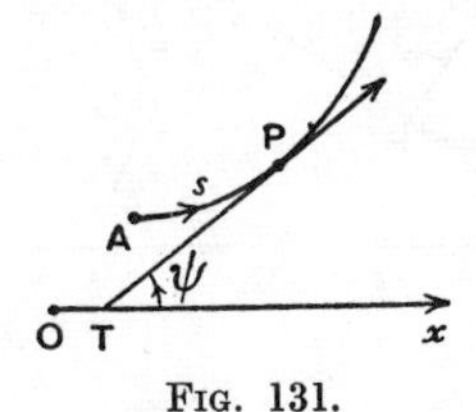

FIG. 131.

is negative. Its value does not depend on the position of the point A or of the line Ox from which s and ψ are measured, but its sign depends on the sense in which s is measured for which conventions have been made on pp. 324, 334; if the equation of the curve is given in a form to which these conventions do not apply it is only possible to evaluate $|\kappa|$, unless new conventions are introduced.

Roughly speaking, the curvature is the " rate at which the curve curves " ; its sign indicates the direction in which the tangent is turning as s increases.

Example 1. Find the curvature at any point of a circle of radius a.

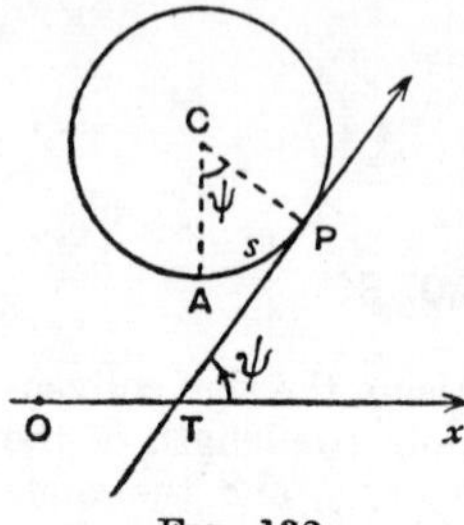

FIG. 132.

Measure s from the point A where $\psi=0$ and in the direction in which ψ increases; denote the centre of the circle by C.

Then $\angle ACP=\psi$ and $s=a\psi$.

$$\therefore \kappa\equiv\frac{d\psi}{ds}=\frac{1}{a}.$$

This is a natural result because a small circle evidently curves more rapidly than a large one.

The direction of the positive tangent at P was defined on p. 324; the line PG which makes an angle $+\frac{1}{2}\pi$ with the positive tangent at P is called the *positive normal* at P. The arrows in Figs. 133 (*a*), (*b*) show the positive tangent and normal in relation to the curve, in the cases where ψ increases with s and where ψ decreases with s, respectively.

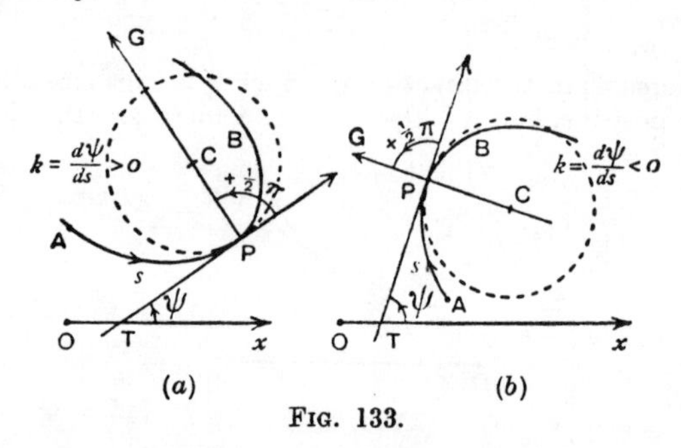

FIG. 133.

The reciprocal of the curvature at any point P on a curve is called the radius of curvature at P because it gives the radius of the circle which has curvature equal to that of the curve at P; the radius of curvature is always denoted by ρ, thus

$$\rho=\frac{1}{\kappa}=\frac{ds}{d\psi}.$$

In Fig. 133*a* the value of ρ at P is positive, and in Fig. 133*b* it is negative.

If a length **PC** equal to ρ is measured from **P** along the positive normal, the point **C** is called the **centre of curvature** at **P**, and the circle, centre **C** and radius **CP**, is called the **circle of curvature** at **P**. Fig. 133*a* and Fig. 133*b* show the positions of **C** where ρ is positive and negative respectively; in each case **C** lies on the "inward drawn" normal, but in Fig. 133*b* the positive normal is outward.

The circle of curvature at **P** not only has the same curvature, both as to magnitude and sign, as the curve at **P**, but also touches the curve at **P**. It is, roughly speaking, the "best fitting" circle at the point.

The circle of curvature usually crosses the curve as well as touching it at **P**. Suppose, as in Fig. 133*a*, that the curvature of the curve increases with s along the arc **APB**. The circle of curvature at **P** curves at the same rate as the curve is curving at **P** and therefore more rapidly than the curve at points between **P** and **A** and less rapidly than the curve at points between **P** and **B**. Hence the circle lies inside the arc **AP** and outside the arc **PB**.

The relation of a curve and its circle of curvature may be compared to that of a curve and an inflexional tangent.

Example 2. Find the radius of curvature of the catenary $s=c\tan\psi$ at its vertex $\psi=0$.

$$\rho\equiv\frac{1}{\kappa}\equiv\frac{ds}{d\psi}=c\sec^2\psi;$$

$$\therefore \text{ at } \psi=0, \quad \rho=c.$$

Note. Since $\kappa=\frac{1}{c}\cos^2\psi$, $\kappa<\frac{1}{c}$ both for $\psi<0$ and for $\psi>0$; therefore the catenary curves more slowly than the circle of curvature on each side of the vertex. Thus the circle of curvature at the vertex does not cross the curve but lies wholly inside it; but at all other points on the curve, the circle of curvature crosses the curve.

Cartesian Equations. Since the intrinsic equation of a curve is not always available in a convenient form, it is necessary to find formulae for the curvature applicable to cartesian and other equations.

(i) To find κ for the curve $y=f(x)$.

For brevity denote $\frac{dy}{dx}, \frac{d^2y}{dx^2}$ by y_1, y_2.

$$\kappa=\frac{d\psi}{ds}=\frac{d\tan^{-1}y_1}{dx}\frac{dx}{ds}.$$

But by the convention on p. 324, dx/ds is positive,

$$\therefore\ \kappa=\frac{y_2}{1+y_1{}^2}\frac{1}{\sqrt{(1+y_1{}^2)}};$$

$$\therefore\ \kappa=\frac{d^2y}{dx^2}\bigg/\sqrt{\left\{1+\left(\frac{dy}{dx}\right)^2\right\}^3}.$$

Thus the sign of κ is the same as that of $\frac{d^2y}{dx^2}$ and determines the direction of concavity; see p. 71 (Vol. I).

In particular at a point on the curve where $\psi=0$, $\kappa=y_2$.

(ii) To find κ for the curve given by $x=f(t)$, $y=g(t)$.

It is convenient to use dots to denote differentiation with respect to t; thus $\dot{x}\equiv\frac{dx}{dt}, \ddot{x}\equiv\frac{d^2x}{dt^2}, \tan\psi=\frac{dy}{dx}=\frac{\dot{y}}{\dot{x}}$, etc.

$$\kappa=\frac{d\psi}{ds}=\frac{d\psi}{dt}\frac{dt}{ds}=\frac{d}{dt}\left(\tan^{-1}\frac{\dot{y}}{\dot{x}}\right)\div\frac{ds}{dt}.$$

But by the convention on p. 324, ds/dt is positive,

$$\therefore\ \kappa=\frac{1}{1+\frac{\dot{y}^2}{\dot{x}^2}}\frac{\dot{x}\ddot{y}-\dot{y}\ddot{x}}{\dot{x}^2}\div\sqrt{(\dot{x}^2+\dot{y}^2)}$$

$$=(\dot{x}\ddot{y}-\dot{y}\ddot{x})/\sqrt{(\dot{x}^2+\dot{y}^2)^3}.$$

(iii) To find κ at the origin for the curve $y=f(x)$.

Suppose that near the origin the equation of the curve can be expressed in the form

$$y=f(x)=ax+bx^2+cx^3g(x),$$

where $g(x)$ and its derivatives are numerically less than some constant k for all values of x near $x=0$.

$$\text{Then } f'(0)=a \quad\text{and}\quad f''(0)=2b.$$

$$\therefore \text{ at the origin, } \kappa=2b/\sqrt{(1+a^2)^3}.$$

If the curve touches Ox at O, $a=f'(0)=0$, and $\kappa=2b$,

$$\therefore \frac{y}{x^2}=b+cx\,g(x)\,; \quad \therefore b=\lim_{x\to 0}\frac{y}{x^2}.$$

$$\therefore \kappa=\lim_{x\to 0}\frac{2y}{x^2}.$$

This is known as *Newton's formula.*

Note. A more detailed investigation of the form of a curve near the origin is made in Chapter XXI.

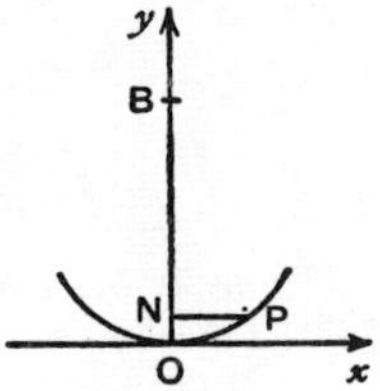

Alternatively, Newton's formula may be obtained as follows :

We shall assume that if a circle is drawn to touch the curve at O and pass through another point P (x, y) on the curve, then the limit of that circle, when $P\to 0$ along the curve, is the circle of curvature at O.

Let OB be the diameter of the circle, and draw PN perpendicular to it ; then

FIG. 134.

$$ON.NB=NP^2, \quad \therefore NB=\frac{x^2}{y}\,; \quad \therefore \rho=\tfrac{1}{2}\lim NB=\lim_{x\to 0}\frac{x^2}{2y}.$$

Example 3. Find the curvature at any point on the parabola $4ax=y^2$.

It may be shown by the method of p. 346 that, for a curve $x=\phi(y)$, if s is measured so that it increases with y,

$$\kappa=-\frac{d^2x}{dy^2}\Big/\sqrt{\left\{1+\left(\frac{dx}{dy}\right)^2\right\}^3}.$$

For the curve $x=\dfrac{y^2}{4a}$, $\dfrac{dx}{dy}=\dfrac{y}{2a}$ and $\dfrac{d^2x}{dy^2}=\dfrac{1}{2a}$;

$$\therefore \kappa=-\frac{1}{2a}\Big/\sqrt{\left(1+\frac{y^2}{4a^2}\right)^3}=-1\Big/\left\{2a\sqrt{\left(1+\frac{x}{a}\right)^3}\right\}.$$

Note. If the curve $x=\phi(y)$ is represented by parametric equations $x=\phi(t)$, $y=t$ the convention adopted in Example 3 for the measurement of s agrees with that given on p. 324 for the curve $x=f(t)$, $y=g(t)$.

Example 4. Find the curvature at any point t of the semi-cubical parabola given by $x=at^2$, $y=at^3$.

Since $\dot{x}=2at$ and $\dot{y}=3at^2$,

$$\dot{s}^2 \equiv \dot{x}^2+\dot{y}^2=a^2t^2(4+9t^2) \text{ and } \tan\psi=\dot{y}/\dot{x}=\tfrac{3}{2}t\,;$$

$$\therefore\ \kappa=\frac{d\psi}{ds}=\dot{\psi}/\dot{s}$$

$$=\left\{\frac{d}{dt}(\tan^{-1}\tfrac{3}{2}t)\right\}\Big/\sqrt{\left\{a^2t^2(4+9t^2)\right\}}$$

$$=\frac{\tfrac{3}{2}}{1+\tfrac{9}{4}t^2}\Big/\sqrt{\left\{a^2t^2(4+9t^2)\right\}}$$

$$=6/\sqrt{\{a^2t^2(4+9t^2)^3\}}.$$

Example 5. Find the value of $|\rho|$ at the origin for the ellipse $3x^2+4y^2=5y$.

$$\lim_{x\to 0}\frac{x^2}{2y}=\lim_{y\to 0}\frac{5y-4y^2}{6y}=\lim_{y\to 0}\left(\tfrac{5}{6}-\tfrac{2}{3}y\right)=\tfrac{5}{6}.$$

$\therefore$ by Newton's formula $|\rho|=\frac{5}{6}$.

EXERCISE XVII a

1. Find the curvature of the cycloid $s=4a\sin\psi$ at $\psi=\alpha$.

2. Find the radius of curvature of the tractrix $\cos\psi=e^{-s/c}$ at $\psi=\alpha$.

3. Find the intrinsic equation of the curve for which $\rho=2(s+a)$.

4. Find ρ in terms of x for the curve $y=a\log\sin(x/a)$.

5. Prove that the curvature of the catenary $y=c\operatorname{ch}(x/c)$ is c/y^2.

6. Prove for the rectangular hyperbola $xy=c^2$ that $2c^2\rho=r^3$.

7. Find ρ at the point $t=\frac{1}{4}\pi$ on the ellipse given by

$$x=a\cos t,\ y=b\sin t.$$

8. Find ρ in terms of t for the cycloid

$$x=a(t+\sin t),\ y=a(1-\cos t).$$

9. Find ρ in terms of ψ for the astroid

$$x=a\cos^3 t,\ y=a\sin^3 t.$$

10. Find $|\rho|$ for the parabola $y = x^2/(4a)$ at the origin.

11. Find the value of $|\kappa|$ at the origin for the ellipse

$$x^2 + 6y^2 + 2x - y = 0.$$

12. Find $|\rho|$ at (1, 0) for the curve $y^2(x^2 - 4) = 4(x^2 - 1)$.

13. Prove that the chord of curvature parallel to Oy at any point of the curve $y = a \log \sec (x/a)$ is of constant length.
[Any chord through P of the circle of curvature at P is called a *chord of curvature.*]

14. Prove that the least radius of curvature of $y = a + a^3/(x^2\sqrt{5})$ is $\frac{9}{10}a$.

15. If C is the centre of curvature at any point P on $y = c \operatorname{ch}(x/c)$ and if CP cuts Ox at G, prove that CP = PG.

16. The radius of curvature at any point of a curve which touches Ox at O is $c \sec \psi$. Prove that $x = c\psi$ and $y = c \log \sec \psi$.

17. For the curve $x^2 = a^2(\sec \psi - \tan \psi)$ prove that

$$\rho = -\tfrac{1}{2}x \sec^2 \psi.$$

18. Prove that for any curve

$$\kappa = -\frac{d^2x}{ds^2}\bigg/\frac{dy}{ds} = +\frac{d^2y}{ds^2}\bigg/\frac{dx}{ds}, \quad \text{and that } \kappa^2 = \left(\frac{d^2x}{ds^2}\right)^2 + \left(\frac{d^2y}{ds^2}\right)^2.$$

p, r Equations.

To find ρ for the curve $r = f(p)$.

$$\kappa = \frac{d\psi}{ds} = \frac{d\theta}{ds} + \frac{d\phi}{ds}, \text{ (p. 334)};$$

but $\sin \phi = r\dfrac{d\theta}{ds}$ and $\cos \phi = \dfrac{dr}{ds}$,

$$\therefore \kappa = \frac{1}{r} \sin \phi + \frac{d\phi}{dr} \cos \phi = \frac{1}{r}\frac{d}{dr}(r \sin \phi);$$

$$\therefore \kappa = \frac{1}{r}\frac{dp}{dr}; \quad \rho = r\frac{dr}{dp}.$$

Alternatively as follows:

From p. 339, $\dfrac{dp}{d\psi} = r \cos \phi = r\dfrac{dr}{ds}$; $\therefore \rho = \dfrac{ds}{d\psi} = r\dfrac{dr}{dp}$.

Polar Equations.

The expression for ρ in terms of r and θ can be obtained from the relations,

$$\frac{1}{\rho}=\frac{d\theta}{ds}+\frac{d\phi}{ds}, \quad \tan\phi=\frac{rd\theta}{dr}, \quad \sin\phi=\frac{rd\theta}{ds};$$

but the result is not easy to remember, see Exercise XVII b, No. 12. The best procedure in practice is to begin by finding the pedal equation.

Example 6. Find the radius of curvature at any point of the cardioid $r=a(1+\cos\theta)$, $(a>0)$.

As in Example 5, p. 336, the pedal equation of the curve is $2ap^2=r^3$.

$$\therefore\ 4ap\,dp=3r^2dr;$$

$$\therefore\ \rho\equiv r\frac{dr}{dp}=r\frac{4ap}{3r^2}=\frac{4a}{3r}\sqrt{\frac{r^3}{2a}};$$

$$\therefore\ \rho=\tfrac{2}{3}\sqrt{(2ar)}.$$

Since $r=2a\cos^2\frac{1}{2}\theta$, this may also be written

$$\rho=\tfrac{4}{3}a\cos\tfrac{1}{2}\theta, \text{ if } -\pi<\theta\leqslant\pi.$$

p, ψ Equations.

To find ρ for the curve $p=f(\psi)$.

Since $p=r\sin\phi$ and $\dfrac{dp}{d\psi}=r\cos\phi$, (p. 339),

$$r^2=p^2+\left(\frac{dp}{d\psi}\right)^2,$$

and regarding each term as a function of ψ,

$$2r\frac{dr}{d\psi}=2p\frac{dp}{d\psi}+2\frac{dp}{d\psi}\frac{d^2p}{d\psi^2};$$

$$\therefore\ rdr=pdp+\frac{d^2p}{d\psi^2}dp;$$

$$\therefore\ \boldsymbol{\rho\equiv r\frac{dr}{dp}=p+\frac{d^2p}{d\psi^2}}.$$

EXERCISE XVII b

[*Assume in this Exercise that a is positive.*]

1. Find ρ in terms of r for the parabola $p^2 = ar$.

2. Find ρ in terms of r for the curve $r^2p^3 = k^5$.

3. Find the curvature at the point $\theta = \frac{1}{2}\pi$ on the parabola $r(1 + \cos\theta) = 2a$.

4. Find the curvature in terms of r at any point of the equi-angular spiral $r = ae^{\theta \cot \alpha}$.

5. Prove for the curve $r^2 = a^2 \cos 2\theta$ that $3r\rho = a^2$.

6. Find ρ in terms of p for the curve $p = a \sin n\psi$. Hence prove that $r^2 + (n^2 - 1)p^2$ is constant for this curve.

7. Prove that the length of the chord of curvature through the pole of the hyperbola $r^2 \cos 2\theta = a^2$ is $2r$. [See note to Exercise XVII a, No. 13.]

8. Prove that the radius of curvature at the point $\theta = 0$ on the curve $r \cos 2\theta = a$ is $-\frac{1}{3}a$.

9. Prove that the curvature of $r^2 = a^2(1 + \cos\theta)$ at the pole is $(2\sqrt{2})/a$.

10. If $\phi = 3\theta$, prove that the circle of curvature bisects the radius vector.

11. For the curve $r^2 = p^2 + a^2$ prove that $\dfrac{d^2p}{d\psi^2} = 0$, and express s in terms of ψ, given that $s = \psi = 0$ when $r = a$.

12. By writing $\dfrac{d\psi}{ds}$ in the form $\dfrac{d\theta}{ds}\left\{1 + \dfrac{d}{d\theta}\left(\tan^{-1}\dfrac{rd\theta}{dr}\right)\right\}$, prove that

$$\kappa = \left\{r^2 + 2\left(\frac{dr}{d\theta}\right)^2 - r\frac{d^2r}{d\theta^2}\right\} \Big/ \sqrt{\left\{r^2 + \left(\frac{dr}{d\theta}\right)^2\right\}^3},$$

where s is measured so that it increases with θ.

Deduce that the curvature at the origin of each branch of the curve $r = a \sin 2\theta$ is $1/a$, and verify this result by using Newton's formula.

Envelopes. Consider the curve whose equation is

$$f(x, y, t) = 0,$$

where t is any constant.

For different values of t we obtain different curves, see Fig. 135, and these are said to form a *system* or *family*; t is called a *parameter*.

A curve, $\alpha\beta\gamma\delta$ . . . . in Fig. 135, which is touched by every member of the family, is called the *envelope* of the family.

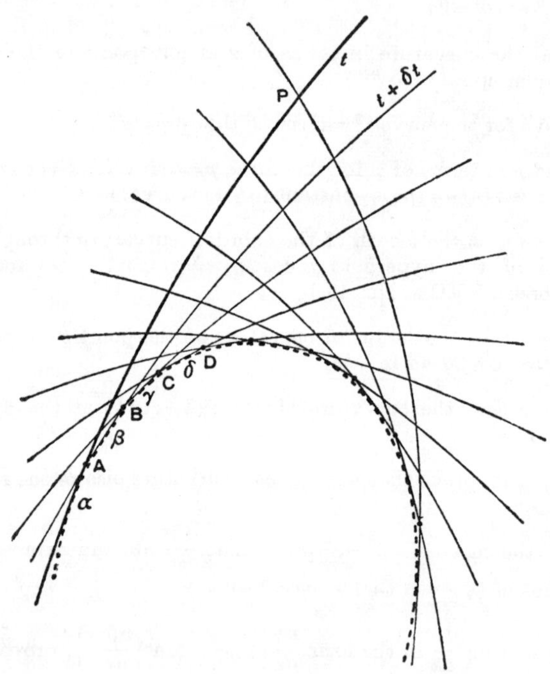

FIG. 135.

In Fig. 135, the curve **PA** whose parameter is t is supposed to touch the envelope at α and to cut the curve whose parameter is $t + \delta t$ at **A**. We shall *assume* that if the curves have continuous curvatures in the neighbourhood of **A**, the limit of **A** when $\delta t \to 0$ is α. This fact is suggested by the diagram but an analytical proof is unsuitable for an elementary course. Hence

the envelope of the system is the locus of the limit, when $\delta t \to 0$, *of the point of intersection of the curves* $f(x, y, t)=0, f(x, y, t+\delta t)=0$.

Since the coordinates of **A** are given by these equations, they satisfy

$$\frac{f(x, y, t+\delta t)-f(x, y, t)}{\delta t}=0,$$

and therefore the coordinates of the limit a of **A** satisfy

$$\lim_{\delta t \to 0} \frac{f(x, y, t+\delta t)-f(x, y, t)}{\delta t}=0,$$

and this is written $$\frac{\partial}{\partial t} f(x, y, t)=0,$$

where the symbol $\frac{\partial}{\partial t}$ is used instead of $\frac{d}{dt}$ to show that x and y remain constant when differentiating with respect to t.

Hence the coordinates of a satisfy

$$\mathbf{f}(\mathbf{x}, \mathbf{y}, \mathbf{t})=0, \quad \frac{\partial}{\partial \mathbf{t}} \mathbf{f}(\mathbf{x}, \mathbf{y}, \mathbf{t})=0,$$

and the equation of the envelope is found by eliminating t from these equations.

Example 7. Find the envelope of the family

$$t^2 f_1(x, y)+2t f_2(x, y)+f_3(x, y)=0$$

and interpret the result.

Points on the envelope are given by

$$t^2 f_1+2t f_2+f_3=0, \quad 2t f_1+2 f_2=0;$$

eliminating t,

$$f_1\left(-\frac{f_2}{f_1}\right)^2+2 f_2\left(-\frac{f_2}{f_1}\right)+f_3=0;$$

$$\therefore f_2^2=f_1 f_3.$$

If (x, y) is any point in the plane, the parameters of the two members of the family which pass through it are the roots of the quadratic, $t^2 f_1+2t f_2+f_3=0$. The condition for this quadratic to have equal roots is $f_2^2=f_1 f_3$. This is the equation of the envelope and it expresses the property that the point lies on the envelope if the two corresponding parameters are equal, a result which is suggested by Fig. 135.

Thus the envelope of the system of straight lines given by $x - ty + at^2 = 0$, where t is the parameter, can be written down by expressing the condition that the roots of the quadratic in t are equal; it is therefore the parabola $y^2 = 4ax$.

Example 8. Find the envelope of the system of curves given by $\frac{x^2}{s} + \frac{y^2}{t} = 1$, where $s + t = c^2$ and c is a given constant.

If (x, y) is any point on the envelope,

$$\frac{x^2}{s^2}ds + \frac{y^2}{t^2}dt = 0 \text{ where } ds + dt = 0;$$

$$\therefore \frac{x^2}{s^2} = \frac{y^2}{t^2};$$

$$\therefore \frac{x}{s} = \frac{y}{t} = \frac{x+y}{c^2} \quad \text{or} \quad \frac{x}{s} = \frac{-y}{t} = \frac{x-y}{c^2};$$

$$\therefore 1 = \frac{x^2}{s} + \frac{y^2}{t} = \frac{(x+y)^2}{c^2} \quad \text{or} \quad \frac{(x-y)^2}{c^2};$$

$\therefore$ the envelope is 4 straight lines $x \pm y = \pm c$.

Alternatively as follows:

Since $s = c^2 - t$, the equation may be written

$$\frac{x^2}{c^2 - t} + \frac{y^2}{t} = 1,$$

that is $$t^2 + t(x^2 - y^2 - c^2) + c^2y^2 = 0;$$

therefore the envelope is

$$(x^2 - y^2 - c^2)^2 = 4c^2y^2,$$

that is $$x^2 - y^2 - c^2 = \pm 2cy \quad \text{or} \quad x^2 = (y \pm c)^2,$$

which may be written $$x = \pm y \pm c.$$

EXERCISE XVII c

Find the envelopes of the systems in Nos. 1-10 where s, t are parameters:

1. $x \cos t + y \sin t = a$.
2. $x \sec t + y \operatorname{cosec} t = a$.
3. $y = tx \pm \sqrt{(a^2t^2 + b^2)}$.
4. $y^2 = t^2(x - t)$.

5. $x \cos 2t + y \sin 2t = a \cos^2 t$. 6. $\dfrac{x}{s} + \dfrac{y}{t} = 1$, where $s + t = a$.

7. $xs + yt = 1$, where $a^2 st = 1$. 8. $\dfrac{x^2}{s^2} + \dfrac{y^2}{t^2} = 1$, where $s + t = a$.

9. The parabolic trajectories $y = x \tan\theta - \frac{1}{2} g x^2 v^{-2} \sec^2\theta$ where the parameter is θ.

10. The normals to the given cycloid $x = a(\theta + \sin\theta)$, $y = a(1 - \cos\theta)$, expressing the result parametrically.

11. Find the envelope of a system of circles whose centres lie on $x^2 + y^2 = a^2$ and which pass through the fixed point $(c, 0)$.

12. Find the envelope of a system of circles whose centres lie on $x^2 - y^2 = a^2$ and which pass through the origin.

Evolutes. The locus of the centre of curvature of a given curve is called the *evolute* of the curve. If the coordinates of any point P on the curve are (x, y), and those of the corresponding centre of curvature C are (ξ, η), we have

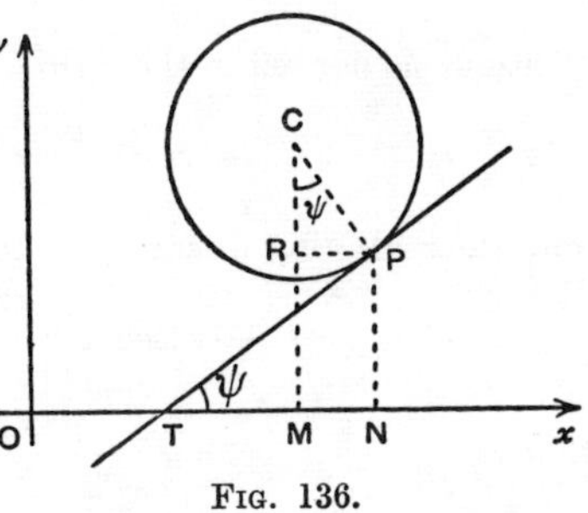

FIG. 136.

$$\xi = x - \rho \sin\psi, \quad \eta = y + \rho\cos\psi.$$

$$\therefore d\xi = dx - \rho\cos\psi\, d\psi - \sin\psi\, d\rho$$

$$= dx - \frac{ds}{d\psi}\frac{dx}{ds} d\psi - \sin\psi\, d\rho;$$

$$\therefore d\xi = -\sin\psi\, d\rho.$$

Similarly,
$$d\eta = dy - \rho\sin\psi\, d\psi + \cos\psi\, d\rho$$

$$= dy - \frac{ds}{d\psi}\frac{dy}{ds} d\psi + \cos\psi\, d\rho;$$

$$\therefore d\eta = \cos\psi\, d\rho.$$

$$\therefore \frac{d\eta}{d\xi} = -\cot\psi = \text{gradient of PC}.$$

Therefore the normal PC to the given curve touches the locus of C, that is the evolute. Hence the evolute is the *envelope of the normals* of the given curve, and the centre of curvature is the point of contact of the normal with the evolute.

Example 9. Find the coordinates of the centre of curvature at the point $(a\cos\phi, b\sin\phi)$ on the ellipse $\dfrac{x^2}{a^2}+\dfrac{y^2}{b^2}=1$, and determine the evolute.

At the point ϕ, $\tan\psi=\dfrac{dy}{d\phi}\div\dfrac{dx}{d\phi}=-\dfrac{b}{a}\cot\phi$;

$\therefore$ the normal at the point ϕ is

$$y-x\frac{a}{b}\tan\phi=b\sin\phi-\frac{a^2}{b}\sin\phi=\frac{b^2-a^2}{b}\sin\phi.$$

For the envelope of the normals, differentiate with respect to ϕ; therefore the centre of curvature is given by

$$-\frac{ax}{b}\sec^2\phi=\frac{b^2-a^2}{b}\cos\phi,$$

that is
$$x=\frac{a^2-b^2}{a}\cos^3\phi.$$

Writing the normal in the form

$$y\cot\phi-\frac{ax}{b}=\frac{b^2-a^2}{b}\cos\phi,$$

and differentiating with respect to ϕ, we have

$$-y\operatorname{cosec}^2\phi=\frac{b^2-a^2}{b}(-\sin\phi),$$

that is
$$y=-\frac{a^2-b^2}{b}\sin^3\phi.$$

Therefore the centre of curvature is

$$\left(\frac{a^2-b^2}{a}\cos^3\phi,\quad -\frac{a^2-b^2}{b}\sin^3\phi\right),$$

and the parametric equations of the evolute are

$$ax:by:(a^2-b^2)=\cos^3\phi:-\sin^3\phi:1.$$

Since this may be written

$$(ax)^{2/3}:(by)^{2/3}:(a^2-b^2)^{2/3}=\cos^2\phi:\sin^2\phi:1,$$

the equation of the evolute can be expressed in the (less convenient) form

$$(ax)^{2/3}+(by)^{2/3}=(a^2-b^2)^{2/3}.$$

Arc of the Evolute. Denote by A the fixed point on the curve AP, $s=f(\psi)$, from which s is measured, and by C_0, C the centres of curvature corresponding to A, P; also denote by σ the length of the arc C_0C of the evolute measured from C_0 in the sense for which s increases.

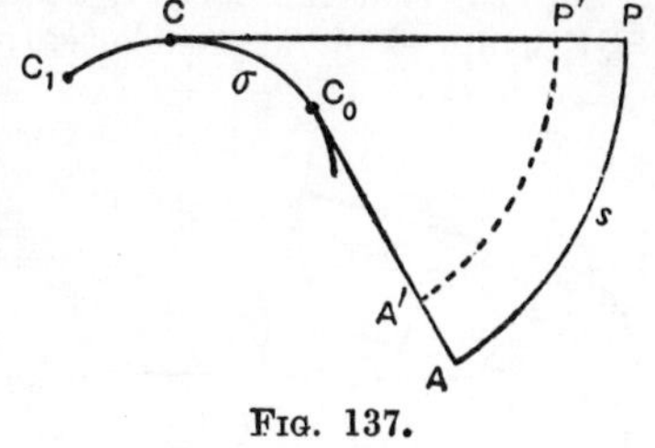

FIG. 137.

From p. 355, if C is the point (ξ, η),

$$d\xi = -\sin\psi\, d\rho, \quad d\eta = \cos\psi\, d\rho;$$

$$\therefore (d\sigma)^2 \equiv (d\xi)^2 + (d\eta)^2 = (d\rho)^2.$$

Therefore if ρ increases steadily with s, that is with σ,

$$d\sigma = d\rho;$$

integrating, $\sigma = \rho - \rho_0$, where ρ_0 is AC_0;

and if ρ decreases steadily as s increases,

$$d\sigma = -d\rho; \quad \sigma = \rho_0 - \rho.$$

Thus the length of an arc of the evolute is equal to the difference between the radii of curvature corresponding to its ends, provided that ρ increases steadily, or decreases steadily, in the interval.

Involutes. This property of the length of arc of the evolute has a physical interpretation :

Suppose that one end C_1 of a string is attached to a point of the evolute, see Fig. 137, and that the string is wrapped round the curve and leaves it tangentially at C_0, continuing as far as A. Then, when the string is unwrapped, the moving end will trace out the original curve AP because, at any moment, the portion unwrapped equals (AC_0 + arc C_0C), which equals PC.

Owing to this property the original curve is called *an involute* of the curve C_0CC_1.

There is an unlimited number of involutes corresponding to a given curve, for any definite point A′ of the string AC_0 traces out a curve A′P′ to which the various positions of the straight part of the string are normal. Thus C_0CC_1 is also the envelope of the normals of A′P′; therefore the curve A′P′ is an involute of C_0CC_1. Further, the tangents at corresponding points of the curves AP, A′P′ are parallel and for this reason the involutes are often called " parallel curves ".

Example 10. Determine the involute of a circle of radius a.

Suppose that the curve is described by one end P of a string PQK wrapped round a circle centre O and radius a, the other end K being fixed; and let KQA be the position of the string when wholly in contact with the circle.

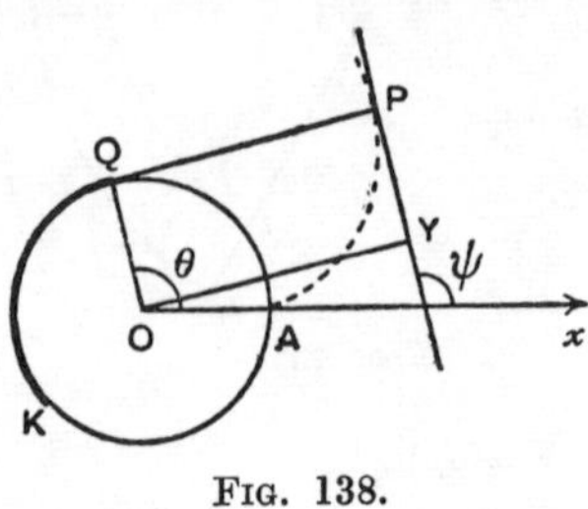

FIG. 138.

Take OA as x-axis and let $\angle x\text{OQ} = \theta$. Then Q is the point

$$(a \cos \theta, a \sin \theta);$$

also $\text{QP} = \text{arc QA} = a\theta$, and QP makes with O$x$ an angle $\theta - \frac{1}{2}\pi$.

$\therefore$ P is the point (x, y) given by

$$x = a \cos \theta + a\theta \cos (\theta - \tfrac{1}{2}\pi),$$
$$y = a \sin \theta + a\theta \sin (\theta - \tfrac{1}{2}\pi);$$

that is, parametric equations of the involute are

$$x = a(\cos \theta + \theta \sin \theta), \quad y = a(\sin \theta - \theta \cos \theta).$$

It follows from these equations, or from the mechanical description of the curve, that all involutes of a given circle are congruent. It is also obvious from the figure that the pedal equation of the involute is $r^2 = p^2 + a^2$, since $p \equiv \text{OY} = \text{QP}$, $\text{OQ} = a$, $\text{OP} = r$. Further for the involute, $\psi = \angle \text{AOQ}$,

$$\therefore \frac{ds}{d\psi} \equiv \rho = \text{PQ} = a\psi;$$

$\therefore$ the intrinsic equation of the involute is $s = \frac{1}{2}a\psi^2$ if s is measured from A, where $\psi = 0$.

Example 11. If $p = f(\psi) > 0$, interpret the equations

$$\text{(i) } x \sin \psi - y \cos \psi = p; \quad \text{(ii) } x \cos \psi + y \sin \psi = \frac{dp}{d\psi};$$

$$\text{(iii) } x \sin \psi - y \cos \psi = -\frac{d^2p}{d\psi^2}.$$

(i) is the equation of a line making an angle ψ with Ox and at a distance p from O. It touches the curve (S) whose p, ψ equation is $p=f(\psi)$. Hence (S) can be found as the envelope of (i) by differentiating with respect to ψ, and this gives (ii). Therefore the coordinates of the point of contact of (i) with (S) are obtained by solving (i) and (ii).

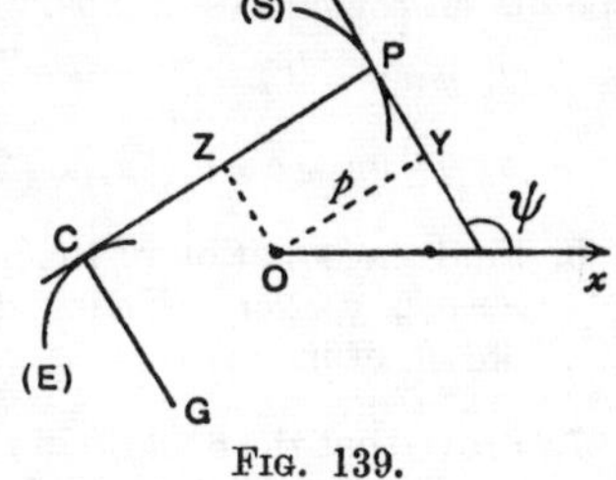

FIG. 139.

But the form of (ii) shows that it is perpendicular to (i); therefore (ii) represents the normal at P to (S).

The evolute (E) of (S) may be found as the envelope of the normal (ii) by differentiating with respect to ψ, and this gives (iii), which therefore passes through the centre of curvature C. But the form of (iii) shows that it is perpendicular to (ii); therefore (iii) represents the normal at C to (E).

If ψ is not confined to values for which $f(\psi)$ is positive the envelope of (i) consists of (S) given by $p=f(\psi)\geqslant 0$ and another curve given by $-p=-f(\psi)>0$; this is because (i) can be written in the form

$$x\sin(\psi+\pi)-y\cos(\psi+\pi)=-p.$$

Some results previously obtained by other methods can be deduced from Example 11:

The perpendicular from O to (ii) is $\dfrac{dp}{d\psi}$;

$$\therefore \text{YP}=\frac{dp}{d\psi}, \text{ (p. 339).}$$

Also, the distance between the parallel lines (i) and (iii) is $p+\dfrac{d^2p}{d\psi^2}$;

$$\therefore \rho=p+\frac{d^2p}{d\psi^2}, \text{ (p. 350).}$$

EXERCISE XVII d

1. Prove that the coordinates (ξ, η) of the centre of curvature at any point (x, y) on a curve are given by

$$\xi=x-\frac{dy}{d\psi}, \quad \eta=y+\frac{dx}{d\psi}.$$

Find the coordinates of the centres of curvature at the named points on the curves in Nos. 2-5 :

2. $y = x^2$; $(\frac{1}{2}, \frac{1}{4})$.
3. $xy = c^2$; (c, c).
4. $y = \log \sec x$; $(\frac{1}{3}\pi, \log 2)$.
5. $y = \sin^2 x$; $(0, 0)$.

6. Find the equation of the normal at the point t of the parabola $x = at^2$, $y = 2at$. Hence determine the evolute and give a sketch of it.

7. Prove that the evolute of the hyperbola $x = \pm a \operatorname{ch} \theta, y = b \operatorname{sh} \theta$, can be expressed in the form $(ax)^{2/3} - (by)^{2/3} = (a^2 + b^2)^{2/3}$.

8. Sketch the evolute of the ellipse $\frac{x^2}{a^2} + \frac{y^2}{b^2} = 1$ and find its total length.

9. Find the length of the portion of the evolute of the cycloid $s = 4a \sin \psi$ corresponding to the arc from $\psi = 0$ to $\psi = \frac{1}{2}\pi$ and prove that the evolute is an equal cycloid.

10. Find the length of the portion of the evolute of the catenary $y = c \operatorname{ch}(x/c)$ which corresponds to the arc from $x = 0$ to $x = c$.

11. Find the intrinsic equation of the evolute of the catenary $s = c \tan \psi$.

12. Find the coordinates of the centre of curvature, **C**, at the point t of the envelope of $x \cos t + y \sin t = a \sec t$. Also find the radius of curvature of the locus of **C**.

13. Prove that the intrinsic equation of the evolute of $s = f(\psi)$ is of the form $s - f'(\psi - \frac{1}{2}\pi) =$ constant, and that, for a suitable origin for s and a suitable initial line, this can be written $s = f'(\psi)$.

14. Prove that the intrinsic equation of an involute of $s = f(\psi)$ can be expressed in the form $s - c\psi = \int f(\psi)\, d\psi$ where c is a constant.

15. Prove that the p, r equation of the evolute of $p = a \sin 2\psi$ is $r^2 + 3p^2 = 16a^2$.

16. Prove that the p, ψ equation of a cardioid can be written in the form $p = 2a \cos^3 \frac{1}{3}\psi$, and find the p, ψ equation of its involute.

CHAPTER XVIII

FURTHER AREAS AND VOLUMES

Area of a Sector. In Fig. 140, AB is an arc of the curve whose polar equation is $r=f(\theta)$, and we shall suppose at first that r increases steadily as θ increases from α at A to β at B.

Take two points P, (r, θ) and Q, $(r+\delta r, \theta+\delta\theta)$ on the arc. Then the area of the sector OPQ lies between the areas of the circular sectors OPP′, OQQ′, that is between $\frac{1}{2}r^2\,\delta\theta$ and $\frac{1}{2}(r+\delta r)^2\,\delta\theta$. The sector OAB may be divided into sectors like OPQ, and by the summation property of a definite integral its area is the common limit of

$$\sum_{\theta=\alpha}^{\theta=\beta} \tfrac{1}{2}r^2\,\delta\theta \quad \text{and} \quad \sum_{\theta=\alpha}^{\theta=\beta} \tfrac{1}{2}(r+\delta r)^2\,\delta\theta,$$

which is $\int_\alpha^\beta \frac{1}{2}\mathbf{r}^2\,\mathbf{d\theta}$, assuming that the limit exists.

The same argument applies if r decreases steadily as θ increases, and the result holds for any arc which can be divided into a finite number of parts for each of which r increases steadily or decreases steadily as θ increases.

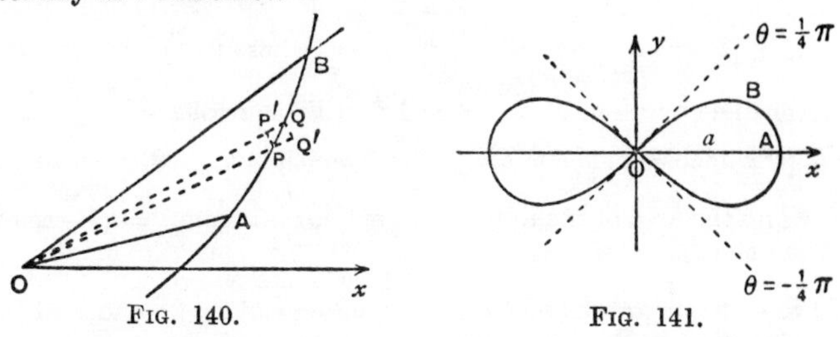

FIG. 140. FIG. 141.

Example 1. Sketch the lemniscate $r^2=a^2\cos 2\theta$ and find the area of one of its loops.

As θ increases from 0 to $\frac{1}{4}\pi$, r decreases from a to 0, giving the arc ABO in Fig. 141. As θ increases from $\frac{1}{4}\pi$ to $\frac{1}{2}\pi$, $\cos 2\theta$ is negative, therefore no value of r exists. Also, since the equation is unaltered when θ is replaced by $\pi-\theta$ or by $-\theta$, the curve is symmetrical about Oy and about Ox and therefore consists of two loops which touch the lines $\theta=\pm\frac{1}{4}\pi$ at the pole.

The area of each loop $=2\int_0^{\pi/4} \frac{1}{2}r^2\,d\theta$

$$=\int_0^{\pi/4} a^2 \cos 2\theta\, d\theta.$$

$$=\left[\tfrac{1}{2}a^2 \sin 2\theta\right]_0^{\pi/4}=\tfrac{1}{2}a^2.$$

Note. To confirm the accuracy of the sketch the reader should prove that $\phi=\frac{1}{2}\pi+2\theta$ for this curve, which shows that the tangent at **A** is perpendicular to **O**x.

EXERCISE XVIII a

Sketch the curves in Nos. 1-3 and find their areas:

1. $r=a\cos\theta$. **2.** $r=a\cos^2\frac{1}{2}\theta$. **3.** $r=a(3+2\cos\theta)$.

Find the areas of the sectors indicated in Nos. 4-6:

4. $r=e^{\theta}$; $\theta=-1$, $\theta=+1$. **5.** $r=a\tan\theta$; $\theta=0$, $\theta=\frac{1}{4}\pi$.

6. $r^2\sin\theta=a^2$; $\theta=\alpha$, $\theta=\pi-\alpha$, where $0<\alpha<\pi$.

Sketch the curves in Nos. 7-9 and find their areas:

7. $r^2=a^2\cos^2\theta+b^2\sin^2\theta$. **8.** $r^2=a^2\sin 2\theta$. **9.** $r=a\sin 2\theta$.

10. Find the area of the minor segment cut off from $r=a(1+\cos\theta)$ by the half-line $\theta=\frac{1}{2}\pi$.

11. Sketch the graphs of $r=a\sin\theta$ and $r^2=a^2\cos 2\theta$ and find the area of the region inside both curves.

12. With the usual notation prove that $pds=r^2d\theta$ and interpret $\int p\,ds$ and $\int p^2\,d\psi$ geometrically.

Sketch the curves in Nos. 13, 14 and find the areas of their loops:

13. $r=a\sin\theta\cos^4\theta$. **14.** $r\cos\theta=a\cos 2\theta$.

15. Find the area of a loop of $r=a\sin 3\theta+b\cos 3\theta$.

From the formula $\frac{1}{2}\int_{\alpha}^{\beta} r^2\, d\theta$ for the area of a sector may be deduced a formula applicable to a curve given by cartesian parametric equations.

Since $\tan\theta = \frac{y}{x}, \quad \sec^2\theta\, d\theta = \frac{x\,dy - y\,dx}{x^2}$;

but $x\sec\theta = r, \quad \therefore r^2 d\theta = x\,dy - y\,dx.$

If a curve is given by the equations

$$x = f(t), \quad y = g(t),$$

$\frac{dx}{dt}$ and $\frac{dy}{dt}$ may be calculated from these equations, and

$$r^2\frac{d\theta}{dt} = x\frac{dy}{dt} - y\frac{dx}{dt}.$$

Thus if in Fig. 140 the points A$(\theta = \alpha)$ and B$(\theta = \beta)$ are given by $t = t_1$ and $t = t_2$,

$$\text{area of sector AOB} = \tfrac{1}{2}\int_{\alpha}^{\beta} r^2\, d\theta = \tfrac{1}{2}\int_{t_1}^{t_2} r^2\frac{d\theta}{dt}\, dt$$

$$= \tfrac{1}{2}\int_{t_1}^{t_2}\left(x\frac{dy}{dt} - y\frac{dx}{dt}\right) dt.$$

Note. This result may be obtained independently from the cartesian formulae $\int x\, dy$, $\int y\, dx$ for areas and may then be used to establish the polar formula for the area of a sector.

Example 2. Find the area of the sector of the half-hyperbola $x = a \operatorname{ch} u$, $y = b \operatorname{sh} u$ bounded by the lines joining the origin to the points $u = 0$, $u = u_1$.

$$x\,dy - y\,dx = ab(\operatorname{ch}^2 u - \operatorname{sh}^2 u)du = ab\, du;$$

$$\therefore \text{area of sector} = \tfrac{1}{2}\int_0^{u_1}\left(x\frac{dy}{du} - y\frac{dx}{du}\right) du$$

$$= \tfrac{1}{2}\int_0^{u_1} ab\, du = \tfrac{1}{2}abu_1.$$

This result for the special case $a = b$ was obtained by a less convenient method on p. 272.

Example 3. Sketch the curve $x^3 + y^3 = 3axy$, $(a > 0)$, and find the area of the loop.

$$\text{Put } y = tx, \text{ then } x^3(1 + t^3) = 3atx^2;$$

$$\therefore x = 0 \text{ or } \frac{3at}{1 + t^3} \quad \text{and} \quad y = tx = 0 \text{ or } \frac{3at^2}{1 + t^3};$$

$\therefore$ the curve is represented parametrically by the equations

$$x = \frac{3at}{1 + t^3}, \quad y = \frac{3at^2}{1 + t^3},$$

and could be plotted by giving numerical values to t.

Unless great accuracy is required it is sufficient to proceed as follows:

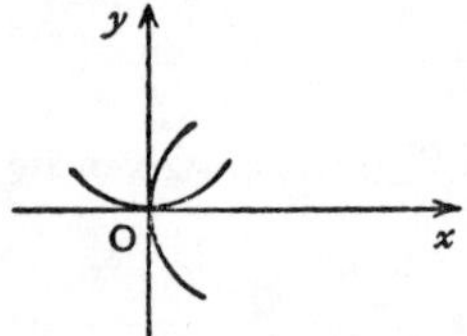

FIG. 142.

If t is small, $\quad x \simeq 3at, \quad y \simeq 3at^2$;
if t is large, $\quad x \simeq 3a/t^2, \quad y \simeq 3a/t$;
these approximations give two branches near the origin like the parabolas in Fig. 142.

If $t \simeq -1$, $1 + t^3$ is small and therefore the values of x and y are numerically large and of opposite signs; but if $t = -1$ there is no value of x and y, and therefore the line $y = -x$ does not cut the curve except at the origin.

$$\text{Since} \qquad y + x = \frac{3at(1 + t)}{1 + t^3} = \frac{3at}{1 - t + t^2}, \; t \neq -1,$$

$$y + x \to \frac{-3a}{3}, = -a, \text{ when } t \to -1.$$

Therefore for values of t near -1 the curve approximates to the line

$$y + x = -a,$$

which is called an *asymptote*; and since

$$x + y + a, \equiv \frac{a(1 + t)^2}{1 - t + t^2},$$

is positive for all values of t, therefore the curve lies wholly on the same side of the asymptote as the origin.

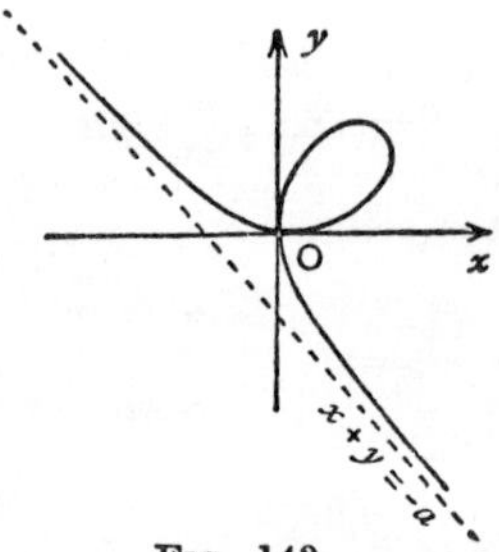

FIG. 143.

There are no very large values of x or y except those given by $t \simeq -1$. The graph is therefore as shown in Fig. 143.

The reader should consider which parts of the curve are given by different values of t; actually the loop is obtained by allowing t to increase from 0 to ∞.

$$\therefore \text{ the area of the loop} = \tfrac{1}{2}\int_0^\infty \left(x\frac{dy}{dt} - y\frac{dx}{dt}\right) dt.$$

$$\text{Since } \frac{y}{x} = t, \quad \frac{xdy - ydx}{x^2} = dt;$$

$$\therefore x\frac{dy}{dt} - y\frac{dx}{dt} = x^2 = \frac{9a^2t^2}{(1+t^3)^2};$$

$$\therefore \text{ the area} = \tfrac{1}{2}\int_0^\infty \frac{9a^2t^2}{(1+t^3)^2}\,dt$$

$$= \tfrac{1}{2}\left[\frac{-3a^2}{1+t^3}\right]_0^\infty = \tfrac{3}{2}a^2.$$

Note. Additional examples of curve tracing and asymptotes will be found in Chapter XXI.

EXERCISE XVIII b

Sketch the curves in Nos. 1-4 and find the areas enclosed by them:

1. $x = 3\cos t$, $y = 2\sin t$.

2. $x = a\cos^2 t$, $y = a\sin t\cos t$.

3. $x = a\sin^2 t$, $y = a\sin^2 t\cos t$.

4. $x = a\cos t(1-\cos t)$, $y = a\sin t(1-\cos t)$.

Sketch the curves in Nos. 5-8 and find the area of a loop:

5. $x = a\sin 2t$, $y = b\cos t$.

6. $x^4 - y^4 = x^6$.

7. $x = t + t^2$, $y = t^2 + t^3$.

8. $x^4 + y^4 = 2a^2xy$.

9. Find the area enclosed by the curve

$$x = a(2\cos t + \cos 2t),\ y = a(2\sin t + \sin 2t).$$

10. Sketch the curve $x = t^3 - t$, $y = t^4 - 1$ and find the area of the loop.

11. Find the area of the ellipse $5x^2 + 4xy + 8y^2 = 1$.

12. Sketch the curve $y^3 = ax(x+y)$ and prove that the area of the loop is $\frac{1}{60}a^2$.

Area of the Surface of a Circular Cone.

The surface of a cone can be "developed" into the sector of a circle, and the area of the surface of a cone may be defined to be the area of this sector. We assume the formula πrl for this area; it is easy to deduce a formula for the area of the curved surface of a frustum.

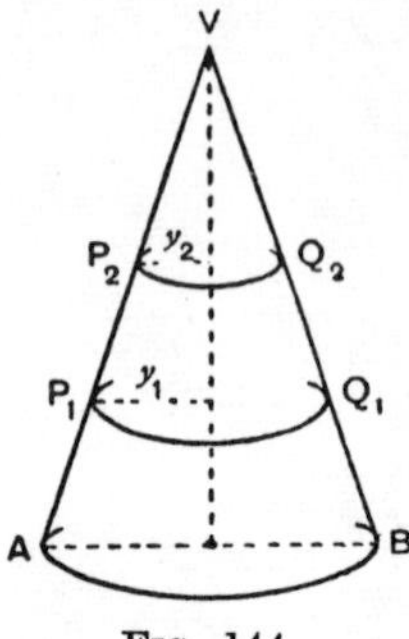

Fig. 144.

Consider the frustum, in Fig. 144, between two circles P_1Q_1, P_2Q_2, of radii y_1, y_2; let $VP_1 = l_1$, $VP_2 = l_2$. Then

$$\text{area of frustum} = \pi y_1 l_1 - \pi y_2 l_2$$
$$= \pi(y_1 + y_2)(l_1 - l_2)$$

since $y_1/y_2 = l_1/l_2$.

Area of the Surface of a Solid of Revolution.

When the word "area" is applied to a curved surface it is used in a new sense which needs to be defined. We have defined the meaning for a conical surface by using the idea of "development", and we shall now deduce a meaning for the area of the surface of any solid of revolution.

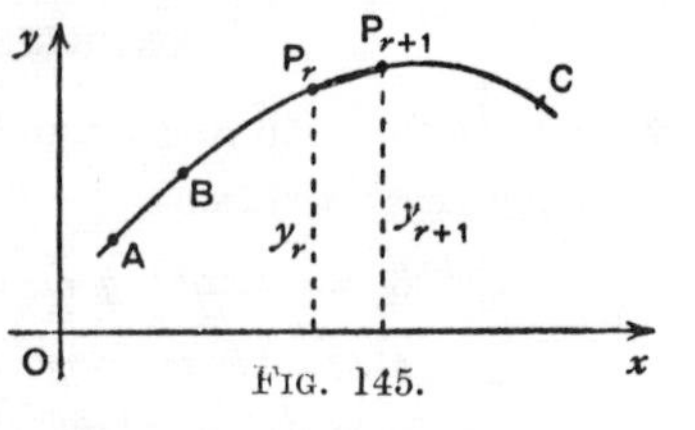

Fig. 145.

Suppose that the surface is generated by the revolution of an arc BC of the curve $y = f(x)$ about Ox, where BC lies above Ox. As in Chapter XVI, let $BP_1P_2 \ldots P_{n-1}C$ be an open n-sided polygon inscribed in this arc, and denote by S_n the sum of the areas of the conical surfaces generated by its sides. Now let the polygon vary in any manner such that $n \to \infty$ and the length of the longest side tends to zero. If then S_n tends to a definite limit S, the surface is said to be of *area* S.

Let the points B, C be given by $s = \beta$, $s = \gamma$ where s is the length of the arc of the given curve measured from a fixed point A and denote the coordinates of P_r by (x_r, y_r).

Then the area of the conical surface generated by P_rP_{r+1} is $\pi(y_r + y_{r+1})P_rP_{r+1}$, and

$$S_n = \sum_{0}^{n-1} \pi(y_r + y_{r+1})P_rP_{r+1}$$

where $r = 0$, $r = n$ correspond to B, C.

The form of S_n suggests that $\lim_{n\to\infty} \mathsf{S}_n = \int_\beta^\gamma 2\pi y\, ds$, and this can be proved by using the fact that the limit of the perimeter of the polygon is the length of the arc. The proof is reserved for the advanced volume. The formula for the area of the surface is

$$\mathsf{S} = \int_\beta^\gamma 2\pi y\, ds.$$

Example 4. Find the area of a zone of a sphere of radius a bounded by parallel planes distant b apart.

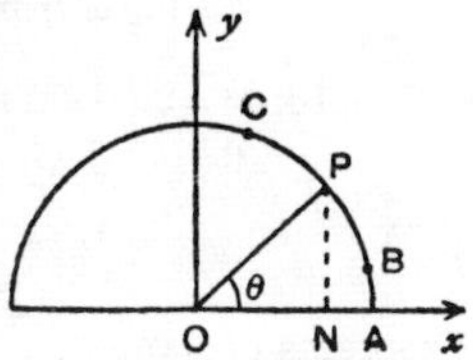

Suppose that the sphere is generated by the revolution about $\mathsf{O}x$ of the semicircle, centre O and radius a, with diameter along $\mathsf{O}x$; then the zone is generated by the revolution of an arc BC.

Any point P on the semicircle is given by $x = a\cos\theta$, $y = a\sin\theta$, where $\angle x\mathsf{OP} = \theta$; let B, C be given by $\theta = \beta$, $\theta = \gamma$, then

$$b = a\cos\beta - a\cos\gamma.$$

FIG. 146.

Also, arc $\mathsf{AP} = s = a\theta$, $\quad \therefore ds = a\,d\theta$.

$$\therefore \text{area of zone} = 2\pi \int_{a\beta}^{a\gamma} y\, ds$$

$$= 2\pi \int_\beta^\gamma (a\sin\theta)\, a\, d\theta$$

$$= 2\pi a^2 \Big[-\cos\theta \Big]_\beta^\gamma$$

$$= 2\pi a^2(\cos\beta - \cos\gamma) = 2\pi ab.$$

Note. Putting $b = 2a$, area of surface of sphere $= 4\pi a^2$.

EXERCISE XVIII c

Find the areas of the surfaces generated by revolution about $\mathsf{O}x$ of the arcs in Nos. 1-8:

1. $4y = 3x$ from $x = 1$ to $x = 2$.
2. $y = c\,\mathrm{ch}\,(x/c)$ from $x = -a$ to $x = +a$.
3. $y = \sqrt{(4ax)}$ from $x = 0$ to $x = b$.
4. $x = t^2$, $y = t - \frac{1}{3}t^3$ from $t = 0$ to $t = \sqrt{3}$.

5. $x = a\cos^3 t$, $y = a\sin^3 t$ from $t = 0$ to $t = \frac{1}{2}\pi$.

6. $x = a(\theta - \sin\theta)$, $y = a(1 - \cos\theta)$ from $\theta = 0$ to $\theta = 2\pi$.

7. $y = \sin x$ from $x = 0$ to $x = \pi$.

8. $r = a(1 + \cos\theta)$ from $\theta = 0$ to $\theta = \pi$.

Find the areas of the surfaces generated by revolution about Oy of the arcs in Nos. 9, 10 :

9. $y = \log x$ from $x = 1$ to $x = 2$.

10. $x = 2\theta - \sin 2\theta$, $y = 8\sin\theta$ from $\theta = 0$ to $\theta = \frac{1}{2}\pi$.

Theorems of Pappus.

I. If an arc of a plane curve revolves about an axis in its plane which does not cross the arc, the area of the surface generated is equal to the length of the arc multiplied by the length of the path traced out by the centre of mass of the arc.

In Fig. 147, G is the centre of mass $(\bar{x}, \bar{y})$ of an arc BC of length l; the axis of revolution is taken as x-axis, and the points B, C are given by $s = \beta$, $s = \gamma$ where s is the length of arc measured from a fixed point A.

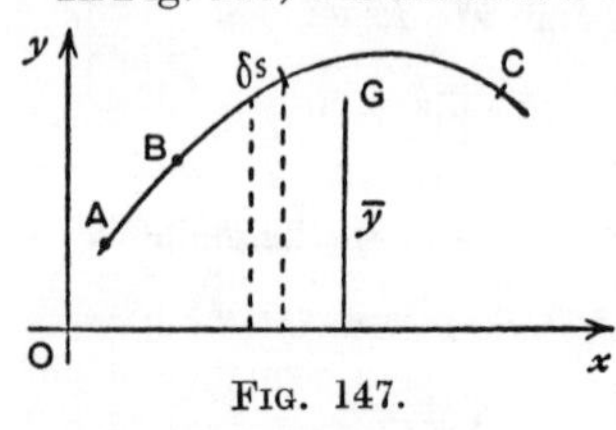

FIG. 147.

$$\text{Then } \bar{y} = \frac{\int_\beta^\gamma y\,ds}{\int_\beta^\gamma ds} = \frac{1}{l}\int_\beta^\gamma y\,ds;$$

$\therefore$ area of surface, $\equiv 2\pi\int_\beta^\gamma y\,ds$, $= 2\pi\bar{y}l$, which proves the theorem.

II. If a plane closed curve revolves about an axis in its own plane which does not cut the curve, the volume of the solid generated is equal to the area enclosed by the curve multiplied by the length of the path traced out by the centre of mass of the area.

Take the axis of revolution as x-axis and suppose that the closed curve has only two tangents parallel to Oy; denote the abscissae of their points of contact B, C by b, c. See Fig. 148.

In Fig. 148, G is the centre of mass $(\bar{x}, \bar{y})$ of an area A bounded by the curve BPCQ, and PQQ′P′ is one of the strips into which A can be divided by lines parallel to Oy; P, Q are the points (x_1, y_1), (x_2, y_2). Then

$$\bar{y} = \frac{\int_b^c \tfrac{1}{2}(y_2 + y_1)(y_2 - y_1)\,dx}{\int_b^c (y_2 - y_1)\,dx} = \frac{\int_b^c (y_2{}^2 - y_1{}^2)\,dx}{2\mathsf{A}};$$

∴ the volume of the solid, $\equiv \int_b^c \pi(y_2{}^2 - y_1{}^2)\,dx, = 2\pi\bar{y}\mathsf{A}$,

which proves the theorem.

This proof can be extended to a closed area which can be divided into a finite number of areas bounded by closed curves each of which has only two tangents parallel to Oy.

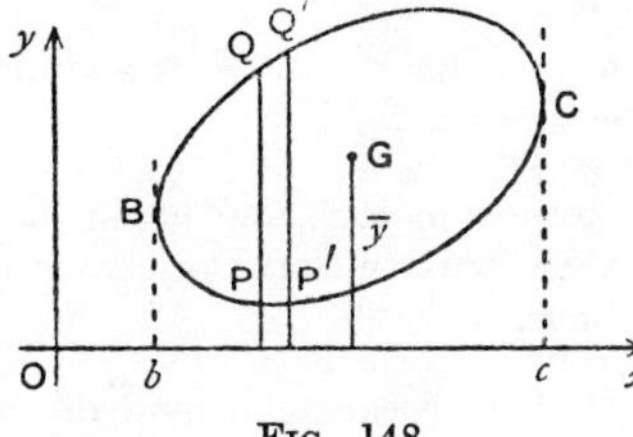

FIG. 148.

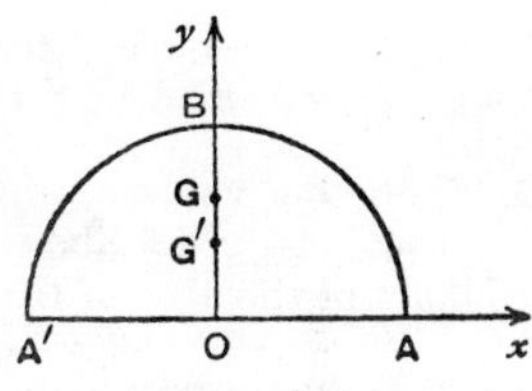

FIG. 149.

Example 5. Find the centres of mass of the arc and area of a semicircle.

Let AA′ be the diameter of the semicircle, centre O, radius a. By symmetry the centres of mass G, G′ lie on the radius OB perpendicular to AA′.

(i) By Pappus' first theorem the area of the surface generated by revolving the arc ABA′ about AA′ is $2\pi\mathsf{OG}\,.\,\pi a$; but the area is that of the surface of a sphere, $4\pi a^2$.

$$\therefore\ 2\pi\,.\,\mathsf{OG}\,.\,\pi a = 4\pi a^2; \quad \therefore\ \mathsf{OG} = \frac{2a}{\pi}.$$

(ii) By Pappus' second theorem the volume of the solid generated by the area ABA′ is $2\pi\mathsf{OG}'\,.\,\tfrac{1}{2}\pi a^2$; but this must equal the volume $\tfrac{4}{3}\pi a^3$ of the sphere.

$$\therefore\ 2\pi\mathsf{OG}'\,.\,\tfrac{1}{2}\pi a^2 = \tfrac{4}{3}\pi a^3; \quad \therefore\ \mathsf{OG}' = \frac{4a}{3\pi}.$$

EXERCISE XVIII d

1. Find the distance of the centre of mass of (i) the arc, (ii) the area of a quadrant of a circle, radius a, from the centre of the circle.

2. A square ABCD, side a, is rotated about a line through A parallel to BD. Describe the form of the solid generated and find (i) its volume, (ii) the area of its surface.

3. A circle, radius a, is rotated about a line in its plane at distance b from the centre, $(b > a)$. Find the surface-area and the volume of the anchor-ring so formed.

4. Give the surface-area and volume of the solid obtained by the revolution of a circle of radius a about one of its tangents.

5. Find the volume of the solid formed by revolving the ellipse $x = a \cos \phi$, $y = b \sin \phi$ about the line $x = 2a$.

6. Find the volume generated by revolving the half-ellipse $x = a\sqrt{(1 - y^2/b^2)}$ about the y-axis. Hence find the centre of mass of the area of the half-ellipse.

7. Prove that the volume generated by rotating a loop of the curve $a^2y^2 = x^2(a^2 - x^2)$ about the y-axis is $\frac{1}{4}\pi^2a^3$ and find the centre of gravity of the area of the loop.

8. Find the centre of gravity of the complete arc of the cardioid $r = a(1 + \cos \theta)$.

9. Find the volume of the solid formed by the revolution of the cardioid $r = a(1 + \cos \theta)$ about the tangent $x = 2a$.

10. Find the centre of gravity of the arc of the cycloid $x = a(t - \sin t)$, $y = a(1 - \cos t)$ from $t = 0$ to $t = 2\pi$ and the area of the surface formed by revolution of that arc about the tangent $y = 2a$.

11. A solid is formed by the revolution of a circle of variable radius r about an axis in its plane at a constant distance b from its centre. Find the total volume of the solid if when the plane of the circle has turned through an angle θ, the value of r is $a(2 + \sin \theta)$. $(a < \frac{1}{3}b)$.

The Notation of Analytical Geometry in Three Dimensions.

Geometrical methods of illustrating the properties of functions of one variable may be extended to functions of two variables.

Take three mutually perpendicular lines Ox, Oy, Oz as axes of reference in three dimensions. Then it is possible to move from

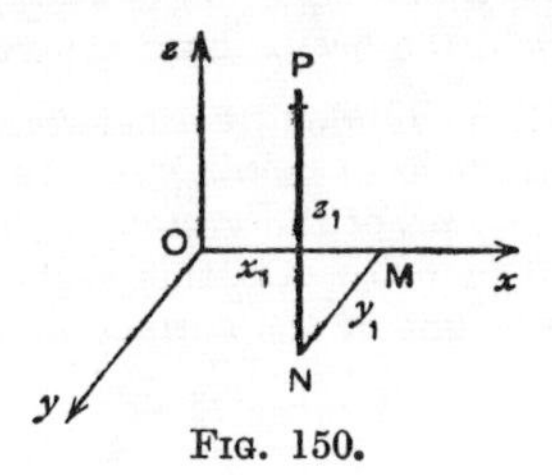

FIG. 150.

O to any point P by successive displacements OM, MN, NP in the directions of these axes, and if the displacements are measured algebraically by x_1, y_1, z_1 respectively, we call (x_1, y_1, z_1) the *cartesian coordinates* of P. To any set of coordinates (x_1, y_1, z_1) for given axes of reference there corresponds one point P, and conversely.

Denote by z any function $f(x, y)$ of two independent variables x and y. Any pair of values, $x=x_1$, $y=y_1$, can be represented by a point N (x_1, y_1) in the plane xOy, see Fig. 150; and if NP is drawn parallel to Oz and equal to $z_1, \equiv f(x_1, y_1)$, its height (positive or negative) represents the value of $f(x, y)$ for $x=x_1$, $y=y_1$.

If x and y vary independently and $z=f(x, y)$, then the point P(x, y, z) moves on a surface whose form illustrates geometrically the proporties of the function $f(x, y)$; we call $z=f(x, y)$ the *equation of the surface.*

More generally, if the coordinates (x, y, z) of a point P satisfy an equation $f(x, y, z)=0$, the point P lies on a surface and $f(x, y, z)=0$ is called the equation of the surface.

The equation of a plane parallel to xOy is evidently of the form $z=c$, and it can be proved that the equation of any plane can be expressed in the form $ax+by+cz+d=0$. The reader will find it easy to prove that the equation of a sphere, centre O, radius a, is $x^2+y^2+z^2=a^2$.

Double Integrals.

We first illustrate the meaning of a *double* integral by solving an example.

Example 6. The base of a solid is a rectangle OADB with edges OA, OB of lengths a, b along the axes Ox, Oy ; its upper face is part of the surface whose equation is $z = x^2y^3$ and its sides are planes perpendicular to the base. Find the volume of the solid.

Let the base OADB be divided by a network of lines parallel to Ox, Oy into a large number of rectangles. Take any one of these and denote the coordinates of the vertex nearest O by (x, y) ; its area may be denoted by $\delta x \delta y$. On this rectangle erect a column whose height is the height of the surface at (x, y), that is x^2y^3.

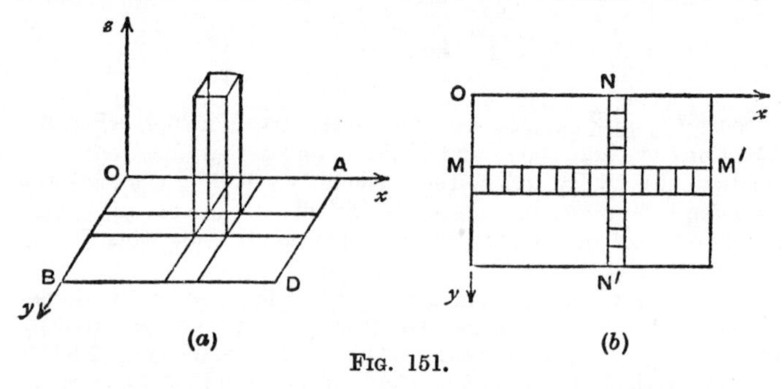

FIG. 151.

If this is done for all the rectangles into which OADB is divided, the volume of the solid may be taken to be the limit of the sum of the volumes of the columns, that is $\lim \Sigma x^2y^3\delta x\delta y$, when the sides of all the rectangles tend to zero.

Note 1. The volume of the column over $\delta x\delta y$ is not exactly $x^2y^3\delta x\delta y$, but is $\xi^2\eta^3\delta x\delta y$ where ξ is between x and $x+\delta x$ and η is between y and $y+\delta y$. This makes no difference to the limit, just as in Chapter X an area is found from $\lim \Sigma y\delta x$ though it is really $\lim \Sigma \eta\delta x$ where η is between y and $y+\delta y$.

Note 2. The sum $\Sigma x^2y^3\delta x\delta y$ is a *double* sum in the sense that x and y both vary. We therefore write

$$\text{volume} = \lim \Sigma\Sigma z\delta x\delta y$$

where, in the example considered, $z = x^2y^3$.

We now proceed to calculate the limit of the double sum.

We begin by finding the sum of the volumes of columns given by rectangles along the strip NN′ in Fig. 151 (b). For this purpose x and δx are treated as constants, thus

$$\Sigma\Sigma x^2 y^3 \delta x \delta y = \Sigma\{x^2 \delta x \Sigma(y^3 \delta y)\}$$

and
$$\lim \Sigma(y^3 \delta y) = \int_0^b y^3\, dy = \tfrac{1}{4}b^4;$$

$$\therefore \lim \Sigma\Sigma x^2 y^3 \delta x \delta y = \lim \Sigma(x^2 \delta x\, \tfrac{1}{4}b^4).$$

We now find the sum of the volumes corresponding to different strips like NN′; this gives

$$\tfrac{1}{4}b^4 \int_0^a x^2\, dx = \tfrac{1}{4}b^4\, \tfrac{1}{3}a^3 = \tfrac{1}{12}a^3 b^4.$$

The process which we have just explained may be represented symbolically by

$$\int_0^a \left\{ \int_0^b z\, dy \right\} dx, \text{ or shortly, } \int_0^a \int_0^b z\, dx\, dy, \text{ or by } \int_0^a dx \int_0^b z\, dy,$$

where in the first integration with respect to y, x was treated as constant.

Alternatively we might have begun by finding the sum of the volumes corresponding to the rectangles which compose MM′, treating y and δy as constant; the process is then completed by the summation for different positions of MM′. The reader should verify that the result is the same.

This second process may be represented by

$$\int_0^b \left\{ \int_0^a z\, dx \right\} dy, \text{ or shortly, } \int_0^b \int_0^a z\, dy\, dx, \text{ or by } \int_0^b dy \int_0^a z\, dx.$$

In the shortened notation the first integral sign refers to the first differential, but the other integration is performed first.

In this example we see that

$$\int_0^a \int_0^b z\, dx\, dy = \int_0^b \int_0^a z\, dy\, dx,$$

but when the limits are not constant the change of order affects their value. This is shown in the next example.

Example 7. The base of a right cylinder is the quadrant of

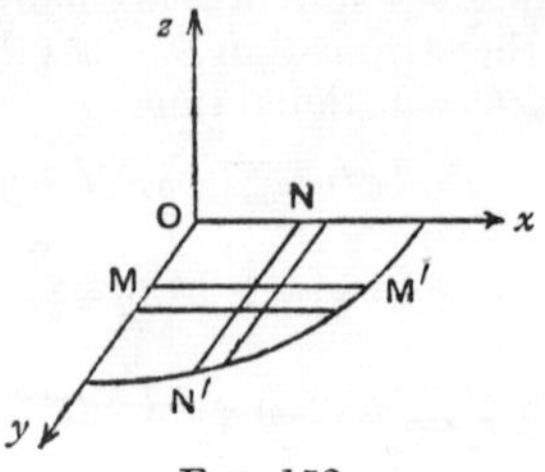

FIG. 152.

the ellipse $\frac{x^2}{a^2}+\frac{y^2}{b^2}=1$ for which $x>0$, $y>0$. Find the volume of the cylinder enclosed between the plane $z=0$ and the surface $z=xy$.

The volume is

$$\lim \Sigma\Sigma\, xy\delta x\delta y.$$

If we begin by summing for a strip NN′ bounded by lines parallel to Oy the contribution from this strip is

$$x\delta x\int_0^{y_1} y\,dy$$

where 0, y_1 are the values of y at N, N′. This gives

$$x\delta x\left[\tfrac{1}{2}y^2\right]_0^{y_1}=\tfrac{1}{2}x\delta x y_1{}^2=\tfrac{1}{2}xb^2(1-x^2/a^2)\delta x,$$

since

$$\frac{x^2}{a^2}+\frac{y_1{}^2}{b^2}=1.$$

To sum for the different strips NN′ we take values of x from 0 to a; this gives

$$\int_0^a \tfrac{1}{2}xb^2(1-x^2/a^2)\,dx=\tfrac{1}{2}b^2\left[\tfrac{1}{2}x^2-\tfrac{1}{4}x^4/a^2\right]_0^a=\tfrac{1}{8}a^2b^2.$$

In this example we have found the volume by evaluating $\int_0^a\int_0^{y_1} xy\,dx\,dy$. If we had first summed for a strip MM′ bounded by lines parallel to Ox we would have evaluated $\int_0^b\int_0^{x_1} xy\,dy\,dx$ where 0, x_1 are the values of x at M, M′. Thus the limits are different in the two methods. The reader should verify that the result is the same.

Cylindrical Coordinates. It is sometimes convenient to replace x and y by polar coordinates ρ, ϕ, in the plane $x\mathrm{O}y$; then $\rho\cos\phi = x$, $\rho\sin\phi = y$, and the position of any point in space is given by the *cylindrical coordinates* (ρ, ϕ, z).

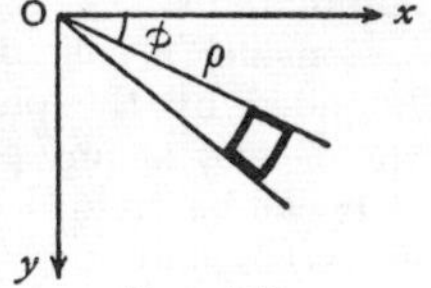

FIG. 153.

If the base, in the plane $x\mathrm{O}y$, of a solid is divided into a network by a series of circles, centre O, (ρ = constant), and a series of radii (ϕ = constant), the area of a mesh may be taken to be $\rho\,\delta\rho\,\delta\phi$, and the formula for the volume becomes

$$\iint z\rho\,d\rho\,d\phi.$$

Spherical Polar Coordinates. If the plane through Oz and any point P makes an angle ϕ with the plane $z\mathrm{O}x$, and if (r, θ) are the polar coordinates of P in this plane with Oz as initial line, (r, θ, ϕ) are called the *spherical polar* coordinates of P.

In Fig. 154, $z = \mathrm{NP} = r\cos\theta$; and $\mathrm{ON} = r\sin\theta$,

$$\therefore\ x = r\sin\theta\cos\phi,\ y = r\sin\theta\sin\phi.$$

Thus cartesian coordinates (x, y, z) are transformed into cylindrical coordinates by replacing x, y by polar coordinates ρ, ϕ given by

$$x = \rho\cos\phi, \quad y = \rho\sin\phi;$$

and spherical polar coordinates are then obtained from the cylindrical coordinates (ρ, ϕ, z) by replacing ρ, z by polar coordinates r, θ given by

$$z = r\cos\theta, \quad \rho = r\sin\theta.$$

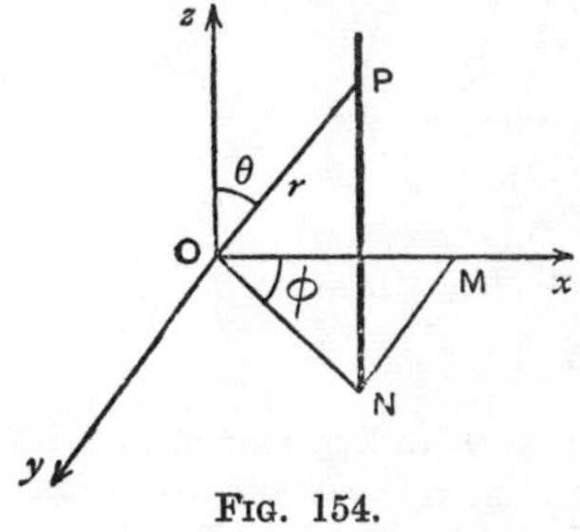

FIG. 154.

In spherical polar coordinates P (x, y, z) becomes

$$(r\sin\theta\cos\phi,\ r\sin\theta\sin\phi,\ r\cos\theta).$$

The use of *double* integrals has been illustrated in Examples 6, 7 by applications to measurements of volume; they apply also to the measurement of any quantity which involves the limit of a double summation, such as the mass of a body of variable density.

Triple Integrals. The method may be extended to *triple* summations.

Consider a solid in the form of a cuboid of variable density bounded by the planes $x=0,\ x=a$; $y=0,\ y=b$; $z=0,\ z=c$; the density at any point (x, y, z) being a given function $f(x, y, z)$.

It can be divided by planes parallel to $y\mathrm{O}z$, $z\mathrm{O}x$, $x\mathrm{O}y$ into small cuboids; and one of these, having (x, y, z), $(x+\delta x, y+\delta y, z+\delta z)$ as opposite corners, would have a mass approximately equal to $f(x, y, z)\delta x\delta y\delta z$, and the mass of the solid can be found as

$$\lim \Sigma\Sigma\Sigma f(x, y, z)\delta x\delta y\delta z,$$

which would be denoted by

$$\int_0^a\int_0^b\int_0^c f(x, y, z)\,dx\,dy\,dz, \text{ or } \int_0^a dx\int_0^b dy\int_0^c f(x, y, z)\,dz.$$

As with double integrals, if the limits are constant the order of integration is immaterial; but for variable limits a change of order involves a change of limits.

Volume in Spherical Polar Coordinates.

Fig. 155 represents an element of surface PQRS of a sphere, centre O, radius r; it is bounded by two cones, axis Oz, semi-vertical angles θ, $\theta+\delta\theta$, and by two planes through Oz, making angles ϕ, $\phi+\delta\phi$ with the plane $z\mathrm{O}x$.

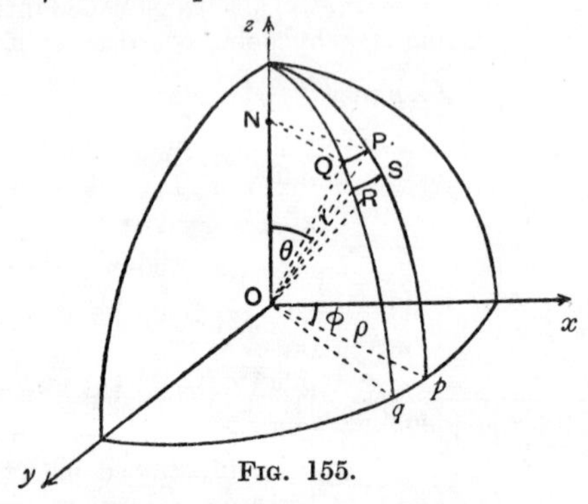

FIG. 155.

PQ is an arc of a *small* circle, centre N, radius $r\sin\theta$, and its length is $r\sin\theta\,.\,\delta\phi$; PS is an arc of a *great circle* and its length is $r\delta\theta$. Therefore the area of the element of surface PQRS may be taken as $r\sin\theta\,\delta\phi\,.\,r\delta\theta$.

If a solid is divided into small cells by spheres with **O** as centre, cones with Oz as axis, and planes through Oz, the volume of the cell which has **PQRS** as base and lies between the spheres of radii r, $r+\delta r$, may be taken as $r\sin\theta\,\delta\phi\,.\,r\delta\theta\,.\,\delta r$, that is $r^2\sin\theta\,\delta r\delta\theta\delta\phi$. Hence the spherical polar formula for a volume is

$$\iiint r^2\sin\theta\,dr\,d\theta\,d\phi.$$

EXERCISE XVIII e

Interpret geometrically the integrals in Nos. 1-4 and evaluate them:

1. $\int_0^2 dx\int_0^3 x\,dy.$

2. $\int_0^1\int_0^{\pi/2} r\,dr\,d\theta.$

3. $\int_0^a\int_0^b\int_0^c x^3\,dx\,dy\,dz.$

4. $\int_0^a dr\int_0^{\pi/2} d\theta\int_0^{\pi} r^2\sin\theta\,d\phi.$

Evaluate the integrals in Nos. 5-8:

5. $\int_0^3\int_1^{1+y}\frac{y}{\sqrt{x}}\,dy\,dx.$

6. $\int_0^{\pi/2}\int_{\sin\theta}^{\sin 2\theta} r\cos\theta\,d\theta\,dr.$

7. $\int_0^{\pi} d\theta\int_0^{\cos\theta}\sqrt{(1-r)}\,dr.$

8. $\int_0^a dx\int_0^b dy\int_0^c (y^2+z^2)\,dz.$

9. Prove that the area common to the parabolas $y=2x^2$, $y^2=4x$ is $\int_0^1\int_{2x^2}^{2\sqrt{x}} 1\,dx\,dy$ and find its value.

10. Find the coordinates of the centre of gravity of the area in the first quadrant enclosed by the curves $y^2=x$, $y^2=x^3$.

11. The thickness of a circular disc, of radius a, at distance r from the centre is $2ac/\sqrt{(4a^2-r^2)}$. Find its volume.

12. Find the area of the portion of the surface of a globe of radius a bounded by the meridians of longitude β, γ and the parallels of latitude λ, μ.

13. The density of the circular lamina $r=2a\cos\theta$ at any point (r,θ) is λr^2 per unit area where λ is constant. Find its mass.

14. The base of a right cylinder is a semicircle, radius a, diameter BC, and a plane is drawn through BC making 45° with the base. Prove that the volume of the cylinder enclosed by this plane and the base is $\frac{2}{3}a^3$.

15. The section of a right cylinder is a circle, radius a, diameter BC ; a sphere is drawn with centre B and radius $2a$. Taking the centre of the sphere as origin and using cylindrical coordinates, prove that the volume of the space inside both the sphere and cylinder is $4\int_0^{\pi/2} d\phi \int_0^{2a\cos\phi} \rho\sqrt{(4a^2-\rho^2)}\, d\rho$ and find its value.

16. Use spherical polar coordinates to prove that $\iiint \frac{1}{\sqrt{(1-x^2-y^2-z^2)}}\, dx\, dy\, dz$ equals $\pi^2/8$ where the summation extends to all positive values of x, y, z for which the integrand exists.

CHAPTER XIX

PARTIAL DIFFERENTIATION

Functions of Two Variables.

A graphical method of representing a function $z, =f(x, y)$, of two *independent* variables x, y has been explained in Chapter XVIII.

There is no single definite rate of change of z analogous to the derivative $\frac{dy}{dx}$ of a function of one variable. For since the variables x and y are independent of one another we may have to consider the effect on z of an increase in x while y remains constant, or of an increase in y with x constant, or of simultaneous increases of x and y. The last case is considered on page 384.

If δz is the increase of z due to an increase δx in x, when y remains constant,

$$\delta z = f(x+\delta x, y) - f(x, y);$$

if $\lim\limits_{\delta x \to 0} \frac{f(x+\delta x, y) - f(x, y)}{\delta x}$ exists, it is the derivative of z with respect to x, y being regarded as constant, and it is called the *partial derivative of z with respect to x*. It is denoted by $\frac{\partial z}{\partial x}$, or $\frac{\partial}{\partial x}f(x, y)$, or f_x.

Similarly, if $\lim\limits_{\delta y \to 0} \frac{f(x, y+\delta y) - f(x, y)}{\delta y}$ exists it is called the *partial derivative of z with respect to y* and is denoted by $\frac{\partial z}{\partial y}$, or $\frac{\partial}{\partial y}f(x, y)$, or f_y.

The partial derivatives may be interpreted in terms of the geometry of the surface $z=f(x, y)$ in the neighbourhood of the point P (x, y, z).

If we move from the point P to a neighbouring point Q of the surface, the gradient at which we climb relative to the plane xOy depends on the direction of the displacement from P to Q on the surface. If y is kept constant we are confined to the plane through P parallel to zOx; $\dfrac{\partial z}{\partial x}$ is the gradient at P of the curve which is

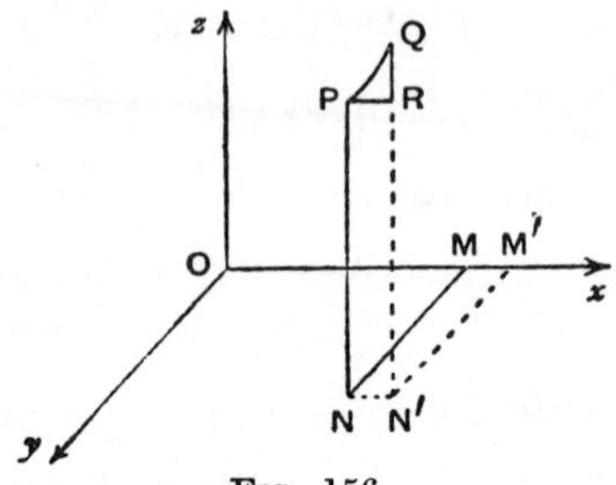

FIG. 156.

the section of the surface by that plane. Similarly $\dfrac{\partial z}{\partial y}$ is the gradient of the curve of section by the plane parallel to zOy.

Successive Partial Derivatives. Since $\dfrac{\partial z}{\partial x}, \dfrac{\partial z}{\partial y}$ are themselves functions of x and y, partial derivatives of $\dfrac{\partial z}{\partial x}, \dfrac{\partial z}{\partial y}$ may be defined. They are called second order partial derivatives of z.

$$\frac{\partial}{\partial x}\left(\frac{\partial z}{\partial x}\right),\quad \frac{\partial}{\partial y}\left(\frac{\partial z}{\partial x}\right),\quad \frac{\partial}{\partial x}\left(\frac{\partial z}{\partial y}\right),\quad \frac{\partial}{\partial y}\left(\frac{\partial z}{\partial y}\right)$$

are written $\dfrac{\partial^2 z}{\partial x^2}, \quad \dfrac{\partial^2 z}{\partial y \partial x}, \quad \dfrac{\partial^2 z}{\partial x \partial y}, \quad \dfrac{\partial^2 z}{\partial y^2}$,

or $f_{xx}, \quad f_{yx}, \quad f_{xy}, \quad f_{yy}$ respectively.

Subject to certain conditions which are usually satisfied by ordinary functions, it can be shown that $\dfrac{\partial^2 z}{\partial y \partial x} = \dfrac{\partial^2 z}{\partial x \partial y}$; but we shall content ourselves in this book with verifying it in special cases; see Example 1, p. 382, and Exercise XIX *a*, Nos. 7-12.

Partial derivatives of higher orders can be defined in succession, but these are not often required in applications.

Functions of Three or More Variables.

Partial derivatives of functions of any number of *independent* variables are defined in the same way.

For example, if $\qquad V = f(x, y, z),$

and if $\qquad \lim_{\delta x \to 0} \dfrac{f(x+\delta x, y, z) - f(x, y, z)}{\delta x}$ exists,

this limit is denoted by $\dfrac{\partial V}{\partial x}$ or $\dfrac{\partial}{\partial x} f(x, y, z)$ or f_x.

Also, $\dfrac{\partial V}{\partial y}, \dfrac{\partial V}{\partial z}$, and partial derivatives of higher orders can be defined as before.

Calculation of Partial Derivatives. No new principle is involved; the partial derivatives are in effect ordinary derivatives of functions of one variable, because the other variables which occur are treated, by definition, as constants.

If $V = f(z)$ where z is a function of a single variable x, we have, by p. 50 (Vol. I),

$$\frac{dV}{dx} = \frac{dV}{dz}\frac{dz}{dx} \equiv f'(z)\frac{dz}{dx};$$

if instead z is a function of two independent variables x and y, and if we wish to calculate the *partial* derivative of V with respect to x, the work is precisely the same as before, because y is treated as constant. But the formula used is now written

$$\frac{\partial V}{\partial x} = \frac{dV}{dz}\frac{\partial z}{\partial x} \equiv f'(z)\frac{\partial z}{\partial x}$$

to indicate that y has been treated as constant. In this relation $\dfrac{dV}{dz}$ is the ordinary derivative of V, $\equiv f(z)$, with respect to z and is therefore not written $\dfrac{\partial V}{\partial z}$.

Partial derivatives of higher orders are calculated in the same way. Thus

$$\frac{\partial^2 V}{\partial x^2}=\frac{\partial}{\partial x}\left(\frac{\partial V}{\partial x}\right)=\frac{\partial}{\partial x}\left\{f'(z)\frac{\partial z}{\partial x}\right\};$$

hence by the rule for differentiating a product

$$\frac{\partial^2 V}{\partial x^2}=\frac{\partial z}{\partial x}\frac{\partial}{\partial x}f'(z)+f'(z)\frac{\partial}{\partial x}\frac{\partial z}{\partial x};$$

but from the formula $\frac{\partial V}{\partial x}=\frac{dV}{dz}\frac{\partial z}{\partial x}$, writing $f'(z)$ in place of V,

$$\frac{\partial}{\partial x}f'(z)=\frac{d}{dz}f'(z)\frac{\partial z}{\partial x}=f''(z)\frac{\partial z}{\partial x};$$

$$\therefore \frac{\partial^2 V}{\partial x^2}=f''(z)\left(\frac{\partial z}{\partial x}\right)^2+f'(z)\frac{\partial^2 z}{\partial x^2}.$$

Example 1. If $z=\frac{x}{y}\log x$, verify the relation

$$\frac{\partial^2 z}{\partial x\partial y}=\frac{\partial^2 z}{\partial y\partial x}.$$

$$\frac{\partial z}{\partial y}\equiv\frac{\partial}{\partial y}\left(\frac{x}{y}\log x\right)=-\frac{x}{y^2}\log x;$$

$$\therefore \frac{\partial^2 z}{\partial x\partial y}\equiv\frac{\partial}{\partial x}\left(-\frac{x}{y^2}\log x\right)=-\frac{1}{y^2}\log x-\frac{x}{y^2}\frac{1}{x}.$$

Also,
$$\frac{\partial z}{\partial x}\equiv\frac{\partial}{\partial x}\left(\frac{x}{y}\log x\right)=\frac{1}{y}\log x+\frac{x}{y}\frac{1}{x};$$

$$\therefore \frac{\partial^2 z}{\partial y\partial x}\equiv\frac{\partial}{\partial y}\left(\frac{1}{y}\log x+\frac{1}{y}\right)=-\frac{1}{y^2}\log x-\frac{1}{y^2}.$$

Example 2. If $z=x^n f\left(\frac{y}{x}\right)$, prove that

$$x\frac{\partial z}{\partial x}+y\frac{\partial z}{\partial y}=nz.$$

Take logarithms, then $\log z=n\log x+\log f\left(\frac{y}{x}\right)$,

$$\therefore \frac{1}{z}\frac{\partial z}{\partial x}=\frac{n}{x}+\frac{1}{f}\frac{\partial f}{\partial x}.$$

$$\text{Put } \frac{y}{x}=t, \text{ then } \frac{\partial f}{\partial x}=\frac{df}{dt}\frac{\partial t}{\partial x}=f'(t)\left(-\frac{y}{x^2}\right),$$

$$\therefore \frac{1}{z}\frac{\partial z}{\partial x}=\frac{n}{x}-\frac{f'(t)}{f(t)}\frac{y}{x^2}.$$

Also,
$$\frac{1}{z}\frac{\partial z}{\partial y}=\frac{1}{f}\frac{\partial f}{\partial y}=\frac{1}{f}\frac{df}{dt}\frac{\partial t}{\partial y}=\frac{f'(t)}{f(t)}\frac{1}{x}.$$

$$\therefore \frac{x}{z}\frac{\partial z}{\partial x}+\frac{y}{z}\frac{\partial z}{\partial y}=\left\{n-\frac{f'(t)}{f(t)}\frac{y}{x}\right\}+\frac{f'(t)}{f(t)}\frac{y}{x};$$

$$\therefore x\frac{\partial z}{\partial x}+y\frac{\partial z}{\partial y}=nz.$$

EXERCISE XIX a

Find f_x, f_y for the functions $f(x, y)$ in Nos. 1-6 :

1. x/y.
2. $ax^2+2hxy+by^2$.
3. $\tan^{-1}(y/x)$.
4. $(x-y)/(x+y)$.
5. $\sin^{-1}(x/y)$.
6. $1/\sqrt{(x^2+y^2)}$.

Find f_{xx}, f_{xy}, f_{yx}, f_{yy} for the functions $f(x, y)$ in Nos. 7-12 :

7. xy.
8. $ax^3+3bx^2y+cy^3$.
9. $\log(xy)$.
10. e^{x+y}.
11. $x\cos y+y\cos x$.
12. $\operatorname{sh} x \operatorname{ch} 2y$.

13. If $z=f(ax+by)$, find $\dfrac{\partial z}{\partial x}\Big/\dfrac{\partial z}{\partial y}$.

14. If $z=f(x/y)$, prove that $xf_x+yf_y=0$.

15. If $z=\log(x^2+y^2)$, prove that $\dfrac{\partial^2 z}{\partial x^2}+\dfrac{\partial^2 z}{\partial y^2}=0$.

16. If $\tan\theta=\dfrac{y}{x}$, prove that $\dfrac{\partial^2\theta}{\partial x^2}+\dfrac{\partial^2\theta}{\partial y^2}=0$.

17. If $r^2=x^2+y^2$, prove that $\dfrac{\partial^2 r}{\partial x^2}\dfrac{\partial^2 r}{\partial y^2}=\left(\dfrac{\partial^2 r}{\partial x\partial y}\right)^2$.

18. If $V=x^2+y^2+z^2$, prove that $xV_x+yV_y+zV_z=2V$.

19. If $V=\tan^{-1}\dfrac{x}{y+z}$, prove that $xV_x+yV_y+zV_z=0$.

20. If $V = f(x + cy) + g(x - cy)$, prove that $\dfrac{\partial^2 V}{\partial y^2} = c^2 \dfrac{\partial^2 V}{\partial x^2}$.

21. If $V = 1/\sqrt{(x^2 + y^2 + z^2)}$, prove that $\dfrac{\partial^2 V}{\partial x^2} + \dfrac{\partial^2 V}{\partial y^2} + \dfrac{\partial^2 V}{\partial z^2} = 0$.

Total Variation of a Function of Two Variables.

The rest of this chapter needs a stricter theory than is suitable for Elementary Calculus.

If $z = f(x, y)$, and δz is the increase of z due to a simultaneous increase δx in x and δy in y, then

$$\delta z = f(x + \delta x, y + \delta y) - f(x, y).$$

By the mean-value theorem, see p. 129 (Vol. I),

$$F(x + \delta x) - F(x) = \delta x F'(x + \theta \delta x), \quad (0 < \theta < 1);$$

therefore, assuming that the conditions of that theorem are satisfied,

$$f(x + \delta x, y + \delta y) - f(x, y + \delta y) = \delta x \frac{\partial}{\partial x} f(x + \theta_1 \delta x, y + \delta y)$$

and
$$f(x, y + \delta y) - f(x, y) = \delta y \frac{\partial}{\partial y} f(x, y + \theta_2 \delta y)$$

where θ_1, θ_2 have values between 0 and 1. Adding,

$$\delta z = \delta x \frac{\partial}{\partial x} f(x + \theta_1 \delta x, y + \delta y) + \delta y \frac{\partial}{\partial y} f(x, y + \theta_2 \delta y);$$

hence, assuming the continuity of $\dfrac{\partial f}{\partial x}, \dfrac{\partial f}{\partial y}$,

$$\delta z = \delta x \left\{ \frac{\partial}{\partial x} f(x, y) + \epsilon_1 \right\} + \delta y \left\{ \frac{\partial}{\partial y} f(x, y) + \epsilon_2 \right\}$$

$$\text{i.e. } \delta z = \frac{\partial z}{\partial x} \delta x + \frac{\partial z}{\partial y} \delta y + \epsilon_1 \delta x + \epsilon_2 \delta y \quad . \quad . \quad (1)$$

where $\epsilon_1 \to 0$ and $\epsilon_2 \to 0$ when $\delta x \to 0$ and $\delta y \to 0$.
Therefore if δx and δy are small, and $f_x \neq 0$, $f_y \neq 0$,

$$\delta z \simeq \frac{\partial z}{\partial x} \delta x + \frac{\partial z}{\partial y} \delta y \quad . \quad . \quad . \quad . \quad (2)$$

A similar result holds for a function of any finite number of variables.

Derivative of a Function of Two Functions of t.

Suppose that $V, =f(x, y)$, is a function of x and y, where x and y are functions of an independent variable t given by, say, $x=\phi(t)$, $y=\psi(t)$. Then V also is a function $f\{\phi(t), \psi(t)\}$ of the single independent variable t, and the value of $\frac{dV}{dt}$ might be found by the ordinary rules of differentiation; but it is usually simpler to avoid the substitution for x and y.

Let δt denote a small increase in t, and let

$$\delta x \equiv \phi(t+\delta t)-\phi(t) \quad \text{and} \quad \delta y \equiv \psi(t+\delta t)-\psi(t)$$

denote the corresponding increases in x and y; these will be small if the functions ϕ, ψ are continuous. The corresponding increase in V, by equation (1) on p. 384, is

$$\delta V, =\frac{\partial V}{\partial x}\delta x+\frac{\partial V}{\partial y}\delta y+(\epsilon_1\delta x+\epsilon_2\delta y);$$

$$\therefore \frac{\delta V}{\delta t}=\frac{\partial V}{\partial x}\frac{\delta x}{\delta t}+\frac{\partial V}{\partial y}\frac{\delta y}{\delta t}+\left(\epsilon_1\frac{\delta x}{\delta t}+\epsilon_2\frac{\delta y}{\delta t}\right);$$

taking the limit when $\delta t \to 0$,

$$\frac{dV}{dt}=\frac{\partial V}{\partial x}\frac{dx}{dt}+\frac{\partial V}{\partial y}\frac{dy}{dt} \quad . \quad . \quad . \quad (3)$$

This may also be written, in the form of a relation between the differentials, as

$$dV=\frac{\partial V}{\partial x}dx+\frac{\partial V}{\partial y}dy \quad . \quad . \quad . \quad (4)$$

And this *exact* relation may be compared with the *approximate* relation (2) on p. 384.

A similar result holds if V is a function of three variables x, y, z given by $x=f_1(t)$, $y=f_2(t)$, $z=f_3(t)$; the differentials are connected by

$$dV=\frac{\partial V}{\partial x}dx+\frac{\partial V}{\partial y}dy+\frac{\partial V}{\partial z}dz \quad . \quad . \quad . \quad (5)$$

Euler's Theorem for Homogeneous Functions.

If $V \equiv f(x, y, z)$ is a homogeneous function of the nth degree in x, y, z, then

$$x\frac{\partial V}{\partial x} + y\frac{\partial V}{\partial y} + z\frac{\partial V}{\partial z} = nV. \qquad (6)$$

It is easy to verify this result for the function $x^p y^q z^r$ where $p+q+r=n$; for if $V = x^p y^q z^r$, by logarithmic differentiation,

$$\frac{1}{V}\frac{\partial V}{\partial x} = \frac{p}{x}, \quad \frac{1}{V}\frac{\partial V}{\partial y} = \frac{q}{y}, \quad \frac{1}{V}\frac{\partial V}{\partial z} = \frac{r}{z};$$

$$\therefore x\frac{\partial V}{\partial x} + y\frac{\partial V}{\partial y} + z\frac{\partial V}{\partial z} = (p+q+r)V = nV.$$

And it follows that it holds also for any function which is the sum of terms of the form $x^p y^q z^r$, that is for a polynomial of degree n.

The general proof is as follows:

Take *any* set of *constant* values x_1, y_1, z_1 of x, y, z and denote $f(x_1, y_1, z_1)$ by V_1.

Then, if $x = tx_1$, $y = ty_1$, $z = tz_1$,

$$V \equiv f(x, y, z) = f(tx_1, ty_1, tz_1) = t^n f(x_1, y_1, z_1)$$

since the function is homogeneous of degree n;

$$\therefore V = t^n V_1.$$

By logarithmic differentiation with respect to t,

$$\frac{1}{V}\frac{dV}{dt} = \frac{n}{t}, \quad \therefore t\frac{dV}{dt} = nV.$$

But from p. 385,

$$\frac{dV}{dt} = \frac{\partial V}{\partial x}\frac{dx}{dt} + \frac{\partial V}{\partial y}\frac{dy}{dt} + \frac{\partial V}{\partial z}\frac{dz}{dt}$$

$$= x_1\frac{\partial V}{\partial x} + y_1\frac{\partial V}{\partial y} + z_1\frac{\partial V}{\partial z}$$

$$= \frac{1}{t}\left(x\frac{\partial V}{\partial x} + y\frac{\partial V}{\partial y} + z\frac{\partial V}{\partial z}\right);$$

$$\therefore x\frac{\partial V}{\partial x} + y\frac{\partial V}{\partial y} + z\frac{\partial V}{\partial z} = t\frac{dV}{dt} = nV.$$

Equation of a Tangent to a Curve.

Suppose that $\mathsf{V}=f(x, y)$ where $y=\phi(x)$; then by equation (3), p. 385, with $t=x$,

$$\frac{d\mathsf{V}}{dx}=\frac{\partial\mathsf{V}}{\partial x}+\frac{\partial\mathsf{V}}{\partial y}\frac{dy}{dx}.$$

$$=f_x+f_y\frac{dy}{dx}.$$

If (x, y) are the cartesian coordinates of any point on a given curve we may regard y as a function $\phi(x)$ of x, although this fact may only be expressed implicitly by the equation

$$f(x, y)=0$$

of the curve. The above result, with $\mathsf{V}=0$, then gives

$$0=f_x+f_y\frac{dy}{dx}.$$

Now the equation of the tangent is sometimes written in the form

$$y-y_1=m(x-x_1)$$

where m is the value of $\frac{dy}{dx}$ at (x_1, y_1). Hence it may also be written

$$y-y_1=-\frac{f_{x_1}}{f_{y_1}}(x-x_1),$$

that is

$$(x-x_1)f_{x_1}+(y-y_1)f_{y_1}=0 \tag{7}$$

where f_{x_1}, f_{y_1} are the values of f_x, f_y at (x_1, y_1).
This result may be further transformed by means of Euler's Theorem.

Suppose that the equation $f(x, y)=0$ of the curve is made homogeneous by the introduction of z and becomes $\phi(x, y, z)=0$. where ϕ is homogeneous of degree n and reduces to $f(x, y)$ for $z=1$; then

$$x_1\phi_{x_1}+y_1\phi_{y_1}+z_1\phi_{z_1}\equiv n\phi(x_1, y_1, z_1),$$

$$\therefore \text{ for } z_1=1,\ x_1\phi_{x_1}+y_1\phi_{y_1}+z_1\phi_{z_1}=nf(x_1, y_1)=0.$$

But when $z_1=1$, $\phi_{x_1}=f_{x_1}$ and $\phi_{y_1}=f_{y_1}$,

$$\therefore \text{ for } z_1=1,\ x_1f_{x_1}+y_1f_{y_1}=-z_1\phi_{z_1};$$

the equation

$$(x-x_1)f_{x_1}+(y-y_1)f_{y_1}=0$$

therefore reduces for $z_1=1$ to

$$xf_{x_1}+yf_{y_1}+z_1\phi_{z_1}=0,$$

and may be expressed in the form

$$x\phi_{x_1}+y\phi_{y_1}+z\phi_{z_1}=0 \tag{8}$$

where z, z_1 are put equal to 1 after differentiation.

Example 3. The length of the side a of a triangle ABC is calculated from measurements of b, c, A. Find approximately the error in the value of a due to small errors y, z, α in the three measurements, if $A \neq \frac{1}{2}\pi$.

$$a^2 = b^2 + c^2 - 2bc \cos A,$$

$$\therefore\ \delta(a^2) \simeq \frac{\partial a^2}{\partial b}\delta b + \frac{\partial a^2}{\partial c}\delta c + \frac{\partial a^2}{\partial A}\delta A,$$

$$\therefore\ 2a\delta a \simeq 2(b - c\cos A)\delta b + 2(c - b\cos A)\delta c + 2bc\sin A\,\delta A\,;$$

$\therefore$ the approximate error is

$$\frac{1}{a}\{(b - c\cos A)y + (c - b\cos A)z + \alpha bc\sin A\}.$$

Note. It is assumed that α is measured in radians. If $A = \frac{1}{2}\pi$, it is necessary to proceed to a closer approximation.

Example 4. Find the equation of the tangent at (x_1, y_1) to the conic $ax^2 + 2hxy + by^2 + 2gx + 2fy + c = 0$.

Make the equation homogeneous in x, y, z, thus :

$$\phi(x, y, z) \equiv ax^2 + 2hxy + by^2 + 2gzx + 2fyz + cz^2\,;$$

$$\therefore\ \phi_{x_1} = 2(ax_1 + hy_1 + gz_1),$$

$$\phi_{y_1} = 2(hx_1 + by_1 + fz_1),$$

$$\phi_{z_1} = 2(gx_1 + fy_1 + cz_1)\,;$$

$\therefore$ the equation of the tangent at (x_1, y_1) is

$$x(ax_1 + hy_1 + g) + y(hx_1 + by_1 + f) + (gx_1 + fy_1 + c) = 0.$$

EXERCISE XIX b

1. If $V = x^2 + y^2$ where $x = t^2 + 1$, $y = t - 1$, find $\frac{dV}{dt}$ by first substituting for x and y and verify it by equation (3), p. 385.

2. If x and y are functions of t, find relations between dV, dx, dy given that

   (i) $V = x^2y^3$; (ii) $V = x/y$; (iii) $V = \tan^{-1}(y/x) - y/x$.

3. If $V = \tan x \cot y$, prove that $\frac{\delta V}{V} \simeq 2(\operatorname{cosec} 2x\,\delta x - \operatorname{cosec} 2y\,\delta y)$.

Find $\frac{dy}{dx}$ at any point (x, y) of the curves in Nos. 4-6:

4. $x^3 + y^3 = a^3$. 5. $\frac{1}{x} - \frac{1}{y} = \frac{1}{c}$. 6. $x \sin y = y \sin x$.

Find the equation of the tangent at (x_1, y_1) to the curves in Nos. 7-9:

7. $ax^2 + by^2 = 1$. **8.** $x^{2/3} + y^{2/3} = a^{2/3}$. **9.** $(x + y)^3 = 3xy$.

10. Find the lengths of the subtangent and subnormal at the point (x, y) of the curve $f(x, y) = 0$.

11. Find $\frac{d^2y}{dx^2}$ at the point (x, y) of the curve

(i) $x^4 + y^4 = a^4$; (ii) $\tan x = \operatorname{sh} y$.

12. If the area Δ of the triangle ABC is calculated from measurements of a, b, C, prove that if $\mathrm{C} \neq \frac{1}{2}\pi$,

$$\frac{\delta\Delta}{\Delta} \simeq \frac{\delta a}{a} + \frac{\delta b}{b} + \cot \mathrm{C}\, \delta \mathrm{C}$$

where δa, δb, $\delta\mathrm{C}$ are the errors of measurement and $\delta\Delta$ the consequent error in Δ.

13. For the triangle ABC, $2\Delta(\cot \mathrm{B} + \cot \mathrm{C}) = a^2$; prove that

$$\delta\Delta \simeq b \sin \mathrm{C}\, \delta a + \tfrac{1}{2}c^2 \delta\mathrm{B} + \tfrac{1}{2}b^2 \delta\mathrm{C}.$$

14. If the area Δ of the triangle ABC is expressed in the form $f(a, b, c)$, prove that $f_a = \frac{1}{2}a \cot \mathrm{A}$.

15. If the side c of the triangle ABC is expressed in terms of a, b and the area Δ, prove that

$$\frac{\partial c}{\partial a} = -\cos \mathrm{A} \sec \mathrm{C} \text{ and find } \frac{\partial c}{\partial b}.$$

16. Write down the value of $x\frac{\partial \mathrm{V}}{\partial x} + y\frac{\partial \mathrm{V}}{\partial y} + z\frac{\partial \mathrm{V}}{\partial z}$ if

$$\text{(i) } \mathrm{V} = \cos^{-1}\frac{x}{\sqrt{(x^2 + y^2 + z^2)}}; \quad \text{(ii) } \mathrm{V} = \frac{\sqrt{(x + y - z)}}{x^2 + y^2 + z^2}.$$

17. If V is a homogeneous function of x and y of degree n, prove that

$$x^2\frac{\partial^2 \mathrm{V}}{\partial x^2} + 2xy\frac{\partial^2 \mathrm{V}}{\partial x \partial y} + y^2\frac{\partial^2 \mathrm{V}}{\partial y^2} = n(n - 1)\mathrm{V}.$$

Differential of a Function of Two Independent Variables.

If V is a function of a single variable x, the differential $d\mathrm{V}$ is defined by the relation $d\mathrm{V}=f'(x)dx$, where dx has the same meaning as δx, see p. 25 (Vol. I); and $\delta\mathrm{V} \simeq f'(x)\delta x$.

If V is a function of *two independent variables* x, y we have from p. 384

$$\delta\mathrm{V} \simeq \frac{\partial\mathrm{V}}{\partial x}\delta x + \frac{\partial\mathrm{V}}{\partial y}\delta y\ ;$$

and we now define the differential $d\mathrm{V}$ by the relation

$$d\mathrm{V} = \frac{\partial\mathrm{V}}{\partial x}\delta x + \frac{\partial\mathrm{V}}{\partial y}\delta y \quad . \quad . \quad . \quad . \quad (9)$$

This definition assigns meanings to dx and dy:

if $\mathrm{V} \equiv x$, $dx = \frac{\partial x}{\partial x}\delta x + 0 = \delta x$; and similarly if $\mathrm{V} \equiv y$, $dy = \delta y$.

We therefore modify the relation (9) into

$$d\mathrm{V} = \frac{\partial\mathrm{V}}{\partial x}dx + \frac{\partial\mathrm{V}}{\partial y}dy \quad . \quad . \quad . \quad . \quad (10)$$

A comparison of relation (4), p. 385, with (10) shows that the expression for the differential $d\mathrm{V}$ of a function of two variables is the same whether the variables are independent or not. The convenience of the differential notation is largely due to this fact.

In general $\delta\mathrm{V}$ is not equal to $d\mathrm{V}$, but when $\delta x \to 0$ and $\delta y \to 0$ in any manner, $\frac{\delta\mathrm{V}}{d\mathrm{V}} \to 1$, and for small values of δx, δy, the value of $d\mathrm{V}$ is a good approximation to that of $\delta\mathrm{V}$. Hence it is easy to write down an *approximate* expression for $\delta\mathrm{V}$ by thinking of the *accurate* expression for $d\mathrm{V}$.

The differential of a function of any number of independent variables is defined in the same way. Thus if V is a function of three independent variables x, y, z,

$$d\mathrm{V} = \frac{\partial\mathrm{V}}{\partial x}dx + \frac{\partial\mathrm{V}}{\partial y}dy + \frac{\partial\mathrm{V}}{\partial z}dz \quad . \quad . \quad . \quad (11)$$

Partial Derivatives of a Function of Two Functions.

Let $\mathrm{V} = \phi(\xi, \eta)$ where $\xi = f(x, y)$, $\eta = g(x, y)$ and x, y are independent variables.

If we write V in the form $\phi\{f(x, y), g(x, y)\}$ we can obtain $\frac{\partial\mathrm{V}}{\partial x}$, $\frac{\partial\mathrm{V}}{\partial y}$ by the ordinary rules of partial differentiation; but it is usually simpler to avoid the substitution.

By definition, since x, y are independent,

$$dV = \frac{\partial V}{\partial x}dx + \frac{\partial V}{\partial y}dy.$$

Also for the same reason

$$d\xi = \frac{\partial \xi}{\partial x}dx + \frac{\partial \xi}{\partial y}dy \quad \text{and} \quad d\eta = \frac{\partial \eta}{\partial x}dx + \frac{\partial \eta}{\partial y}dy.$$

But these values of dV, $d\xi$, $d\eta$ satisfy the relation

$$dV = \frac{\partial V}{\partial \xi}d\xi + \frac{\partial V}{\partial \eta}d\eta;$$

$$\therefore dV = \frac{\partial V}{\partial \xi}\left(\frac{\partial \xi}{\partial x}dx + \frac{\partial \xi}{\partial y}dy\right) + \frac{\partial V}{\partial \eta}\left(\frac{\partial \eta}{\partial x}dx + \frac{\partial \eta}{\partial y}dy\right)$$

$$= \left(\frac{\partial V}{\partial \xi}\frac{\partial \xi}{\partial x} + \frac{\partial V}{\partial \eta}\frac{\partial \eta}{\partial x}\right)dx + \left(\frac{\partial V}{\partial \xi}\frac{\partial \xi}{\partial y} + \frac{\partial V}{\partial \eta}\frac{\partial \eta}{\partial y}\right)dy.$$

Since dx, dy are independent and arbitrary we have, by comparing the two expressions for dV in terms of dx and dy,

$$\frac{\partial V}{\partial x} = \frac{\partial V}{\partial \xi}\frac{\partial \xi}{\partial x} + \frac{\partial V}{\partial \eta}\frac{\partial \eta}{\partial x} \quad . \quad . \quad . \quad (12)$$

$$\frac{\partial V}{\partial y} = \frac{\partial V}{\partial \xi}\frac{\partial \xi}{\partial y} + \frac{\partial V}{\partial \eta}\frac{\partial \eta}{\partial y} \quad . \quad . \quad . \quad (13)$$

These results may be expressed by saying that the operators $\frac{\partial}{\partial x}$ and $\frac{\partial \xi}{\partial x}\frac{\partial}{\partial \xi} + \frac{\partial \eta}{\partial x}\frac{\partial}{\partial \eta}$ are equivalent, and similarly with y instead of x. To find a higher partial derivative of V, say $\frac{\partial^2 V}{\partial x^2}$, we apply the operator $\frac{\partial}{\partial x}$ to $\frac{\partial V}{\partial x}$, that is we apply the operator $\frac{\partial \xi}{\partial x}\frac{\partial}{\partial \xi} + \frac{\partial \eta}{\partial x}\frac{\partial}{\partial \eta}$ to $\frac{\partial V}{\partial \xi}\frac{\partial \xi}{\partial x} + \frac{\partial V}{\partial \eta}\frac{\partial \eta}{\partial x}$. In this way it is possible to express $\frac{\partial^2 V}{\partial x^2}$, $\frac{\partial^2 V}{\partial x \partial y}$, $\frac{\partial^2 V}{\partial y^2}$ in terms of $\frac{\partial^2 V}{\partial \xi^2}$, $\frac{\partial^2 V}{\partial \xi \partial \eta}$, $\frac{\partial^2 V}{\partial \eta^2}$, $\frac{\partial V}{\partial \xi}$, $\frac{\partial V}{\partial \eta}$. The method is illustrated in the following example.

Example 5. Transform the expression $\frac{\partial^2 V}{\partial x^2}+\frac{\partial^2 V}{\partial y^2}$ from cartesian to polar coordinates.

We have $$x=r\cos\theta,\quad y=r\sin\theta,$$

and $$r^2=x^2+y^2,\quad \tan\theta=\frac{y}{x};$$

$$\therefore\ 2r\frac{\partial r}{\partial x}=2x,\quad \therefore\ \frac{\partial r}{\partial x}=\frac{x}{r}=\cos\theta;$$

and $$\sec^2\theta\frac{\partial\theta}{\partial x}=-\frac{y}{x^2},\quad \therefore\ \frac{\partial\theta}{\partial x}=-\frac{y}{r^2}=-\frac{\sin\theta}{r}.$$

$$\therefore\ \frac{\partial V}{\partial x}\equiv\frac{\partial V}{\partial r}\frac{\partial r}{\partial x}+\frac{\partial V}{\partial\theta}\frac{\partial\theta}{\partial x}=\cos\theta\frac{\partial V}{\partial r}-\frac{\sin\theta}{r}\frac{\partial V}{\partial\theta};$$

thus $\frac{\partial}{\partial x}$ and $\cos\theta\frac{\partial}{\partial r}-\frac{\sin\theta}{r}\frac{\partial}{\partial\theta}$ are equivalent operators,

$$\therefore\ \frac{\partial^2 V}{\partial x^2}=\left(\cos\theta\frac{\partial}{\partial r}-\frac{\sin\theta}{r}\frac{\partial}{\partial\theta}\right)\left(\cos\theta\frac{\partial V}{\partial r}-\frac{\sin\theta}{r}\frac{\partial V}{\partial\theta}\right)$$

$$=\cos\theta\left(\cos\theta\frac{\partial^2 V}{\partial r^2}+\frac{\sin\theta}{r^2}\frac{\partial V}{\partial\theta}-\frac{\sin\theta}{r}\frac{\partial^2 V}{\partial r\partial\theta}\right)$$

$$-\frac{\sin\theta}{r}\left(-\sin\theta\frac{\partial V}{\partial r}+\cos\theta\frac{\partial^2 V}{\partial\theta\partial r}-\frac{\cos\theta}{r}\frac{\partial V}{\partial\theta}-\frac{\sin\theta}{r}\frac{\partial^2 V}{\partial\theta^2}\right);$$

$$\therefore\ \frac{\partial^2 V}{\partial x^2}=\cos^2\theta\frac{\partial^2 V}{\partial r^2}-\frac{2\sin\theta\cos\theta}{r}\frac{\partial^2 V}{\partial r\partial\theta}$$

$$+\frac{\sin^2\theta}{r^2}\frac{\partial^2 V}{\partial\theta^2}+\frac{\sin^2\theta}{r}\frac{\partial V}{\partial r}+\frac{2\sin\theta\cos\theta}{r^2}\frac{\partial V}{\partial\theta},$$

assuming that $\frac{\partial^2 V}{\partial r\partial\theta}=\frac{\partial^2 V}{\partial\theta\partial r}$, see p. 380.

Similarly, $$\frac{\partial}{\partial y}=\sin\theta\frac{\partial}{\partial r}+\frac{\cos\theta}{r}\frac{\partial}{\partial\theta}$$

$$=\cos\theta_1\frac{\partial}{\partial r}-\frac{\sin\theta_1}{r}\frac{\partial}{\partial\theta_1}$$

where $\theta_1=\theta-\frac{1}{2}\pi$, which also makes $d\theta_1=d\theta$.

Hence $\frac{\partial V}{\partial y}, \frac{\partial^2 V}{\partial y^2}$ can be found from $\frac{\partial V}{\partial x}, \frac{\partial^2 V}{\partial x^2}$ by replacing θ by $\theta - \frac{1}{2}\pi$. This gives

$$\frac{\partial^2 V}{\partial y^2} = \sin^2\theta \frac{\partial^2 V}{\partial r^2} + \frac{2\sin\theta\cos\theta}{r}\frac{\partial^2 V}{\partial r \partial \theta}$$

$$+ \frac{\cos^2\theta}{r^2}\frac{\partial^2 V}{\partial \theta^2} + \frac{\cos^2\theta}{r}\frac{\partial V}{\partial r} - \frac{2\sin\theta\cos\theta}{r^2}\frac{\partial V}{\partial \theta}.$$

$$\therefore \frac{\partial^2 V}{\partial x^2} + \frac{\partial^2 V}{\partial y^2} = \frac{\partial^2 V}{\partial r^2} + \frac{1}{r^2}\frac{\partial^2 V}{\partial \theta^2} + \frac{1}{r}\frac{\partial V}{\partial r}.$$

EXERCISE XIX c

1. If z is a function of x and y and if $x = u - v$, $y = uv$, express du and dv in terms of dx, dy; hence prove that

$$\text{(i) } (u+v)\frac{\partial z}{\partial x} = u\frac{\partial z}{\partial u} - v\frac{\partial z}{\partial v}; \quad \text{(ii) } (u+v)\frac{\partial z}{\partial y} = \frac{\partial z}{\partial u} + \frac{\partial z}{\partial v}.$$

2. If z is a function of x and y and if $x = e^u \sin v$, $y = e^u \cos v$, prove that

$$\text{(i) } \frac{\partial z}{\partial u} = x\frac{\partial z}{\partial x} + y\frac{\partial z}{\partial y}; \quad \text{(ii) } \frac{\partial z}{\partial x} = e^{-u}\left(\sin v \frac{\partial z}{\partial u} + \cos v \frac{\partial z}{\partial v}\right).$$

3. The cartesian and polar coordinates of a point are (x, y) and (r, θ). Illustrate geometrically the meanings of

(i) $\frac{\partial x}{\partial r}$ where x is a function of r, θ;

(ii) $\frac{\partial r}{\partial x}$ where r is a function of x, y.

What are their values? Find also the values of $\frac{\partial x}{\partial \theta}$ and of $\frac{\partial \theta}{\partial x}$.

4. If $V = f(x, y)$ and if $x = r\cos\theta$, $y = r\sin\theta$, prove that $r\frac{\partial V}{\partial r} = x\frac{\partial V}{\partial x} + y\frac{\partial V}{\partial y}$ and express $\frac{\partial V}{\partial \theta}$ in terms of x and y.

5. If $f(x, y, z)=0$, explain the meanings of the symbols $\frac{\partial y}{\partial x}$ and $\frac{\partial x}{\partial y}$. Prove that

$$\text{(i)}\ \frac{\partial x}{\partial y}\frac{\partial y}{\partial x}=1\,;\quad \text{(ii)}\ \frac{\partial x}{\partial y}\frac{\partial y}{\partial z}=-\frac{\partial x}{\partial z}.$$

6. If $V=f(x, y)$ and if $x=r\cos\theta$, $y=r\sin\theta$, express $\left(\frac{\partial V}{\partial x}\right)^2+\left(\frac{\partial V}{\partial y}\right)^2$ in terms of r, θ.

7. If $x=r\cos\theta$, $y=r\sin\theta$ where r and θ are functions of t and dots denote differentiation with respect to t, express $\dot{x}$ and $\dot{y}$ in terms of $\dot{r}$, $\dot{\theta}$, r, θ; also prove that

$$\text{(i)}\ \ddot{x}\cos\theta+\ddot{y}\sin\theta=\ddot{r}-r\dot{\theta}^2\,;$$

$$\text{(ii)}\ \ddot{y}\cos\theta-\ddot{x}\sin\theta=\frac{1}{r}\frac{d}{dt}(r^2\dot{\theta}).$$

Interpret these results if (x, y) is the position of a moving particle at time t.

8. If $V=f(x, y)$ and if $x=\xi^2-\eta^2$, $y=2\xi\eta$, prove that

$$\text{(i)}\ 2(\xi^2+\eta^2)\frac{\partial V}{\partial x}=\xi\frac{\partial V}{\partial \xi}-\eta\frac{\partial V}{\partial \eta}\,;$$

$$\text{(ii)}\ \frac{1}{4(\xi^2+\eta^2)}\left(\frac{\partial^2 V}{\partial \xi^2}+\frac{\partial^2 V}{\partial \eta^2}\right)=\frac{\partial^2 V}{\partial x^2}+\frac{\partial^2 V}{\partial y^2}.$$

9. If $V=f(x+ht, y+kt)$ where h, k are constants and x, y are independent of t, prove that

$$\frac{d^2V}{dt^2}=h^2\frac{\partial^2 V}{\partial x^2}+2hk\frac{\partial^2 V}{\partial x\partial y}+k^2\frac{\partial^2 V}{\partial y^2}.$$

10. If $V=f\left(\frac{x}{z}, \frac{y}{z}\right)$, prove that

$$x\frac{\partial V}{\partial x}+y\frac{\partial V}{\partial y}+z\frac{\partial V}{\partial z}=0.$$

11. If $V=f(x-y, y-z, z-x)$, prove that

$$\frac{\partial V}{\partial x}+\frac{\partial V}{\partial y}+\frac{\partial V}{\partial z}=0.$$

12. If $y^3+3x^2y+1=0$, prove that

$$(x^2+y^2)^3\frac{d^2y}{dx^2}+2(x^2-y^2)=0.$$

13. If $z = f(x, y)$ and if $u = x + y$ and $v = y(x + y)$, express $\dfrac{\partial^2 z}{\partial x^2} + \dfrac{\partial^2 z}{\partial y^2}$ in terms of u, v.

14. If $x = r\sin\theta\cos\phi, y = r\sin\theta\sin\phi, z = r\cos\theta$, and if $V = f(x, y, z)$, prove that $\dfrac{\partial^2 V}{\partial x^2} + \dfrac{\partial^2 V}{\partial y^2} + \dfrac{\partial^2 V}{\partial z^2}$ equals

$$\frac{\partial^2 V}{\partial r^2} + \frac{1}{r^2}\frac{\partial^2 V}{\partial \theta^2} + \frac{1}{r^2\sin^2\theta}\frac{\partial^2 V}{\partial \phi^2} + \frac{2}{r}\frac{\partial V}{\partial r} + \frac{\cot\theta}{r^2}\frac{\partial V}{\partial \theta}.$$

[Begin by using Example 5, p. 392, to express $\dfrac{\partial^2 V}{\partial x^2} + \dfrac{\partial^2 V}{\partial y^2}$ in terms of ρ, ϕ where $\rho = r\sin\theta$.]

15. If V is a function of two independent variables x, y, prove that a necessary (though not sufficient) condition for a maximum or minimum value of V is that $\dfrac{\partial V}{\partial x} = 0$ and $\dfrac{\partial V}{\partial y} = 0$.

The sum of the volumes of a sphere and a right circular cylinder is constant. If the total surface has a stationary value, prove that the radius of the sphere is equal to the base-radius of the cylinder and is half the height of the cylinder.

16. For what values of x and y are there stationary values of $x^4 + y^4 - 2(x - y)^2$?

CHAPTER XX

DIFFERENTIAL EQUATIONS

ANY relation between the variables x, y and the derivatives $\frac{dy}{dx}, \frac{d^2y}{dx^2}$, etc., is called an *ordinary differential equation*. If the highest derivative that occurs in the equation is $\frac{d^ny}{dx^n}$ the equation is said to be of *order n*; the *degree* of an algebraic differential equation is the power to which that highest derivative occurs when the equation is expressed in a rational integral form.

Construction of Differential Equations. From an equation involving x, y, a which may represent a family of curves there can be obtained by differentiation another equation involving x, y, $\frac{dy}{dx}$, a; if the constant a is eliminated between the two equations we obtain what is called the differential equation of the family.

Example 1. Find the differential equation of the system of parabolas $y^2 = 4ax$ where the constant a may have any value.

By differentiation, $$2y\frac{dy}{dx} = 4a;$$

eliminating a, $$2xy\frac{dy}{dx} = 4ax = y^2,$$

$$\therefore\ 2x\frac{dy}{dx} = y, \quad (y \neq 0).$$

Alternatively, $$2\log y = \log 4a + \log x,$$

$$\therefore\ \frac{2}{y}\frac{dy}{dx} = \frac{1}{x}.$$

The differential equation of a system of curves expresses a property common to all the members.

In Example 1, since $y \Big/ \frac{dy}{dx}$ equals the subtangent, the differential equation asserts that the subtangent is twice the abscissa for any point of any parabola of the system. (This is the property **TA = AN** of Geometrical Conics.)

The differential equation obtained for the system of curves $y^2 = 4ax$ was of the *first order*. Starting from an equation involving x, y, a, b where a and b are both constants, it would be necessary to differentiate twice before the elimination could be performed.

Example 2. Find the differential equation for the space-time relation $x = a\cos t + b\sin t$ where a and b are arbitrary constants.

Here the variables are x and t instead of y and x respectively.

Differentiating twice in succession with respect to t

$$\frac{dx}{dt} = -a\sin t + b\cos t,$$

and
$$\frac{d^2x}{dt^2} = -a\cos t - b\sin t\,;$$

$\therefore$ the differential equation is $\dfrac{d^2x}{dt^2} = -x$.

The differential equation in Example 2 expresses a property of the acceleration of any particle which moves according to the given law. It is an equation of the *second order*.

The method suggests that the differential equation corresponding to a relation between x, y, and n arbitrary constants will be of *order n*.

EXERCISE XX a

1. If $y = ae^x$, verify that $\dfrac{dy}{dx} = y$.

2. If $xy = c^2$, prove that $\dfrac{dy}{dx} = -\dfrac{y}{x}$ and interpret the result geometrically.

3. If $x = a\cos nt + b\sin nt$ where a, b are arbitrary constants and n is fixed, prove that $\dfrac{d^2x}{dt^2} = -n^2x$. (The acceleration is directed towards the origin $x = 0$ and is proportional to the distance from that point. The motion is called *simple harmonic*.)

4. Eliminate a, ϵ from $x = a\cos(nt + \epsilon)$ and interpret the result.

5. Find the differential equation of all straight lines through the origin.

6. Find the differential equation of all straight lines $lx + my + 1 = 0$ and interpret the result.

7 Eliminate a, b from (i) $y^2 = 4a(x + b)$; (ii) $y = a\tan^{-1}x + b$.

8. If $y = ae^{px} + be^{qx}$, prove that

$$\frac{d^2y}{dx^2} - (p+q)\frac{dy}{dx} + pqy = 0.$$

9. If $x = e^{-pt}(a \cos nt + b \sin nt)$, prove that

$$\frac{d^2x}{dt^2} + 2p\frac{dx}{dt} + (n^2 + p^2)x = 0.$$

Solutions of Differential Equations. In practical applications it is often necessary to proceed from a known differential equation to find the relation involving x, y, and constants. This is called *solving* the differential equation. Any relation between x and y which satisfies the equation is called a *particular* solution or particular integral of the equation, and a general solution is called the *complete primitive.* Examples 1 and 2 suggest that the complete primitive of a differential equation of order n involves n arbitrary constants.

We proceed to give methods of solution for those types of equation which arise most frequently in mechanics, physics, and geometry. As regards mechanics it may be said that the fundamental reason why differential equations arise is that forces, and hence accelerations, $\left(\frac{d^2x}{dt^2}\right)$, are usually known and the problem often consists in finding positions (x); for this reason the equations that arise are usually of the second order.

Equations of the First Order.

(1) *One variable absent.*

If y is absent the equation takes the form

$$\frac{dy}{dx} = f(x);$$

the solution is equivalent to a mere integration and may be written

$$y = \int f(x)\,dx.$$

If x is absent the equation may be written

$$\frac{dy}{dx} = f(y) \quad \text{or} \quad \frac{dx}{dy} = \frac{1}{f(y)}$$

and the solution is

$$x = \int \frac{1}{f(y)}\,dy.$$

(2) *Separable variables.*

If the equation can be expressed in the form

$$f(y)\frac{dy}{dx}+g(x)=0$$

we write

$$f(y)dy+g(x)dx=0;$$

$$\therefore \int f(y)\,dy+\int g(x)\,dx=0.$$

Note. The indefinite integrals in these solutions involve an arbitrary constant.

Example 3. Solve $\sin 2x\dfrac{dy}{dx}+y^2=1.$

$$\sin 2x\,dy=(1-y^2)dx, \quad \therefore \frac{dy}{1-y^2}=\frac{dx}{\sin 2x};$$

$$\therefore \int\frac{dy}{1-y^2}=\int\frac{dx}{\sin 2x};$$

$$\therefore \tfrac{1}{2}\log\frac{1+y}{1-y}=\tfrac{1}{2}\log\tan x+a,$$

$$\therefore \log\frac{1+y}{1-y}=\log(b\tan x),$$

$$\therefore \frac{1+y}{1-y}=b\tan x.$$

Example 4. Solve $\dfrac{dy}{dx}+\mathsf{P}y=0$ where P is a function of x; e.g. $\mathsf{P}\equiv x^2$.

$$\frac{dy}{y}+\mathsf{P}dx=0; \quad \therefore \log y+\int\mathsf{P}dx=0$$

$$\therefore y=e^{-\int\mathsf{P}dx}.$$

Note that the arbitrary constant is involved in $\int\mathsf{P}dx$; there is no increase of generality in writing $\log y+\int\mathsf{P}dx=\mathsf{C}$.

$$\text{If } \mathsf{P}\equiv x^2, \quad \int\mathsf{P}dx=\tfrac{1}{3}x^3+a,$$

$$\therefore y=e^{-\frac{1}{3}x^3-a},$$

and this may be written $y=be^{-\frac{1}{3}x^3}$.

Example 5. Find the curves for which the subnormal is constant.

For the curve $y = f(x)$ the subnormal is $y\dfrac{dy}{dx}$ (p. 40, Vol. I).

$$\therefore\ y\frac{dy}{dx} = k\ ; \quad \therefore\ y\,dy = k\,dx\ ;$$

$$\therefore\ \tfrac{1}{2}y^2 = kx + c.$$

This is the equation of a parabola with axis Ox and latus rectum $2k$.

(3) *Homogeneous equations.*

If the equation is of the form

$$f(x, y)\frac{dy}{dx} + g(x, y) = 0,$$

where $f(x, y)$ and $g(x, y)$ are homogeneous and of the same degree, it can be reduced to the form,

$$\frac{dy}{dx} = \mathrm{F}\left(\frac{y}{x}\right)$$

and solved by the substitution $\dfrac{y}{x} = z$.

This is illustrated in Example 6.

Example 6. Solve $xy\dfrac{dy}{dx} = x^2 + y^2$.

$$\text{Put } y = xz, \text{ then } \frac{dy}{dx} = x\frac{dz}{dx} + z,$$

$$\therefore\ x^2z\left(x\frac{dz}{dx} + z\right) = x^2 + x^2z^2\ ;$$

$$\therefore\ xz\frac{dz}{dx} + z^2 = 1 + z^2\ ;\ (x \neq 0)$$

$$\therefore\ xz\frac{dz}{dx} = 1\ ; \quad \therefore\ zdz = \frac{1}{x}dx\ ;$$

$$\therefore\ \tfrac{1}{2}z^2 = \log x + c = \log(ax)\ ;$$

$$\therefore\ y^2 = 2x^2 \log(ax).$$

EXERCISE XX b

Solve the differential equations in Nos. 1-16 :

1. $\dfrac{dy}{dx}=3x^5.$
2. $\dfrac{dy}{dx}=2y^7.$
3. $x^4\dfrac{dy}{dx}=1-x^2.$
4. $\dfrac{dy}{dx}=\tan y.$
5. $r^2\dfrac{d\theta}{dr}=k.$
6. $\left(\dfrac{dy}{dx}\right)^2=x.$
7. $x\dfrac{dy}{dx}+x+y=0.$
8. $(x+y)\dfrac{dy}{dx}=x-y.$
9. $xy\dfrac{dy}{dx}=x^2-y^2.$
10. $(1+x^2)\dfrac{dy}{dx}=y^2.$
11. $y\dfrac{dy}{dx}+x=a.$
12. $x\dfrac{dy}{dx}+y=a.$
13. $x\dfrac{dy}{dx}+x=a.$
14. $\sin^2 x\dfrac{dy}{dx}=\cos^2 y.$
15. $\left(\dfrac{dy}{dx}\right)^2+y^2=1.$
16. $(x^2+y^2)\dfrac{dy}{dx}=xy.$

17. Find the curves for which $\tan\psi=x^3+1/x$.

18. Find the curves for which the subtangent is constant.

19. Find the curves $r=f(\theta)$ for which ϕ is constant.

20. Use the substitution $x-y=z$ to solve $\dfrac{dy}{dx}=(x-y)^2$.

21. Solve $\dfrac{dy}{dx}=\cos(x+y)$ by means of a substitution.

22. Express $d(x^n y)$ in terms of dx and dy. Hence solve

$$x\frac{dy}{dx}+ny=x^k.$$

23. If the perpendicular from the foot of the ordinate to the tangent is of constant length k, find the equation of the curve.

Equations of the First Order (*continued*).

(4) *Linear equations.*

A differential equation is said to be *linear* in y if y and its derivatives occur in the equation to the first degree only, and in separate terms. Thus any linear equation of the first order is of the form

$$\frac{dy}{dx} + \mathsf{P}y = \mathsf{Q}$$

where P and Q are functions of x (or constants).

The solution for the special case $\mathsf{Q} = 0$ has been obtained in Example 4, p. 399, and may be written

$$ye^{\int \mathsf{P}dx} = 1.$$

This solution is verified by the fact that

$$\frac{d}{dx}(ye^{\int \mathsf{P}dx}) = e^{\int \mathsf{P}dx}\left(\frac{dy}{dx} + \mathsf{P}y\right),$$

and this suggests the method of solving the general equation $\dfrac{dy}{dx} + \mathsf{P}y = \mathsf{Q}$.

Multiply both sides by $e^{\int \mathsf{P}dx}$, then

$$e^{\int \mathsf{P}dx}\left(\frac{dy}{dx} + \mathsf{P}y\right) = \mathsf{Q}e^{\int \mathsf{P}dx};$$

$\therefore$ integrating,

$$ye^{\int \mathsf{P}dx} = \int \mathsf{Q}e^{\int \mathsf{P}dx}\,dx.$$

This process is best understood by examining an example.

Example 7. Solve $\dfrac{dy}{dx} + 2y\cot x = \cos x$.

Here $\int \mathsf{P}dx = \int 2\cot x\,dx = 2\log \sin x$,

$$\therefore e^{\int \mathsf{P}dx} = e^{\log \sin^2 x} = \sin^2 x;$$

we therefore multiply both sides by $\sin^2 x$.

$$\sin^2 x\frac{dy}{dx} + 2y\cos x \sin x = \cos x \sin^2 x;$$

$$\therefore \frac{d}{dx}(y\sin^2 x) = \cos x \sin^2 x;$$

$$\therefore y\sin^2 x = \tfrac{1}{3}\sin^3 x + \mathsf{c};$$

$$\therefore y = \tfrac{1}{3}\sin x + \mathsf{c}\operatorname{cosec}^2 x.$$

(5) *Exact equations.*

If the equation

$$f(x, y)dx + g(x, y)dy = 0,$$

can be written in the form

$$\frac{\partial}{\partial x}\mathsf{F}(x, y)dx + \frac{\partial}{\partial y}\mathsf{F}(x, y)dy = 0,$$

it is called an *exact* equation ; the integral, given by

$$d\mathsf{F}(x, y) = 0,$$

is

$$\mathsf{F}(x, y) = c.$$

Equations which are not themselves exact can sometimes be made so by multiplying both sides by some function of x and y ; such a function is called an *integrating factor.*

For example, $\sin^2 x$ is an integrating factor for the equation in Example 7 and $e^{\int \mathsf{P}dx}$ is an integrating factor for the general linear equation $\frac{dy}{dx} + \mathsf{P}y = \mathsf{Q}$.

Example 8. If a current i is flowing at time t in a circuit of self-induction L and resistance R with external E.M.F. equal to $a \cos pt$, it can be proved that

$$\mathsf{L}\frac{di}{dt} + \mathsf{R}i = a \cos pt.$$

Find an expression for the current if L, R, a, p are constants.

For brevity put $\mathsf{R}/\mathsf{L} = b$, $a/\mathsf{L} = c$, then

$$\frac{di}{dt} + bi = c \cos pt.$$

The integrating factor is e^{bt},

$$\therefore\ e^{bt}\frac{di}{dt} + bie^{bt} = ce^{bt} \cos pt\ ;$$

$$\therefore\ ie^{bt} = \int ce^{bt} \cos pt\, dt$$

$$= \frac{ce^{bt}}{b^2 + p^2}(p \sin pt + b \cos pt) + k,$$ see p. 263 ;

$$\therefore\ i = ke^{-\mathsf{R}t/\mathsf{L}} + \frac{a}{\mathsf{R}^2 + \mathsf{L}^2p^2}(\mathsf{L}p \sin pt + \mathsf{R} \cos pt).$$

The more general equation

$$\frac{dy}{dx} + \mathsf{P}y = \mathsf{Q}y^n$$

where P and Q are functions of x may be reduced to the linear form by dividing both sides by y^n and using the substitution $y^{1-n} = z$. This is illustrated in Example 9.

Example 9. Solve $4\frac{dy}{dx} + xy + xy^5 = 0.$

This may be written

$$\frac{4}{y^5}\frac{dy}{dx} + \frac{x}{y^4} = -x.$$

$$\text{Put } y^{-4} = z, \quad \therefore \ -\frac{4}{y^5}\frac{dy}{dx} = \frac{dz}{dx};$$

$$\therefore \ \frac{dz}{dx} - xz = x.$$

The integrating factor is $e^{-\int x\,dx}$, that is $e^{-x^2/2}$;

hence $$\frac{d}{dx}(ze^{-x^2/2}) = xe^{-x^2/2};$$

$$\therefore \ ze^{-x^2/2} = -e^{-x^2/2} + c;$$

$$\therefore \ \frac{1}{y^4} = ce^{x^2/2} - 1.$$

Alternatively, the solution may be completed as follows

$$\frac{dz}{dx} = x(z+1), \quad \therefore \ \frac{dz}{z+1} = x\,dx$$

$$\therefore \ \log(z+1) = \tfrac{1}{2}x^2 + a, \quad \therefore \ y^{-4} + 1 = ce^{x^2/2}.$$

EXERCISE XX c

Solve the differential equations in Nos. 1-10:

1. $\frac{dy}{dx} - y = 1.$

2. $\frac{dy}{dx} + y = e^x.$

3. $\frac{dy}{dx} + y \cot x = \cos x.$

4. $\cos x \frac{dy}{dx} + y \sin x = \sin^3 x \cos x.$

5. $\frac{dy}{dx} - y = x.$

6. $\frac{dy}{dx} - y \tan x = x.$

7. $\frac{dy}{dx} + y = y^4.$

8. $2x\frac{dy}{dx} - y = x^3 y^3.$

9. $3\frac{dy}{dx} + y + xy^4 = 0.$

10. $\sin x \frac{dy}{dx} + y \cos x = y^2.$

11. (Clairaut's equation.) If $y = x\frac{dy}{dx} + f\left(\frac{dy}{dx}\right)$, prove by differentiation with respect to x that either $\frac{d^2y}{dx^2} = 0$ or else $x + f'(p) = 0$ where $p \equiv \frac{dy}{dx}$. Hence show that there is a complete primitive $y = cx + f(c)$ and also another solution not included therein which is the envelope of the system of straight lines $y = cx + f(c)$. (This is called a *singular solution.*)

12. Find a complete primitive and a singular solution for

$$\text{(i) } y = x\frac{dy}{dx} + \frac{dx}{dy}, \quad \text{(ii) } \left(y - x\frac{dy}{dx}\right)^2 = a^2\left(\frac{dy}{dx}\right)^2 + b^2.$$

Orthogonal Systems of Curves.

Suppose the differential equation of a system of curves C, $F(x, y, a)=0$, is

$$f_1(x, y)\frac{dy}{dx}+f_2(x, y)=0.$$

Then if $P(x, y)$ is any point in the plane, the equation $\frac{dy}{dx}=-\frac{f_2}{f_1}$ gives the gradient of the tangent at P to the member of the system C which passes through P. If a second system of curves C′ is such that each member of C′ intersects every member of C at right angles, the gradient of the tangent at P to the member of C′ which passes through P is given by the equation $\frac{dy}{dx}=\frac{f_1}{f_2}$. Hence the differential equation of the system C′ is

$$f_2(x, y)\frac{dy}{dx}-f_1(x, y)=0.$$

The systems C and C′ are called *orthogonal*, and the curves of either system are called *orthogonal trajectories* of the other.

More generally, if the differential equation of a system is $f\left(x, y, \frac{dy}{dx}\right)=0$, the differential equation of the orthogonal system is $f\left(x, y, -\frac{dx}{dy}\right)=0$.

Orthogonal systems are of great importance in physics; for example, in electricity the lines of force are orthogonal to the equipotential curves.

Example 10. Find the system orthogonal to the circles $x^2+y^2=a^2$.

The differential equation of the given system is

$$xdx+ydy=0;$$

∴ that of the orthogonal system is

$$xdy-ydx=0,$$

that is

$$d\left(\frac{y}{x}\right)=0;$$

∴ the orthogonal system is $y/x=b$, which represents a system of straight lines through the origin (the centre of the circles).

Example 11. Find the orthogonal trajectories of the system of circles given by $x^2 + y^2 = ay$.

Here it is best to use polar coordinates. The equation $x^2 + y^2 = ay$ is equivalent to $r^2 = ar \sin \theta$, that is $r = a \sin \theta$.

Hence $$\cot \phi \equiv \frac{1}{r}\frac{dr}{d\theta} = \cot \theta.$$

But the ϕ of the orthogonal system must differ at any point from that of the given system by $\frac{1}{2}\pi$; therefore the differential equation of the orthogonal system is

$$\frac{1}{r}\frac{dr}{d\theta} = -\tan \theta;$$

$$\therefore \frac{dr}{r} = -\tan \theta \, d\theta; \quad \therefore \log r = \log \cos \theta + \log b;$$

$$\therefore r = b \cos \theta.$$

This is equivalent to $x^2 + y^2 = bx$.

EXERCISE XX d

Find the orthogonal trajectories of the systems in Nos. 1-9:

1. $x - y = c$. **2.** $x^2 - y^2 = a^2$. **3.** $xy = c^2$.

4. $y^2 = 4ax$. **5.** $y = cx^5$. **6.** $r = a\theta$.

7. $r = a(1 + \cos \theta)$. **8.** $r^2 = a^2 \sin 2\theta$. **9.** $r^2 = a^2 \cos \theta$.

10. Prove that the system of confocal parabolas $y^2 = 4a(x + a)$ is identical with the orthogonal system.

11. Prove that the system of confocal conics given by $x^2/(a^2 + \lambda) + y^2/(b^2 + \lambda) = 1$, where λ varies, is identical with the orthogonal system.

Equations of the Second Order.

(1) *$\frac{dy}{dx}$ and y absent.*

The equation is $\frac{d^2y}{dx^2}=f(x)$ and is solved by two successive integrations with respect to x.

(2) *$\frac{dy}{dx}$ and x absent.*

The equation is $\frac{d^2y}{dx^2}=f(y)$ and may be solved by the substitution $\frac{dy}{dx}=p$. Then

$$\frac{d^2y}{dx^2}=\frac{dp}{dx}=\frac{dp}{dy}\frac{dy}{dx}=p\frac{dp}{dy},$$

$$\therefore\ pdp=f(y)dy\ ;$$

$$\therefore\ \tfrac{1}{2}p^2=\int f(y)\,dy=\mathsf{F}(y)+a,\ \text{say}\ ;$$

$$\therefore\ \frac{dy}{dx}=\{2\mathsf{F}(y)+2a\}^{1/2}.$$

Hence $$x=\int\{2\mathsf{F}(y)+2a\}^{-1/2}\,dy.$$

Note. This method amounts to multiplying both sides of the equation by $\frac{dy}{dx}$ and then integrating ; see Example 12.

(3) *x or y absent.*

The equation can be reduced to the first order by the substitution $\frac{dy}{dx}=p$:

$$f\left(\frac{d^2y}{dx^2}, \frac{dy}{dx}, x\right)=0 \text{ becomes } f\left(\frac{dp}{dx}, p, x\right)=0\ ;$$

$$f\left(\frac{d^2y}{dx^2}, \frac{dy}{dx}, y\right)=0 \text{ becomes } f\left(p\frac{dp}{dy}, p, y\right)=0.$$

Example 12. Solve $\frac{d^2x}{dt^2} = -n^2x.$

$$\frac{dx}{dt}\frac{d^2x}{dt^2} = -n^2x\frac{dx}{dt};$$

$$\therefore \frac{d}{dt}\left\{\tfrac{1}{2}\left(\frac{dx}{dt}\right)^2\right\} = -n^2\frac{d}{dt}(\tfrac{1}{2}x^2);$$

$$\therefore \tfrac{1}{2}\left(\frac{dx}{dt}\right)^2 = -\tfrac{1}{2}n^2x^2 + \text{constant},$$

and this may be written

$$\left(\frac{dx}{dt}\right)^2 = n^2(a^2 - x^2);$$

$$\therefore \frac{dt}{dx} = \pm\frac{1}{n\sqrt{(a^2 - x^2)}};$$

$$\therefore \pm nt = \int\frac{dx}{\sqrt{(a^2 - x^2)}} = \sin^{-1}\frac{x}{a} + b;$$

$$\therefore x = a\sin(\pm nt - b).$$

Since a and b are arbitrary the ambiguous sign may be taken to be + and the solution can be expressed in any of the forms

$$x = a\sin(nt + \epsilon);\quad x = a\cos(nt + \epsilon_1);\quad x = \mathsf{A}\sin nt + \mathsf{B}\cos nt.$$

This equation represents simple harmonic motion.

EXERCISE XX e

Solve the differential equations in Nos. 1-14 :

For brevity y_1 is written for $\frac{dy}{dx}$ and y_2 for $\frac{d^2y}{dx^2}$.

1. $x^2y_2 = a.$
2. $y_2 = \sin nx.$
3. $y_2 = a\cos^2 x.$
4. $e^xy_2 = 1.$
5. $y_2 = n^2y.$
6. $y^3y_2 = k^2.$
7. $y_2 = ay_1.$
8. $xy_2 = 2y_1.$
9. $y_2 = 2yy_1^3.$
10. $xy_2 + y_1 = 0.$
11. $y_1y_2 = 2c^2.$
12. $y_2 + y_1^2 = -1.$
13. $(1 + x^2)y_2 + 2xy_1 = 0.$
14. $y_2 + y = a^4y^{-3}.$

15. The motion of a particle along Ox is given by $\frac{d^2x}{dt^2} = -n^2x$. If it starts from rest where $x=a$, find the times when its position is given by $x=\frac{1}{2}a$, $x=-\frac{1}{2}a$.

16. A uniform chain of length l hangs over a smooth peg and is initially at rest with its middle point at a depth a below the peg. If at time t the depth of the middle point is x, it can be proved that $l\frac{d^2x}{dt^2}=2gx$. Find x in terms of a, g, l and t.

17. Find the system of curves for which the radius of curvature varies as the cube of the length of the normal.

Equations of the Second Order (*continued*).

(4) *Linear equations.*

The general linear equation of the second order is

$$\frac{d^2y}{dx^2}+P\frac{dy}{dx}+Qy=R \quad . \quad . \quad . \quad . \quad \text{(i)}$$

where P, Q, R are functions of x (or constants).

The solution of this equation depends on that of

$$\frac{d^2y}{dx^2}+P\frac{dy}{dx}+Qy=0 \quad . \quad . \quad . \quad . \quad \text{(ii)}$$

If $y=f(x)$ and $y=g(x)$ are any two particular solutions of (ii), it is easy to show that

$$y=Af(x)+Bg(x),$$

where A, B are arbitrary constants, is also a solution.

For if $\quad f''+Pf'+Qf=0$

and $\quad g''+Pg'+Qg=0,$

then $\quad A(f''+Pf'+Qf)+B(g''+Pg'+Qg)=0,$

that is $\frac{d^2}{dx^2}(Af+Bg)+P\frac{d}{dx}(Af+Bg)+Q(Af+Bg)=0.$

Hence if we know two solutions $y=f(x)$ and $y=g(x)$ of (ii), where $f(x)/g(x)$ is not a mere constant, we can write down a complete primitive in the form $y=Af(x)+Bg(x)$, since this contains two arbitrary constants.

If $y = \phi(x)$ is any particular solution of (i), the substitution $y = \phi(x) + z$ reduces (i) to the form of (ii).

The substitution gives

$$\phi'' + \frac{d^2z}{dx^2} + \mathsf{P}\left(\phi' + \frac{dz}{dx}\right) + \mathsf{Q}(\phi + z) = \mathsf{R} ;$$

and if $\phi'' + \mathsf{P}\phi' + \mathsf{Q}\phi = \mathsf{R}$, this becomes

$$\frac{d^2z}{dx^2} + \mathsf{P}\frac{dz}{dx} + \mathsf{Q}z = 0.$$

If then $z = f(x)$, $z = g(x)$ are two solutions of (ii), where $f(x)/g(x)$ is not a mere constant, the general solution of (ii) is

$$z = \mathsf{A}f(x) + \mathsf{B}g(x),$$

and therefore a complete primitive of (i) is

$$y = \phi(x) + \mathsf{A}f(x) + \mathsf{B}g(x).$$

$y = \phi(x)$ is called a *particular integral* of (i) and $\mathsf{A}f(x) + \mathsf{B}g(x)$ is called the *complementary function* (C.F.).

Example 13. Solve $\dfrac{d^2y}{dx^2} + 4y = 3e^x$.

For a particular integral assume $y = ae^x$;

by substitution, $ae^x + 4ae^x = 3e^x$; $\therefore a = \frac{3}{5}$.

Thus a particular integral is $y = \frac{3}{5}e^x$.

To find the C.F. we solve the equation

$$\frac{d^2y}{dx^2} + 4y = 0.$$

By Example 12, p. 409, the solution is $y = \mathsf{A}\sin 2x + \mathsf{B}\cos 2x$. Hence a complete primitive of the given equation is

$$y = \tfrac{3}{5}e^x + \mathsf{A}\sin 2x + \mathsf{B}\cos 2x.$$

It is outside the scope of this book to deal with linear equations in general, but we shall show by examples how to proceed with equations in which the coefficients P, Q are *constant*; methods of solution of a few more general equations are indicated in Exercise XX f, Nos. 33-38.

Example 14. Solve $\frac{d^2y}{dx^2} - 8\frac{dy}{dx} + 15y = 0.$

This equation is satisfied by $y = e^{nx}$ if

$$n^2e^{nx} - 8ne^{nx} + 15e^{nx} = 0,$$

that is if
$$n^2 - 8n + 15 = 0,$$

or
$$(n-3)(n-5) = 0, \text{ i.e. } n = 3 \text{ or } 5.$$

$\therefore$ a complete primitive is $y = \mathsf{A}e^{3x} + \mathsf{B}e^{5x}$. This may also be written in the form $e^{4x}(\mathsf{F}\text{ sh } x + \mathsf{G}\text{ ch } x)$.

Example 15. Solve $\frac{d^2y}{dx^2} - 8\frac{dy}{dx} + 16y = 0.$

This equation is satisfied by $y = e^{nx}$ if

$$n^2 - 8n + 16 = 0,$$

that is if
$$(n-4)^2 = 0, \text{ i.e. } n = 4.$$

Hence one solution is $y = e^{4x}$; we can find another by the substitution $y = e^{4x}z$; then

$$\frac{dy}{dx} = e^{4x}\left(\frac{dz}{dx} + 4z\right); \quad \frac{d^2y}{dx^2} = e^{4x}\left(\frac{d^2z}{dx^2} + 8\frac{dz}{dx} + 16z\right);$$

$$\therefore \frac{d^2y}{dx^2} - 8\frac{dy}{dx} + 16y = e^{4x}\frac{d^2z}{dx^2};$$

$$\therefore \frac{d^2z}{dx^2} = 0; \quad \therefore z = \mathsf{A}x + \mathsf{B}.$$

$\therefore$ a complete primitive is $y = (\mathsf{A}x + \mathsf{B})e^{4x}$.

Example 16. Solve $\frac{d^2y}{dx^2} - 8\frac{dy}{dx} + 25y = 0.$

The equation is satisfied by $y = e^{nx}$ if

$$n^2 - 8n + 25 = 0,$$

that is if
$$(n-4)^2 + 9 = 0;$$

but this has no solutions (in real algebra).

Make the substitution $y = e^{4x}z$; then as in Example 15 we get

$$\frac{d^2y}{dx^2} - 8\frac{dy}{dx} + 25y = e^{4x}\left(\frac{d^2z}{dx^2} + 9z\right),$$

$$\therefore \frac{d^2z}{dx^2} = -9z.$$

Hence by Example 12, p. 409

$$z = \mathsf{A} \sin 3x + \mathsf{B} \cos 3x.$$

$\therefore$ a complete primitive is $y = e^{4x}(\mathsf{A} \sin 3x + \mathsf{B} \cos 3x)$. This may be compared with the second form of the result of Example 14.

Note. The reader who is familiar with the theory of complex number will recognise that this solution may be inferred from the roots $4 \pm 3i$ of the equation in n, since $\exp(4x \pm 3xi)$ equals $e^{4x}(\cos 3x \pm i \sin 3x)$.

Examples 14-16 illustrate the procedure to be followed in solving the equation

$$\frac{d^2y}{dx^2} + 2b\frac{dy}{dx} + cy = 0$$

where b, c are constants, in the three cases $b^2 > c$, $b^2 = c$, $b^2 < c$.

The reader should now work Exercise XX f, Nos. 1-10.

To complete the solution of the more general equation

$$\frac{d^2y}{dx^2} + 2b\frac{dy}{dx} + cy = \mathsf{R}$$

where b, c are constants and R is a function of x we need to be able to find a *particular integral* of this equation.

We give in Exercise XX f, Nos. 11-20, indications of how to find particular integrals for those forms of R which are of common occurrence; Nos. 11-15 deal with the types kx^n, ke^{px}, $k \sin px + l \cos px$; Nos. 16-20 deal with exceptional cases. The reader should now work these examples; he will then be in a position to appreciate the following remarks:

(i) If R is a polynomial of degree n, a particular integral is also a polynomial, usually of degree n but of degree $n+1$ if $c = 0$.

(ii) If $\mathsf{R} = ke^{px}$, a particular integral is usually $\mathsf{A}e^{px}$; but if e^{px} is a term of the C.F. a particular integral is $\mathsf{A}xe^{px}$ unless xe^{px} belongs to the C.F., in which case a particular integral is $\mathsf{A}x^2e^{px}$.

(iii) If $\mathsf{R} = k \sin px + l \cos px$ a particular integral is usually $\mathsf{A} \sin px + \mathsf{B} \cos px$; but if these terms belong to the C.F. it is $x(\mathsf{A} \sin px + \mathsf{B} \cos px)$.

Further, if R is the sum of two or more terms, say $R_1 + R_2$, it is sufficient to find particular integrals of the separate equations

$$\frac{d^2y}{dx^2} + 2b\frac{dy}{dx} + cy = R_1, \quad \frac{d^2y}{dx^2} + 2b\frac{dy}{dx} + cy = R_2,$$

and to take their sum, since it is obvious by substitution that the sum is a solution of the equation

$$\frac{d^2y}{dx^2} + 2b\frac{dy}{dx} + cy = R_1 + R_2.$$

Example 17. Solve $\dfrac{d^2y}{dx^2} - 8\dfrac{dy}{dx} + 15y = 5e^{2x}$.

From Example 14, p. 412, the C.F. is $Ae^{3x} + Be^{5x}$.

For a particular integral assume $y = Ce^{2x}$;

$\therefore$ by substitution, $Ce^{2x}(2^2 - 8 \, . \, 2 + 15) = 5e^{2x}$; $\quad \therefore C = \frac{5}{3}$.

$\therefore$ a complete primitive is $y = \frac{5}{3}e^{2x} + Ae^{3x} + Be^{5x}$.

Example 18. Solve $\dfrac{d^2y}{dx^2} - 8\dfrac{dy}{dx} + 15y = e^{3x}$.

As in Example 17, the C.F. is $Ae^{3x} + Be^{5x}$.

For a particular integral assume $y = Cxe^{3x}$;

then $$\frac{dy}{dx} = Ce^{3x}(3x + 1); \quad \frac{d^2y}{dx^2} = Ce^{3x}(9x + 6);$$

$\therefore$ by substitution, $Ce^{3x}\{(9x + 6) - 8(3x + 1) + 15x\} = e^{3x}$;

$$\therefore -2C = 1; \quad \therefore C = -\tfrac{1}{2}.$$

$\therefore$ a complete primitive is $y = e^{3x}(A - \frac{1}{2}x) + Be^{5x}$.

Example 19. Solve $\dfrac{d^2y}{dx^2} - 8\dfrac{dy}{dx} + 16y = e^{4x}$.

From Example 15, p. 412, the C.F. is $(Ax + B)e^{4x}$.

For a particular integral assume $y = Cx^2e^{4x}$;

substituting and simplifying, this gives $C = \frac{1}{2}$;

$\therefore$ a complete primitive is $y = e^{4x}(\frac{1}{2}x^2 + Ax + B)$.

From Examples 17, 18 it follows that a complete primitive of

$$\frac{d^2y}{dx^2} - 8\frac{dy}{dx} + 15y = 5e^{2x} + e^{3x}$$

is $$y = \tfrac{5}{3}e^{2x} + e^{3x}(A - \tfrac{1}{2}x) + Be^{5x}.$$

EXERCISE XX f

Find complete primitives of the equations in Nos. 1-10:

1. $y_2 - 3y_1 + 2y = 0.$
2. $y_2 + 2y_1 - 3y = 0.$
3. $y_2 - 4y = 0.$
4. $y_2 + 9y = 0.$
5. $y_2 - 6y_1 + 9y = 0.$
6. $y_2 + 4y_1 + 4y = 0.$
7. $y_2 + 2y_1 + 5y = 0.$
8. $y_2 - 2y_1 + 2y = 0.$
9. $y_2 - 6y_1 + 13y = 0.$
10. $y_2 + y_1 + y = 0.$

Find by substitution values of the constants A, B, C such that the equations in Nos. 11-20 have the particular integrals indicated:

11. $y_2 + 5y_1 + 2y = 7$; $y = \mathsf{A}$.
12. $y_2 + 3y_1 + 2y = 8x$; $y = \mathsf{A}x + \mathsf{B}$.
13. $y_2 + 3y_1 - 2y = 9x^2$; $y = \mathsf{A}x^2 + \mathsf{B}x + \mathsf{C}$.
14. $y_2 - y_1 - 2y = 5e^{3x}$; $y = \mathsf{A}e^{3x}$.
15. $y_2 + y_1 + 5y = 3 \sin 5x + 2 \cos 5x$; $y = \mathsf{A} \sin 5x + \mathsf{B} \cos 5x$.
16. $y_2 + 4y_1 = 2x$; $y = \mathsf{A}x^2 + \mathsf{B}x$.
17. $y_2 - 3y_1 = 5x^2$; $y = \mathsf{A}x^3 + \mathsf{B}x^2 + \mathsf{C}x$.
18. $y_2 - y_1 - 2y = 5e^{2x}$; $y = \mathsf{A}xe^{2x}$.
19. $y_2 + 4y = 3 \sin 2x$; $y = x(\mathsf{A} \sin 2x + \mathsf{B} \cos 2x)$.
20. $y_2 - 4y_1 + 4y = e^{2x}$; $y = \mathsf{A}x^2e^{2x}$.

Solve the equations in Nos. 21-32:

21. $y_2 - 4y_1 + 3y = 12.$
22. $y_2 + 3y_1 + 2y = 4x.$
23. $y_2 - 3y_1 = 5 - 24x.$
24. $y_2 - y = e^{2x}.$
25. $y_2 - 9y = e^{3x}.$
26. $y_2 - 9y = \sin x.$
27. $y_2 + 9y = \sin 3x.$
28. $y_2 - 3y_1 + 2y = \cos x.$
29. $y_2 - 2y_1 + y = e^x.$
30. $y_2 - 5y_1 + 6y = e^{2x}.$
31. $y_4 = 16y.$
32. $y_2 - 4y = e^x - \sin x.$

33. Use the substitution $x = e^z$ to solve the equations:

(i) $x^2y_2 + xy_1 + y = 0$; (ii) $x^2y_2 - 5xy_1 + 9y = 8x$.

34. If $y = \phi(x)$ is a solution of the equation

$$\frac{d^2y}{dx^2} + \mathsf{P}\frac{dy}{dx} + \mathsf{Q}y = 0,$$

where P, Q are functions of x, prove that the substitution $y = z\phi$ reduces the equation $\frac{d^2y}{dx^2} + \mathsf{P}\frac{dy}{dx} + \mathsf{Q}y = \mathsf{R}$, where R is a function of x, to

$$\phi\frac{d^2z}{dx^2} + (2\phi' + \mathsf{P}\phi)\frac{dz}{dx} = \mathsf{R}.$$

What is the integrating factor of the linear equation obtained by putting $\frac{dz}{dx}$ equal to u ?

35. Solve $(1 + x^2)y_2 + xy_1 = y$ given that one solution is $y = x$. [Make the substitution indicated in No. 34.]

36. Solve $(1 - 2x)y_2 + 2y_1 - (3 - 2x)y = 0$ given that one solution is $y = e^x$.

37. Prove that the equation $\frac{d^2y}{dx^2} + \mathsf{P}\frac{dy}{dx} + \mathsf{Q}y = \mathsf{R}$, where P, Q, R are functions of x, can be reduced to the form $\frac{d^2z}{dx^2} + \mathsf{Q}_1z = \mathsf{R}_1$, where Q_1, R_1 are functions of x, by the substitution $y = ze^{-\frac{1}{2}\int \mathsf{P}dx}$.

38. Use the substitution indicated in No. 37 to solve the equation

$$y_2 + \frac{2}{x}y_1 + k^2y = 0.$$

CHAPTER XXI

APPROXIMATIONS: POWER SERIES AND CURVE TRACING

Power Series. An approximation to the value of a function $f(x)$ can often be expressed in the convenient form of a polynomial in x in such a way that the error can be decreased by increasing the degree of the polynomial.

For example, $1+x+x^2+\ldots\ldots+x^{n-1}=\dfrac{1-x^n}{1-x}$, $(x\neq 1)$;

$\therefore$ the function $\dfrac{1}{1-x}$ can be expressed in the form $1+x+x^2+\ldots\ldots+x^{n-1}$, with an error of $x^n/(1-x)$. But if $-1<x<+1$ this error can be made as small as desired by taking n sufficiently large; thus the polynomial $1+x+x^2+\ldots\ldots+x^{n-1}$, if it is of high enough degree, will be a good approximation to $1/(1-x)$ for the range of values $-1<x<+1$. The infinite series

$$1+x+x^2+\ldots\ldots+x^r+\ldots\ldots, \quad (-1<x<+1)$$

is called the *expansion* of the function $1/(1-x)$ in powers of x, and the error $x^n/(1-x)$ is called the *remainder after n terms.* $1/(1-x)$ is called the *sum to infinity* of the series when $-1<x<+1$, and we write

$$\frac{1}{1-x}=1+x+x^2+x^3+\ldots\ldots, \quad (-1<x<+1).$$

The practical value of an expansion depends on the rapidity with which the remainder after n terms decreases as n increases. For $1/(1-x)$, if $x=0{\cdot}01$, the remainder decreases very rapidly and for many purposes it would be sufficient to take the first 3 terms of the expansion, $1+x+x^2$, as an approximation for $1/(1-x)$. On the other hand, if $x\simeq 0{\cdot}9$ an inconvenient number of terms would have to be taken, so the expansion would be less valuable.

Most expansions are valid only for a certain range of values of x.

Maclaurin's Expansion. We shall now discuss a general method of obtaining polynomial approximations to a function.

Consider for example the function $f(x)$, $\equiv e^x$, for values of x near $x=0$ and construct the polynomial $\mathsf{P}_1(x) \equiv a_0 + a_1 x$, which is so related to $f(x)$ that $\mathsf{P}_1(0)=f(0)$, $\mathsf{P}_1'(0)=f'(0)$;

then $a_0 = e^0 = 1$, $a_1 = e^0 = 1$; $\therefore \mathsf{P}_1(x) \equiv 1 + x$.

Geometrically these conditions mean that the graph of $1+x$ is the tangent to the graph of e^x at $x=0$. Thus near $x=0$ we may regard $1+x$ as an approximation to e^x.

Next construct $\mathsf{P}_2(x) \equiv a_0 + a_1 x + a_2 x^2$ so that

$$\mathsf{P}_2(0)=f(0), \quad \mathsf{P}_2'(0)=f'(0), \quad \mathsf{P}_2''(0)=f''(0);$$

then $a_0=1$, $a_1=1$, $2a_2=1$; $\therefore \mathsf{P}_2(x) \equiv 1 + x + \frac{1}{2}x^2$.

Geometrically these conditions mean that the graphs of $1+x+\frac{1}{2}x^2$ and e^x meet where $x=0$ and have not only the same gradient but also the same curvature at this point; they are said to have *contact of the second order* at $x=0$. Thus near $x=0$ we may regard $1+x+\frac{1}{2}x^2$ as a closer approximation to e^x.

Similarly, if we construct $\mathsf{P}_3(x)$, $\equiv a_0 + a_1 x + a_2 x^2 + a_3 x^3$, so that

$$\mathsf{P}_3(0)=f(0), \quad \mathsf{P}_3'(0)=f'(0), \quad \mathsf{P}_3''(0)=f''(0), \quad \mathsf{P}_3'''(0)=f'''(0),$$

we find $a_0=1$, $a_1=1$, $2a_2=1$, $2.3a_3=1$;

$$\therefore \mathsf{P}_3(x) \equiv 1 + x + \frac{x^2}{2!} + \frac{x^3}{3!};$$

and we say that the graph of this polynomial has *contact of the third order* with that of e^x at $x=0$, and we may regard it as a still closer approximation to e^x near $x=0$.

This process can be repeated as often as desired, and the polynomial of degree n whose graph has contact of the nth order with that of e^x is

$$1 + x + \frac{x^2}{2!} + \frac{x^3}{3!} + \ldots\ldots + \frac{x^n}{n!}.$$

The following table illustrates the character of the approximations for the range of values $x = -1{\cdot}5$ to $x = +1{\cdot}5$

x	$-1{\cdot}5$	-1	$-0{\cdot}5$	0	$0{\cdot}5$	1	$1{\cdot}5$
$\mathsf{P}_2(x)$	0·625	0·5	0·625	1	1·625	2·5	3·625
$\mathsf{P}_3(x)$	0·062	0·333	0·604	1	1·646	2·667	4·187
$\mathsf{P}_4(x)$	0·273	0·375	0·607	1	1·648	2·708	4·398
e^x	0·223	0·368	0·607	1	1·649	2·718	4·482

The reader should represent these results graphically. It is evident that for this range of values the graphs of $P_4(x)$ and e^x are almost identical, and for larger values of n the graph of $P_n(x)$ will nearly coincide with that of e^x over a wider range of values of x.

The same process can be applied to any function $f(x)$ which possesses differential coefficients of order n at $x=0$. Let the polynomial $P_n(x) \equiv a_0 + a_1 x + a_2 x^2 + \ldots\ldots + a_n x^n$ be such that

$$P_n(0) = f(0), \quad P_n'(0) = f'(0), \ldots, P_n^n(0) = f^n(0).$$

By successive differentiation,

$$P_n'(x) = a_1 + 2a_2 x + 3a_3 x^2 + 4a_4 x^3 + \ldots\ldots + na_n x^{n-1},$$
$$P_n''(x) = 1.2a_2 + 2.3a_3 x + 3.4a_4 x^2 + \ldots\ldots + (n-1)na_n x^{n-2},$$
$$P_n'''(x) = 1.2.3a_3 + 2.3.4a_4 x + \ldots\ldots + (n-2)(n-1)na_n x^{n-3},$$

and so on.

Then $\quad P_n(0) = a_0, \quad P_n'(0) = a_1, \quad P_n''(0) = 1.2a_2,$
$\quad P_n'''(0) = 1.2.3a_3, \ldots$

$$\therefore a_0 = f(0),\ a_1 = f'(0),\ a_2 = \frac{1}{2!}f''(0), \quad a_3 = \frac{1}{3!}f'''(0), \ldots$$

and in general $a_r = \dfrac{1}{r!}f^r(0), \quad (r \leqslant n).$

Hence $\quad P_n(x) \equiv f(0) + xf'(0) + \dfrac{x^2}{2!}f''(0) + \ldots\ldots + \dfrac{x^n}{n!}f^n(0).$

The graph of this polynomial has contact of the nth order with that of $f(x)$ at $x=0$ and may be regarded as an approximation to $f(x)$ for values of x near $x=0$. It is beyond the scope of this book to determine the error in this approximation, but it can be proved that

$$f(x) - P_n(x) = \frac{x^{n+1}}{(n+1)!}f^{n+1}(\theta x)$$

where θ lies between 0 and 1.

If this expression for the error in the approximation tends to zero when $n \to \infty$,

$$\mathbf{f(0) + xf'(0) + \frac{x^2}{2!}f''(0) + \ldots\ldots + \frac{x^r}{r!}f^r(0) + \ldots}$$

is called *Maclaurin's expansion* of $f(x)$, and the *remainder after n terms* can be expressed in the form $\dfrac{x^n}{n!}f^n(\theta x), \quad 0 < \theta < 1.$

Thus for the function e^x the remainder after n terms of Maclaurin's expansion is $\frac{x^n}{n!}e^{\theta x}$ where θ lies between 0 and 1, and it can be proved that this tends to zero when $n \to \infty$ for all values of x. Hence we write

$$e^x = 1 + x + \frac{x^2}{2!} + \frac{x^3}{3!} + \ldots\ldots + \frac{x^r}{r!} + \ldots.$$

for all values of x.

Example 1. Expand $\log(1+x)$ in powers of x.

If $f(x) = \log(1+x)$,

$$f'(x) = \frac{1}{1+x}, \quad f''(x) = \frac{-1}{(1+x)^2}, \quad f'''(x) = \frac{(-1)(-2)}{(1+x)^3}, \ldots$$

$$\therefore f(0) = 0, \quad f'(0) = 1, \quad f''(0) = -1,$$

$$f'''(0) = (-1)^2\, 2!, \quad f''''(0) = (-1)^3 3!, \ldots$$

$\therefore$ the expansion for $\log(1+x)$ is

$$0 + \frac{1}{1!}x - \frac{1}{2!}x^2 + \frac{2!}{3!}x^3 - \frac{3!}{4!}x^4 + \ldots.$$

that is $$x - \tfrac{1}{2}x^2 + \tfrac{1}{3}x^3 - \tfrac{1}{4}x^4 + \ldots.$$

This series is convergent only if $-1 < x \leqslant 1$; the expansion is meaningless outside this range; it can be [proved to be valid for $-1 < x \leqslant 1$.

Example 2. Obtain an approximation for $\operatorname{ch} x \log(1+x^2)$ when x is small and evaluate $\lim\limits_{x\to 0} \{\operatorname{ch} x \log(1+x^2) - x^2\}/x^6$.

For small values of x,

$$\operatorname{ch} x = \tfrac{1}{2}(e^x + e^{-x})$$

$$\simeq \tfrac{1}{2}\left\{\left(1 + \frac{x}{1!} + \frac{x^2}{2!} + \frac{x^3}{3!} + \frac{x^4}{4!}\right) + \left(1 - \frac{x}{1!} + \frac{x^2}{2!} - \frac{x^3}{3!} + \frac{x^4}{4!}\right)\right\}$$

$$\therefore \operatorname{ch} x \simeq 1 + \frac{x^2}{2!} + \frac{x^4}{4!}.$$

From Example 1, $\log(1+x^2) \simeq x^2 - \tfrac{1}{2}x^4 + \tfrac{1}{3}x^6$;

$$\therefore \operatorname{ch} x \log(1+x^2) \simeq x^2(1 + \tfrac{1}{2}x^2 + \tfrac{1}{24}x^4)(1 - \tfrac{1}{2}x^2 + \tfrac{1}{3}x^4)$$

$$\simeq x^2 + \tfrac{1}{8}x^6.$$

Also, $\operatorname{ch} x \log(1+x^2) - x^2 = \frac{1}{8}x^6 + a_1x^8 + a_2x^{10} + \ldots$,

where it can be shown that $a_1, a_2, a_3, \ldots$ are all less than 1.

Hence $\{\operatorname{ch} x \log(1+x^2) - x^2\}/x^6 - \frac{1}{8} = x^2(a_1 + a_2x^2 + a_3x^4 + \ldots)$,

which, for $x^2 < \frac{1}{2}$, is less than

$$x^2(1 + \tfrac{1}{2} + \tfrac{1}{4} + \ldots), \text{ i.e. } 2x^2.$$

Hence when $x \to 0$ the function $\to \frac{1}{8}$.

Note. The approximation for $\operatorname{ch} x$ may be obtained directly from Maclaurin's expansion.

Taylor's Expansion.

If $g(x) \equiv f(x+a)$, and if the successive derivatives of $g(x)$ at $x=0$ exist, we have from Maclaurin's expansion

$$g(x) = g(0) + \frac{x}{1!}g'(0) + \frac{x^2}{2!}g''(0) + \frac{x^3}{3!}g'''(0) + \ldots$$

and the remainder after n terms is $\frac{x^n}{n!}g^n(\theta x)$, $(0<\theta<1)$.

But $g(0) = f(a)$; also $g'(x) = f'(x+a)$, $\therefore g'(0) = f'(a)$; similarly $g''(0) = f''(a)$, etc.

$$\therefore f(x+a) = f(a) + \frac{x}{1!}f'(a) + \frac{x^2}{2!}f''(a) + \frac{x^3}{3!}f'''(a) + \ldots$$

and the remainder after n terms is $\frac{x^n}{n!}f^n(a+\theta x)$, $(0<\theta<1)$.

This is called *Taylor's expansion*.

Example 3. Expand $\sin(x+h)$ in powers of h.

If $f(x) = \sin x$, $f'(x) = \cos x = \sin(x + \frac{1}{2}\pi)$;

$\therefore f''(x) = \sin(x + \frac{1}{2}\pi + \frac{1}{2}\pi)$; $f'''(x) = \sin(x + \frac{1}{2}\pi + \frac{1}{2}\pi + \frac{1}{2}\pi)$; ...;

hence $$f^r(x) = \sin(x + \tfrac{1}{2}r\pi).$$

$$\therefore \sin(x+h) = \sin x + \frac{h}{1!}\sin(x + \tfrac{1}{2}\pi) + \ldots + \frac{h^r}{r!}\sin(x + \tfrac{1}{2}r\pi) + \ldots$$

The remainder after n terms is $\frac{h^n}{n!}\sin(x + \theta h + \frac{1}{2}n\pi)$ and therefore its numerical value cannot exceed $\frac{h^n}{n!}$; this tends to zero when $n \to \infty$ for all values of h. Thus the expansion is always valid.

A Theorem on Limits.

If $f(x)$ and $g(x)$ are functions which possess derivatives at $x=a$ and if $f(a)=g(a)=0$ and $g'(a)\neq 0$, then

$$\lim_{x\to a}\frac{f(x)}{g(x)}=\frac{f'(a)}{g'(a)}.$$

$$\frac{f(x)}{g(x)}=\frac{f(x)-f(a)}{x-a}\div\frac{g(x)-g(a)}{x-a},\quad x\neq a.$$

But when $x\to a$, $\quad\dfrac{f(x)-f(a)}{x-a}\to f'(a)$

and $\quad\dfrac{g(x)-g(a)}{x-a}\to g'(a),\ \neq 0,$

$$\therefore\ \frac{f(x)}{g(x)}\to\frac{f'(a)}{g'(a)}\text{ when } x\to a.$$

A more useful theorem, which, however, we shall not prove, is as follows :

If $f(x)$, $g(x)$ are functions which possess derivatives near $x=a$, and if $f(a)=g(a)=0$,

$$\lim_{x\to a}\frac{f(x)}{g(x)}=\lim_{x\to a}\frac{f'(x)}{g'(x)}$$

provided that the second limit exists.

The advantage of this theorem is that it does not require $g'(a)\neq 0$.

Example 4. Find $\displaystyle\lim_{x\to 0}\frac{e^x-e^{\sin x}}{x-\sin x}$.

By the theorem just stated

$$\lim_{x\to 0}\frac{e^x-e^{\sin x}}{x-\sin x}=\lim_{x\to 0}\frac{e^x-e^{\sin x}\cos x}{1-\cos x}$$

provided the second limits exists, and by the same theorem

$$\lim_{x\to 0}\frac{e^x-e^{\sin x}\cos x}{1-\cos x}=\lim_{x\to 0}\frac{e^x-e^{\sin x}\cos^2 x+e^{\sin x}\sin x}{\sin x}$$

$$=\lim_{x\to 0}\frac{e^x-e^{\sin x}\cos^3 x+3e^{\sin x}\cos x\sin x+e^{\sin x}\cos x}{\cos x}$$

provided the last limit exists.

But the last limit evidently equals $\dfrac{1-1+1}{1}$ or 1.

Hence the required limit is 1.

EXERCISE XXI a

Use Maclaurin's series to expand the functions in Nos. 1-4 in powers of x and write down expressions for the *remainders after n terms* :

1. e^{-2x}. **2.** $\cos x$. **3.** $\operatorname{sh} x$. **4.** $\log(1-x)$.

5. Use Taylor's series to expand $\cos(x+h)$ in powers of h.

6. Use Taylor's series to expand $\log(x+a)$ in powers of x, $(a>0)$.

Obtain the approximations in Nos. 7-14 where x is small :

7. $\sqrt{(1+x)} \simeq 1+\frac{1}{2}x-\frac{1}{8}x^2$. **8.** $\tan x \simeq x+\frac{1}{3}x^3$.

9. $\cos^2 x \simeq 1-x^2+\frac{1}{3}x^4$. **10.** $\sin^{-1} x \simeq x+\frac{1}{6}x^3$.

11. $e^{\sin x} \simeq 1+x+\frac{1}{2}x^2$. **12.** $\log(1+\sin x) \simeq x-\frac{1}{2}x^2+\frac{1}{6}x^3$.

13. $\sec x \simeq 1+\frac{1}{2}x^2+\frac{5}{24}x^4$. **14.** $x/(e^x-1) \simeq 1-\frac{1}{2}x+\frac{1}{12}x^2$.

Evaluate the limits when $x \to 0$ in Nos. 15-20 :

15. $\dfrac{1-\cos x}{x^2}$. **16.** $\dfrac{x-\sin x}{x^3}$. **17.** $\dfrac{e^x-1}{x}$.

18. $\dfrac{\log(1+x)-x}{\sin^2 x}$. **19.** $\dfrac{\sin x-x}{\sin x-x\cos x}$. **20.** $\dfrac{\sec x-\operatorname{ch} x}{e^{x^2}+2\cos x-3}$.

21. Describe the graph of $y=\lim\limits_{p\to 0}\dfrac{\sin(\pi x)}{p+\sin(\pi x)}$

22. Prove that $\lim\limits_{x\to 0}\left(\dfrac{1}{x^2}-\cot^2 x\right)=\frac{2}{3}$.

nth Derivatives. The general term of a Maclaurin's or Taylor's series involves the nth derivative of the function which is being expanded. Hence the calculation of such derivatives is of some importance. It is not usually possible to express the result in a simple form. We give some examples in which the nth derivative is easily found.

(i) If $f(x)=x^m$, $f^n(x)=m(m-1)(m-2)\ldots(m-n+1)x^{m-n}$, and if m is a positive integer,

$$f^n(x)=\frac{m!\,x^{m-n}}{(m-n)!},\quad m>n;\quad f^m(x)=m!;\quad f^n(x)=0,\quad m<n.$$

(ii) If $f(x)=(ax+b)^m$,

$$f^n(x)=a^n m(m-1)\ldots(m-n+1)(ax+b)^{m-n}.$$

(iii) If $f(x)=e^{ax}$, $\qquad f^n(x)=a^n e^{ax}$.

(iv) If $f(x)=\sin(ax+b)$, $f^n(x)=a^n \sin(ax+b+\frac{1}{2}n\pi)$;
if $f(x)=\cos(ax+b)$, $f^n(x)=a^n \cos(ax+b+\frac{1}{2}n\pi)$.

(v) If $f(x)=e^{ax}\sin bx$, $\qquad f'(x)=e^{ax}(a\sin bx+b\cos bx)$;

$\therefore$ if $r=\sqrt{(a^2+b^2)}$ and $\cos\alpha:\sin\alpha:1=a:b:r$

$$f'(x)=e^{ax}r\sin(bx+\alpha);$$

hence repeating the process

$$f''(x)=e^{ax}r^2\sin(bx+2\alpha);$$
$$f^n(x)=e^{ax}r^n\sin(bx+n\alpha).$$

Similarly if $f(x)=e^{ax}\cos bx$,

$$f^n(x)=e^{ax}r^n\cos(bx+n\alpha),$$

where r, α have the same values as before.

(vi) If $f(x)=uv$, where u, v are functions of x whose successive derivatives are known, $f^n(x)$ can be found by a theorem which will now be proved.

Leibniz' Theorem. If u and v are functions of x and if suffixes are used to denote differentiations with respect to x, so that

$$u_r \equiv \frac{d^r u}{dx^r}, \quad v_r \equiv \frac{d^r v}{dx^r}, \quad (uv)_r \equiv \frac{d^r(uv)}{dx^r},$$

then $$(uv)_n = u_n v + \frac{n}{1}u_{n-1}v_1 + \frac{n(n-1)}{1\,.\,2}u_{n-2}v_2 + \;.\;.\;.\;.\;.\; + uv_n,$$

where the coefficients $1, \frac{n}{1}, \frac{n(n-1)}{1\,.\,2}, \ldots$ are the same as those in the binomial expansion of $(u+v)^n$ and may be denoted by $c_0, c_1, c_2, \ldots, c_n$.

It is easy to verify the result for $n=2$, $n=3$, etc.

$$(uv)_1 = u_1v + uv_1;$$
$$(uv)_2 = (u_2v+u_1v_1)+(u_1v_1+uv_2)$$
$$= u_2v + 2u_1v_1 + uv_2;$$
$$(uv)_3 = (u_3v+u_2v_1)+2(u_2v_1+u_1v_2)+(u_1v_2+uv_3)$$
$$= u_3v + 3u_2v_1 + 3u_1v_2 + uv_3;$$

and it is evident that the coefficients arise in the same way as they do in the successive multiplication of factors $u+v$. (*Pascal's Triangle.*)

The general result may be proved by induction or as follows: From successive differentiation it follows that

$$(uv)_n = p_0 u_n v + p_1 u_{n-1} v_1 + p_2 u_{n-2} v_2 + \dots\dots + p_n u v_n$$

where the values of $p_0, p_1, \dots, p_n$ are independent of u, v and may therefore be obtained from the special case when

$$u = e^{ax}, \quad v = e^{bx}, \quad uv = e^{(a+b)x}.$$

We then have

$$(a+b)^n e^{(a+b)x} = \Sigma(p_r a^{n-r} e^{ax} b^r e^{bx}) = e^{(a+b)x} \Sigma(p_r a^{n-r} b^r);$$

$$\therefore (a+b)^n \equiv \Sigma(p_r a^{n-r} b^r); \quad \therefore c_r = p_r.$$

Example 5. Find $\dfrac{d^5}{dx^5}(x^3 e^x)$.

With the notation of Leibniz' theorem,

$$\text{if } u = e^x, \quad u_1 = u_2 = u_3 = u_4 = u_5 = e^x;$$

$$\text{if } v = x^3, \quad v_1 = 3x^2, \quad v_2 = 6x, \quad v_3 = 6, \quad v_4 = v_5 = 0.$$

$$\therefore (uv)_5 = e^x\left(x^3 + \frac{5}{1}3x^2 + \frac{5 \cdot 4}{1 \cdot 2}6x + \frac{5 \cdot 4 \cdot 3}{1 \cdot 2 \cdot 3}6\right)$$

$$= e^x(x^3 + 15x^2 + 60x + 60).$$

In Maclaurin's expansion of $f(x)$ it is actually $f^n(0)$, not $f^n(x)$, that is required and advantage can sometimes be taken of this fact.

Example 6. Expand $e^{\sin^{-1} x}$ in powers of x, where $|x| \leqslant 1$.

Let $y = e^{\sin^{-1} x}$ and denote its successive derivatives by y_1, $y_2, y_3, \dots$.

By logarithmic differentiation $\dfrac{y_1}{y} = \dfrac{1}{\sqrt{(1-x^2)}}$;

$$\therefore (1-x^2){y_1}^2 = y^2.$$

Differentiating, $(1-x^2)2y_1y_2 - 2x{y_1}^2 = 2yy_1$;

$$\therefore (1-x^2)y_2 - xy_1 = y.$$

Differentiate both sides n times by Leibniz' theorem:

$$\{(1-x^2)y_{n+2} + n(-2x)y_{n+1} + \tfrac{1}{2}n(n-1)(-2)y_n\} - (xy_{n+1} + ny_n) = y_n.$$

Put $x = 0$ and denote the value of y_r for $x = 0$ by a_r, then

$$\{a_{n+2} - n(n-1)a_n\} - na_n = a_n;$$

$$\therefore a_{n+2} = (n^2+1)a_n,$$

and this is true for $n \geqslant 1$, and it also holds for $n = 0$ if a_0 denotes the value of y when $x = 0$.

Now $a_0 = e^{\sin^{-1} 0} = 1$ and $a_1 = a_0/\sqrt{(1-0)} = 1$, hence the formula

$$a_{n+2} = (n^2+1)a_n,$$

for a_{n+2} proves that

$$a_2 = 1, \quad a_4 = (2^2+1)a_2 = 2^2+1, \quad a_6 = (4^2+1)a_4 = (4^2+1)(2^2+1), \text{ etc.}$$

$$a_3 = (1^2+1)a_1 = 1^2+1, \quad a_5 = (3^2+1)a_3 = (3^2+1)(1^2+1), \text{ etc.}$$

Therefore, from Maclaurin's expansion,

$$e^{\sin^{-1} x} = a_0 + \frac{a_1}{1!}x + \frac{a_2}{2!}x^2 + \frac{a_3}{3!}x^3 + \dots$$

$$= 1 + x + \frac{x^2}{2!} + \frac{x^3}{3!}(1^2+1) + \frac{x^4}{4!}(2^2+1) + \frac{x^5}{5!}(1^2+1)(3^2+1) + \dots$$

We have not investigated the validity of the expansion ; the result does in fact hold for $|x| \leqslant 1$.

EXERCISE XXI b

Find the nth derivatives of the functions in Nos. 1-20 :

1. x^{n+2}.
2. x^{-1}.
3. $\log(2x+3)$.
4. $1/\sqrt{(3x+5)}$.
5. $\sin 3x$.
6. $\sin^2 x$.
7. $e^x \sin x$.
8. $e^{3x} \cos 4x$.
9. $\text{ch}\, x$.
10. $\cos^3 x$; use the relation $\cos 3x = 4\cos^3 x - 3\cos x$.
11. $1/(x^2 - a^2)$; express the function in partial fractions.
12. $1/(x^2+3x+2)$.
13. $\sin^3 x$.
14. $e^{2x} \text{ch}\, 3x$.
15. xe^{2x}.
16. $x^3 e^x$.
17. x^2/e^x.
18. $x^2 \sin 3x$.
19. $x^3 \log x$.
20. $(1+x^2)\,\text{sh}\, x$.
21. If $f(x) = e^x \sin x$ find $f^n(x)$ and expand $f(x)$ in a series of powers of x.
22. If $y = \cos(\log x)$ prove that $x^2 y_2 + xy_1 + y = 0$ and deduce $x^2 y_{n+2} + (2n+1)xy_{n+1} + (n^2+1)y_n = 0$.

23. If $y = f(x) = (\sin^{-1} x)/\sqrt{(1-x^2)}$ prove that

$$\text{(i)}\ (1-x^2)y_1 = xy + 1\ ; \quad \text{(ii)}\ f^{n+1}(0) = n^2 f^{n-1}(0).$$

Hence expand y in a series of powers of x.

24. Expand $(\sin^{-1} x)^2$ in a series of powers of x.

25. If $y = \text{ch}\,(p \sin^{-1} x)$ prove that

$$(1-x^2)y_{n+2} - (2n+1)xy_{n+1} - (n^2+p^2)y_n = 0.$$

Hence expand y in a series of powers of x.

Curve Tracing : Form of a Curve near the Origin.

It is often necessary to determine the nature of a curve in the neighbourhood of a particular point. For that purpose it is usually best to take the origin at the point. We therefore give examples to show how the form of a curve near the origin is determined; the method often amounts to finding approximate values of y in terms of x where x, y are the coordinates of a point near the origin.

Example 7. Examine the form of the curve $x^2y^3 + x - y = 0$ near the origin.

The equation is
$$y = x + x^2y^3.$$

Near the origin O, x and y are small, therefore x^2y^3 is small compared to x and y. Hence for points near O a first approximation is

$$y \simeq x.$$

Therefore for points near O, $x^2y^3 \simeq x^5$, and a second approximation is

$$y \simeq x + x^5.$$

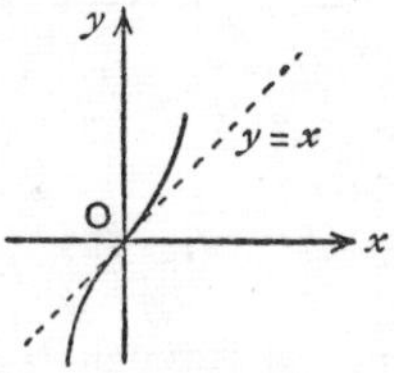

Thus near O the form of the given curve approximates to that of the curve $y = x + x^5$. The gradient of $y = x + x^5$ at $x = 0$ is 1; therefore this curve touches $y = x$ at the origin, and evidently it lies above $y = x$ for $x > 0$ and below $y = x$ for $x < 0$. Hence the form of the given curve near O is roughly as in Fig. 157. It touches and crosses the line $y = x$ at O.

FIG. 157.

Example 8. Examine the form of the curve $x^2y^3 = (x - y)(2x - y)$ near the origin.

When x and y are small, x^2y^3 is small compared to the separate terms $2x^2$, $-3xy$, y^2 of $(x - y)(2x - y)$ and can only equal $(x - y)(2x - y)$ if $y \simeq x$ or $y \simeq 2x$. This means that there are two branches of the curve which in the neighbourhood of the origin approximate to the straight lines $y = x$, $y = 2x$.

The equation of the curve may be written

$$(a)\ y = x + \frac{x^2y^3}{y - 2x} \quad \text{or} \quad (b)\ y = 2x + \frac{x^2y^3}{y - x}.$$

For the branch of the curve to which a first approximation is $y \simeq x$, the second approximation is, from (a),

$$y \simeq x + \frac{x^2x^3}{x - 2x}, \text{ that is } y \simeq x - x^4.$$

For the other branch to which the first approximation is $y \simeq 2x$, the second approximation is, from (b),

$$y \simeq 2x + \frac{x^2(2x)^3}{2x - x}, \text{ that is } y \simeq 2x + 8x^4.$$

Thus near O the form of the given curve approximates to that of the two curves $y = x - x^4$, $y = 2x + 8x^4$. The first touches $y = x$ at O and lies below $y = x$ for $x \neq 0$; the second touches $y = 2x$ at O and lies above $y = 2x$ for $x \neq 0$. Hence the form of the given curve near O is roughly as in Fig. 158.

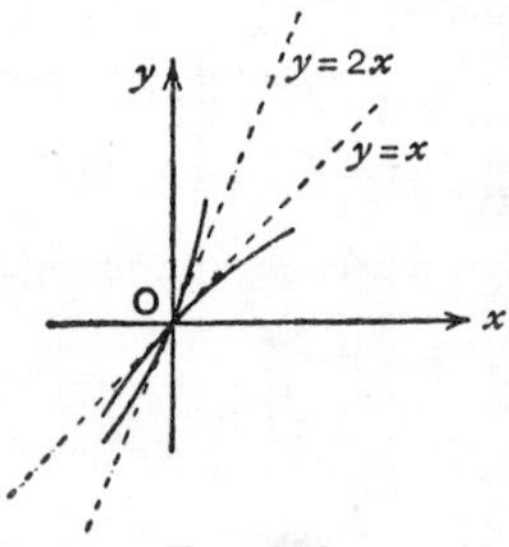

Fig. 158.

The origin is called a *double point* of the curve; $y = x$, $y = 2x$ are tangents to the two branches of the curve at O, and the second approximations show on which sides of the tangents the curve lies.

Examples 7, 8 illustrate a general theorem that if an algebraic curve passes through the origin the term or terms of lowest degree (in the rationalised form of the equation) equated to zero usually represent a tangent or tangents at the origin; but see Exercise XXI c, No. 25.

When the terms of lowest degree are quadratic and have distinct factors, as in Example 8, the origin is an ordinary *double point*. When the terms of lowest degree are quadratic and have no factors, as for the curve $x^3 = x^2 + y^2$, the origin is called an *isolated double point*, because no other point in its neighbourhood lies on the curve. We shall indicate by three examples different forms which a curve can assume near the origin when the terms of lowest degree are of the form $(lx + my)^2$, and for simplicity we take $l = 0$.

$$\text{(i) } y^2 = x^3.$$

This curve consists of two parts given by $y = x\sqrt{x}$, $y = -x\sqrt{x}$ which lie on opposite sides of $\mathsf{O}x$. Since y is undefined for $x < 0$, both parts of the curve lie on the right of $\mathsf{O}y$. $\mathsf{O}x$ is the tangent at O to the curve and to each part of it separately; it can be regarded as two coincident tangents. A point such as the origin on this curve is called a *cusp of the first species* or, shortly, an ordinary *cusp*.

Since $\lim (x^2/2y) = 0$ the radius of curvature of $y^2 = x^3$ at O is zero.

$$\text{(ii) } (y - x^2)^2 + x^5 = 0.$$

This curve consists of two parts given by $y = x^2 + x^2\sqrt{(-x)}$, $y = x^2 - x^2\sqrt{(-x)}$ which lie on the same side of $\mathsf{O}x$. Since y is undefined for $x > 0$, both parts of the curve lie on the left of $\mathsf{O}y$. $\mathsf{O}x$ is the tangent at O; it touches each branch and may be regarded as two coincident tangents. A point such as O is called a *cusp of the second species* or a *rhamphoid cusp*.

Since $\lim (x^2/2y) = \frac{1}{2}$, the radius of curvature at O is $\frac{1}{2}$; the two parts of the curve have equal curvatures at O.

$$\text{(iii) } (y - x^2)^2 = 4x^4(1 + x).$$

This curve consists of two parts given by $y = x^2 \pm 2x^2\sqrt{(1 + x)}$ for which first approximations near O are the parabolas $y = 3x^2$, $y = -x^2$. The two parts touch $\mathsf{O}x$ at O on opposite sides. A point such as O is called a *double point with coincident tangents*.

From Newton's formula, applied to each part of the curve separately, it follows that the curvatures at O of the two parts are $+6$ and -2.

If a curve is described by a moving point an ordinary double point arises where the moving point happens to cross its previous path, a double point with coincident tangents is a place where one part of the track touches another part, and a cusp is a place where the moving point stops and moves back in the directly opposite direction.

Triple points and multiple points of different kinds arise when the terms of lowest degree are cubic or of higher order.

Example 9. Find the curvature at the origin for the two branches, $y=f(x)$, of the curve $y(y+x)=2x^2y+2x^3(y+2x)$.

The two tangents at the origin are $y=0$, $y+x=0$, and since the equation of the curve can be written

$$(a)\ y=\frac{2x^2y+2x^3(y+2x)}{y+x} \quad \text{or} \quad (b)\ y=-x+2x^2+\frac{2x^3(y+2x)}{y}$$

the second approximations are

$$(a)\ y\simeq\frac{0+2x^3(0+2x)}{0+x}, \quad (b)\ y\simeq -x+2x^2,$$

that is $\qquad y\simeq 4x^3 \quad \text{and} \quad y\simeq -x+2x^2.$

For the first branch, at the origin, $\dfrac{dy}{dx}=0$, $\dfrac{d^2y}{dx^2}=0$,

$\therefore$ the curvature is zero.

For the second branch, at the origin, $\dfrac{dy}{dx}=-1$, $\dfrac{d^2y}{dx^2}=4$,

$\therefore$ the curvature is, by p. 346, $4/\sqrt{(1+1)^3}, =\sqrt{2}$.

Note. The first branch is inflexional at the origin, which explains why the curvature is zero.

EXERCISE XXI c

Sketch the forms near the origin of the curves in Nos. 1-8 :

1. $y=x^3$. **2.** $y=x^4$. **3.** $y^2=x^3$.

4. $y^2=-x^5$. **5.** $y=x-x^4$. **6.** $y=x^2-x^3$.

7. $y=x-x^{2/3}$. **8.** $y=x^2\pm x^2\sqrt{x}$.

Find approximations for y in terms of x (or for x in terms of y) when x and y are small for the relations in Nos. 9-20 and sketch the portions of the graphs near the origin :

9. $x^3+y^3=y$. **10.** $x^3+y^3=x-y$.

11. $x^3+y^3=y^2$. **12.** $x^3+y^3=x(x-y)$.

13. $y^2(1-x)=x^3$. **14.** $x^3=(y-x^2)^2$.

15. $x^4+y^4=y(y+x)$. **16.** $x^3+y^3=x^4+y^4$.

17. $2y = \sin(x+y)$.   18. $xy(x^2+y^2) = x^2 - y^2$.

19. $x(x+y)^2 = (x-y)^2$.   20. $(x-y)^2(x+y) = x^4$.

Find the radii of curvature at the origin of the branches $y = f(x)$ of the curves in Nos. 21-24:

21. $y = 3x^2 + 2xy - 5y^2$.   22. $y^2 = x^4(1+x)$.

23. $x^3 + y^3 + x^2 = y - 2x$.   24. $x^2(y^2 + x) = (3x - y)(x+y)$.

25. Are the axes tangents at the origin to the curves:

$$\text{(i) } y^2 + x^4 = x^5; \quad \text{(ii) } y^2x + x^5 = x^6 + y^4\,?$$

Curve Tracing: Asymptotes.

An important part of curve tracing consists in finding the form of the curve at a great distance from the origin. This is often done by finding approximations for y in terms of x and $1/x$.

Example 10. Find the form of the curve $y(y-x)(y-3x) = 6x^2$ for large values of x and sketch the curve.

The equation of the curve may be written as

$$(a)\ y = \frac{6x^2}{(y-x)(y-3x)}; \quad (b)\ y = x + \frac{6x^2}{y(y-3x)}; \quad (c)\ y = 3x + \frac{6x^2}{y(y-x)}$$

and second approximations, for large values of x, are

$$(a)\ y \simeq 2; \quad (b)\ y \simeq x + \frac{6x^2}{x(x-3x)}; \quad (c)\ y \simeq 3x + \frac{6x^2}{3x(3x-x)}$$

$$y \simeq x - 3; \qquad y \simeq 3x + 1.$$

The three straight lines $y = 2$, $y = x - 3$, $y = 3x + 1$, to which the form of the curve approximates at a great distance from the origin, are called *asymptotes* of the curve (cf. p. 364).

It is convenient to proceed to third approximations:

For the branch corresponding to the asymptote $y = 2$

$$y = \frac{6x^2}{(y-x)(y-3x)} = 2 + \frac{8xy - 2y^2}{(y-x)(y-3x)} \simeq 2 + \frac{16x}{3x^2}, \text{ when } x \text{ is large};$$

that is $$y \simeq 2 + \frac{16}{3x}.$$

This shows that the curve lies above $y = 2$ when x is large and positive and below $y = 2$ when x is large and negative.

For the branch corresponding to the asymptote $y=x-3$,

$$y=x+\frac{6x^2}{y(y-3x)}=x-3+\frac{3y^2-9yx+6x^2}{y(y-3x)}=x-3+\frac{3(y-x)(y-2x)}{y(y-3x)},$$

$$\therefore \text{ when } x \text{ is large, } y \simeq x-3+\frac{3(-3)(-3-x)}{(x-3)(-2x-3)} \simeq x-3+\frac{9x}{-2x^2},$$

that is $$y \simeq x-3-\frac{9}{2x}.$$

This shows that the curve lies below $y=x-3$ when x is large and positive, and above it when x is large and negative.

The reader should now use the same method to show that for the branch corresponding to the asymptote $y=3x+1$ the third approximation is $y \simeq 3x+1-\dfrac{5}{6x}$. Thus the relation of the curve to each of its asymptotes has been found and is indicated in Fig. 159. To trace the curve it is convenient to find the points where it meets the asymptotes. This may be done in the ordinary

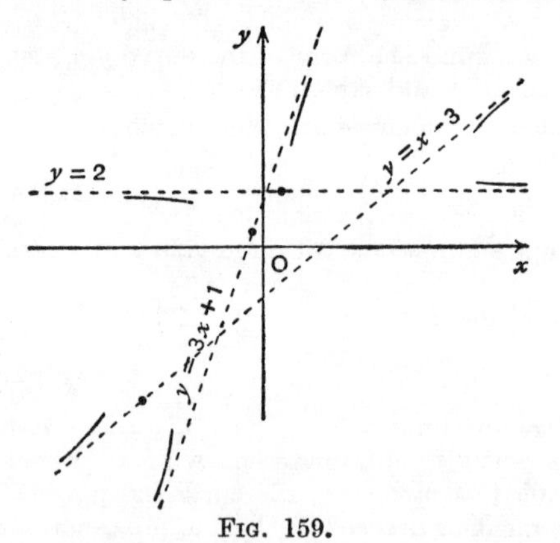

FIG. 159.

way, or *alternatively* we may rearrange the equation of the curve by writing

$$y(y-x)(y-3x)-6x^2 \equiv (y-2)(y-x+3)(y-3x-1)+(\;.\;.\;.\;.\;),$$

where it is easily seen that the contents of the final bracket must be $7y-16x-6$.

Hence the points where the curve meets its asymptotes lie on the line $7y-16x-6=0$. The points are marked in Fig. 159.

Also, near the origin, the curve approximates to $y^3 = 6x^2$. It may therefore be sketched as in Fig. 160, which is, however, not drawn to scale.

The accuracy of the form of the curve in Fig. 160 may be tested by obtaining parametric equations of the curve as on p. 364. Putting $y = tx$ we find

$$x : y : 6 = 1 : t : t(t-1)(t-3)$$

showing that $x \to \infty$ when $t \to 0$, $t \to 1$, $t \to 3$.

Since $\dfrac{dy}{dt} = \dfrac{-12(t-2)}{(t-1)^2(t-3)^2}$ the tangent is parallel to Ox where $t = 2$, that is at $(-3, -6)$.

Also, $\dfrac{dx}{dt} = \dfrac{-6(3t^2 - 8t + 3)}{t^2(t-1)^2(t-3)^2}$; therefore the tangent is parallel to Oy at the two points given by $t = \frac{1}{3}(4 \pm \sqrt{7})$.

Further, any straight line $lx + my + 6 = 0$ cuts the curve where

$$l + mt + t(t-1)(t-3) = 0,$$

that is
$$t^3 - 4t^2 + (m+3)t + l = 0.$$

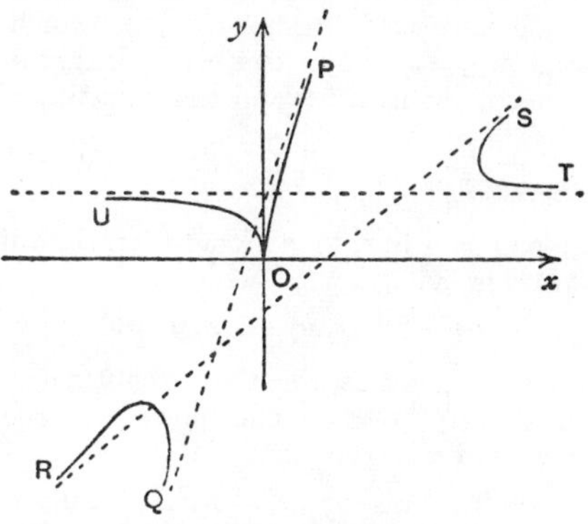

FIG. 160.

Hence if t_1, t_2, t_3 are the parameters of the points of intersection of the line and the curve,

$$t_1 + t_2 + t_3 = 4.$$

For a point of inflexion $t_1 = t_2 = t_3$; hence there is one point of inflexion given by $t = \frac{4}{3}$, namely $(-8{\cdot}1, -10{\cdot}8)$.

Example 10 illustrates the method of obtaining successive approximations to a given algebraic curve when x is large.

Suppose for example that an asymptote of a curve of degree 4 is parallel to $y=7x$; then a first approximation to one branch of the curve when x is large is $y \simeq 7x$, and therefore the equation of the curve must be of the form

$$y=7x+\frac{Q_3(x, y)}{P_3(x, y)}$$

where $P_3(x, y)$ is a polynomial of degree 3 in x, y and $Q_3(x, y)$ is a polynomial whose degree does not exceed 3; thus the equation of the curve is $\quad (y-7x)P_3(x, y)-Q_3(x, y)=0.$

Therefore $y-7x$ is a factor of the terms of highest degree in the equation of the curve. Hence

the lines through the origin parallel to the asymptotes are factors of the terms of highest degree in the equation of the curve.

Thus all the possible directions of the asymptotes can be obtained by factorising the terms of highest degree.

If y is a factor of the terms of highest degree there may be an asymptote parallel to $y=0$. Suppose that $y=3$ is an asymptote of a curve; then an approximation to one branch of the curve when x is large is $y=3$ and a closer approximation is usually $y=3+a/x$ where a is constant; and the equation of the curve is of the form

$$y=3+\frac{Q(x, y)}{P(x, y)}$$

where the degree of x in $P(x, y)$ is higher than that of x in $Q(x, y)$. Thus the equation is

$$(y-3)P(x, y)-Q(x, y)=0$$

and consequently $y-3$ is a factor of the coefficient of the highest power of x in the equation of the curve. If the equation is arranged in powers of x in the form

$$x^{n-1}f_1(y)+x^{n-2}f_2(y)+\ldots\ldots=0,$$

where the degree of $f_r(y)$ in y does not exceed r, and if $f_1(y)$ is not a mere constant, then the first approximation, when x is large, is $f_1(y)=0$ and the second is obtained from $f_1(y)+\dfrac{f_2(y)}{x}=0$. Thus $f_1(y)=0$ is the equation of an asymptote parallel to Ox. Similarly if the term $x^{n-1}f_1(y)$ is absent and $f_2(y)$ is not a mere constant, any asymptotes parallel to Ox are given by $f_2(y)=0$; and so on.

Hence

all asymptotes parallel to Ox *can be obtained by equating to zero the coefficient of the highest power of* x ;

conversely it may be seen that the lines obtained in this way are usually asymptotes; e.g. the asymptote $3y-6=0$ in Example 10, p. 431, can be found from the coefficient of x^2. But it may happen that there is no branch of the curve corresponding to a line so obtained; e.g. $y=0$ is not an asymptote of $x^2y^2=y-1$ because $y \simeq 0$ would imply $x^2y^2 \simeq -1$, which is impossible.

Similarly asymptotes parallel to Oy are usually obtained by equating to zero the coefficient of the highest power of y.

For example in $x(x-y)(x-3y)-6y^2=0$ the coefficient of y^2 is $3x-6$ and therefore $3x-6=0$ is an asymptote (cf. Example 10).

Example 11. Find the asymptotes of the curve

$$x^3y^2-a^2x^3-b^3y^2=0.$$

Since the term of highest degree is x^3y^2 any asymptote must be parallel to $x=0$ or $y=0$.

The coefficient of x^3 is y^2-a^2; hence $y=\pm a$ are asymptotes; these correspond to the branches $y \simeq \pm a(1+\frac{1}{2}b^3/x^3)$. The coefficient of y^2 is x^3-b^3, and $x^3-b^3=0$ gives the single asymptote $x=b$, corresponding to the branch $x \simeq b+\frac{1}{3}a^2b/y^2$.

Asymptotes which are not parallel to either axis may be found by the method of successive approximation illustrated in Example 10. Some difficulties that arise when the terms of highest degree contain a repeated factor are illustrated in Examples 12, 13.

Example 12. Find the asymptotes of $x(y-x)^2=4y$.

Any asymptote is parallel either to $x=0$ or to $y-x=0$.

The coefficient of y^2 is x, and $x=0$ is an asymptote which corresponds to the branch $x \simeq 4/y$.

If the equation is written in the form

$$(y-x)^2=4y/x \quad \text{or} \quad y=x \pm 2\sqrt{(y/x)}$$

we see that there are two branches of the curve for each of which x and y are both large.

For each a first approximation is $y \simeq x$. Hence the second approximations are $y \simeq x \pm 2\sqrt{(x/x)}$, that is $y \simeq x \pm 2$.

Thus the asymptotes are $x=0$, $y=x+2$, $y=x-2$.

Example 13. Find the asymptotes of $x(y-x)^2 = 4y^2$.

The coefficient of y^2 is $x-4$, and $x=4$ is an asymptote, corresponding to the branch $x \simeq 4 + 32/y$.

If the equation is written in the form

$$(y-x)^2 = 4y^2/x \quad \text{or} \quad y = x \pm 2y/\sqrt{x}$$

we see that there are two branches of the curve for which x and y are both large.

For each a first approximation is $y \simeq x$. Hence the second approximations are

$$y \simeq x \pm 2x/\sqrt{x}, \text{ that is } y \simeq x \pm 2\sqrt{x},$$

and this may be written $(y-x)^2 \simeq 4x$.

Thus there is no asymptote in the direction of $y = x$, but the curve approximates to a parabola. The parabola $(y-x)^2 = 4x$ is not, however, the best parabolic approximation that can be obtained. The third approximation is

$$y \simeq x \pm 2(x \pm 2\sqrt{x})/\sqrt{x},$$

that is

$$y \simeq x \pm 2\sqrt{x} + 4.$$

Thus the parabola $(y-x-4)^2 = 4x$ is a closer approximation. If we sketch this parabola we obtain a good approximation to the part of the curve where x and y are large. The reader should now prove that a fourth approximation is $y \simeq x + 4 \pm 2(\sqrt{x} + 4/\sqrt{x})$ and deduce that the parabola $(y-x-4)^2 = 4x + 32$ is a still closer approximation to the given curve. Approximating curves of this kind are sometimes called *curvilinear asymptotes.*

EXERCISE XXI d

Find the asymptotes of the curves in Nos. 1-16 :

1. $x^2y = x + y$.
2. $y^2(a-x) = x^3$.
3. $(y-2x)(y-3x) = 1$.
4. $(y-x)(y-2x) = x$.
5. $(y-x)(y-2x)(y-3x) = x + 5y$.
6. $x^2(y-x) = 4y$.
7. $y^2(y+4x) = x$.
8. $y^2(y+4x) = x^2$.
9. $(y-x)(y-2x)(y-4x) = 6x^2$.
10. $x^3 + y^3 = 6y^2$.
11. $x^4 - y^4 = x(x+2y)$.
12. $xy(x^2+y^2) = x^3 - y^2$.
13. $x(x-y)^2 = x + 3y$.
14. $(y^2-x^2)(y-x) = 8x$.
15. $(x^2-y^2)^2 = 4xy$.
16. $x(x-y)^3 = y - 2x$.

17. Prove that approximations for the branches of the curve $y^2(y+x)=x$ are given by $y=1-\dfrac{1}{2x}$, $y=-1-\dfrac{1}{2x}$, $y=-x+\dfrac{1}{x}$.

18. Prove that approximations for the branches of the curve $y(x-y)^2=x$ are given by $y=\dfrac{1}{x}$, $y=x\pm 1-\dfrac{1}{2x}$.

19. Prove that approximations for the branches of the curve $(y+x)^2(y-3x)+x^2=0$ are given by $y=3x-\frac{1}{16}-\dfrac{1}{512x}$ and the parabola $(2x+2y-\frac{1}{16})^2=x$; also find the closest parabolic approximation.

Sketch the curves in Nos. 20-27; give the equations of the asymptotes and approximations near the origin:

20. $y^2(y+x)=x$.

21. $y(x-y)^2=x$.

22. $(x^2+y^2)(y-2x)=10xy$.

23. $x^4-y^4=2xy$.

24. $x^3+y^3=y^2-4x^2$.

25. $(x^2-y^2)^2+4xy=0$.

26. $x^4=(x^2-1)y^2$.

27. $x(x^3-y^3)=(x+y)^2$.

28. *Write down* the equations of the asymptotes of

$$(y-2x+3)(y+3x-1)(y-5x)+3x-7y=2,$$

and obtain those of

$$(y-x+1)(y-x+2)(2y-x+3)=2y+4x.$$

29. Prove that the three points where the curve

$$x(y^2-x^2)+2x^2+2y=10$$

cuts its asymptotes lie on the straight line $x+2y=10$.

30 Find the equation of the curve of degree 3 which has as asymptotes the lines $y=x+1$, $y+x=2$, $y=0$ and which touches **O**y at **O** and passes through the point (1, 1).

In Nos. 31, 32 sketch the curves and give the equations of any asymptotes:

31. $x^2(xy-1)^2=x-1$.

32. $x^4y^4+x^2+y^2=1$.

CHAPTER XXII

MOMENTS OF INERTIA AND CENTRES OF PRESSURE

Moments of Inertia. If $m_1, m_2, m_3, \ldots$ are the masses of the particles of a system, and $r_1, r_2, r_3, \ldots$ are their distances from a fixed straight line AB, the expression

$$m_1r_1^2 + m_2r_2^2 + m_3r_3^2 + \ldots, \equiv \Sigma mr^2,$$

is called the *moment of inertia* (M.I.) of the system about the axis AB.

A rigid body can be regarded as consisting of particles of mass δm at distance r from AB and its moment of inertia about AB is then the limit of the sum, $\Sigma r^2 \delta m$, taken throughout the body. This limit is $\int r^2\, dm$.

If M is the total mass of the body, and if the M.I., $\int r^2\, dm$, is written in the form Mk^2, k is called the *radius of gyration* of the body about the axis AB. It is the distance from AB at which a particle of mass M must be situated so as to have the same M.I. as the given body about AB.

To see why moments of inertia are required in dealing with the motion of rigid bodies, consider a body of mass M rotating with angular velocity ω about an axis AB. The linear velocity of any particle of mass δm at distance r from the axis is $r\omega$. Hence its kinetic energy is $\frac{1}{2}\delta m(r\omega)^2$ and its angular momentum about AB is $r(\delta m)r\omega$.

$\therefore$ the total kinetic energy of the body

$$= \int \tfrac{1}{2} r^2\omega^2\, dm = \tfrac{1}{2}\omega^2 \int r^2\, dm = \tfrac{1}{2} Mk^2\omega^2;$$

and the total angular momentum about AB

$$= \int r^2\omega\, dm = \omega \int r^2\, dm = Mk^2\omega.$$

Example 1. Find the M.I. of a uniform rod AB of mass M and length $2a$ about an axis Oy bisecting AB at right angles.

Let ρ be the mass per unit of length so that $M = 2a\rho$.

Consider an element PQ where $OP = x$, $PQ = \delta x$, see Fig. 161; its mass is $\rho\delta x$ and its M.I. about Oy is $x^2\rho\delta x$ approximately.

$$\therefore \text{the M.I. of the rod} = \int_{-a}^{+a} \rho x^2\, dx = \tfrac{2}{3}\rho a^3 = \tfrac{1}{3}Ma^2.$$

It follows that the M.I. of a uniform rectangular lamina ABCD of mass M about an axis Oy which bisects AB and CD is $\frac{1}{3}$Ma^2, where AB $=2a$. For the rectangle can be divided into narrow strips by lines parallel to AB and the M.I. of each strip is $\frac{1}{3}$(mass of strip)a^2; therefore the total M.I. is $\frac{1}{3}$(total mass)a^2.

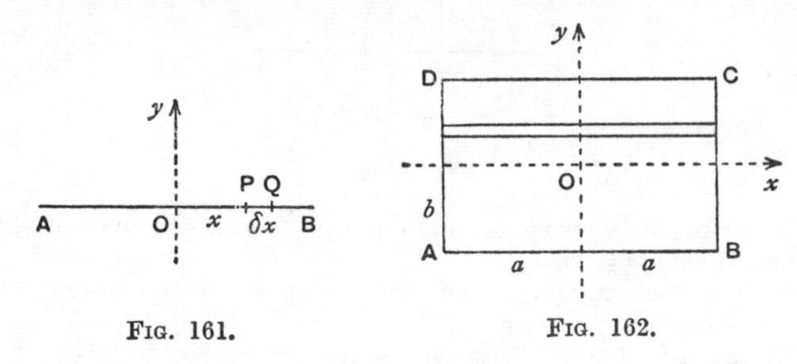

FIG. 161. FIG. 162.

Example 2. Find the M.I. of a uniform circular disc of mass M and radius a about an axis through the centre O perpendicular to the plane of the disc.

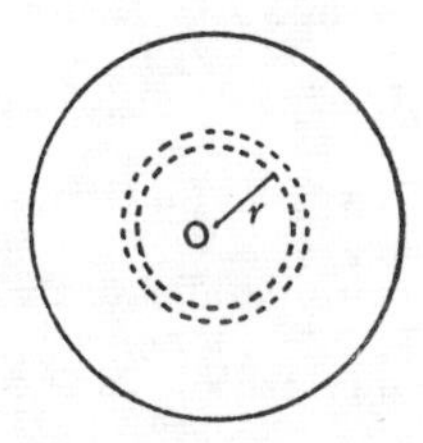

FIG. 163.

Let ρ be the mass per unit of area so that M $=\pi a^2\rho$. Consider the portion of the disc between circles of radii $r, r+\delta r$ concentric with it; the mass of this portion is approximately $\rho 2\pi r\delta r$ and its M.I. can be taken to be $(\rho\, 2\pi r\, \delta r)r^2$.

$$\therefore \text{ the M.I. of the disc} = \int_0^a 2\pi\rho r^3\, dr = \tfrac{1}{2}\pi\rho a^4 = \tfrac{1}{2}\text{M}a^2.$$

Example 3. Find the M.I. of a uniform elliptic lamina $x^2/a^2 + y^2/b^2 = 1$ of mass M about Ox.

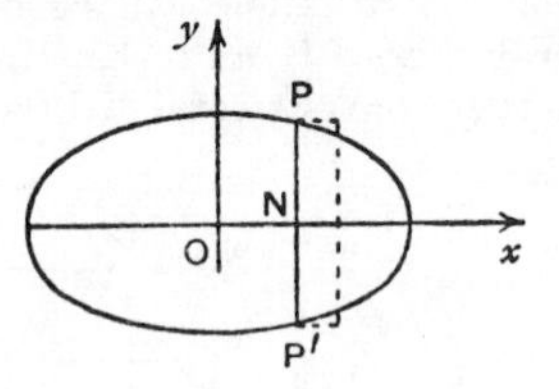

FIG. 164.

Let ρ be the density per unit area; then as the area of the ellipse is πab, $M = \pi ab\rho$.

Denote the coordinates of any point P on the ellipse by $(a \cos \phi, b \sin \phi)$.

Divide the lamina into strips perpendicular to Ox; by Example 1 the M.I. of such a strip is $\frac{1}{3}(\rho 2y\delta x)y^2$. The M.I. of the lamina is therefore

$$2\int_0^a \tfrac{2}{3}\rho y^3\, dx = \tfrac{4}{3}\rho \int_{\pi/2}^0 b^3 \sin^3 \phi(-a \sin \phi)\, d\phi,$$

since $dx = d(a \cos \phi) = -a \sin \phi\, d\phi$. Hence

$$\text{M.I.} = \tfrac{4}{3}\rho ab^3 \int_0^{\pi/2} \sin^4 \phi\, d\phi$$

$$= \tfrac{4}{3}\rho ab^3\, \tfrac{3}{4}\, \tfrac{1}{2}\, \tfrac{1}{2}\pi = \tfrac{1}{4}Mb^2.$$

Alternatively, by using strips parallel to Ox the M.I. about Ox

$$= 2\int_0^b y^2 2\rho x\, dy = 4\rho ab^3 \int_0^{\pi/2} \sin^2 \phi \cos^2 \phi\, d\phi,$$

which gives the same result.

EXERCISE XXII a

[Assume in this Exercise that each body is of mass M and uniform density. The results of Examples 1, 2 may be quoted.]

1. Find the M.I. of a rod AB of length $2a$ about an axis through A perpendicular to AB.

2. Find the M.I. of a rectangular wire ABCD about a line parallel to AB through the centre; $AB = 2a$, $BC = 2b$.

3. Find the M.I. of a rectangular lamina ABCD about AB; AB $=2a$, BC $=2b$.

4. Find the M.I. of a circular cylinder of radius a about its axis.

5. Find the M.I. of an isosceles triangle of height h about a line through the vertex parallel to the base.

6. Find the M.I. of a circular cone of base-radius a about its axis.

7. Find the M.I. about Ox of the segment of the parabola $y^2=4ax$ bounded by the line $x=c$.

8. Find the M.I. about Ox of the solid formed by revolution about Ox of the area bounded by $y^2=4ax$, $y=0$, $x=c$.

9. Find the M.I. of the astroid $x=a\cos^3 t$, $y=a\sin^3 t$ about Ox.

10. Find the M.I. of a wire in the form of a circular arc AB, centre O, radius c, with $\angle$AOB $=2\theta$, about the diameter which bisects AB.

General Theorems about M.I.

The calculation of moments of inertia is often much simplified by the use of two general theorems.

(i) If the M.I. of a lamina about two perpendicular axes Ox, Oy in its plane are I_x, I_y, then its M.I. about the axis Oz perpendicular to its plane is $I_x + I_y$.

Consider first a system of particles in the plane xOy.

If m is the mass of a particle at (x, y) and r its distance from O, its M.I. about Oz is mr^2, that is $m(x^2+y^2)$. Hence the M.I. of the system about Oz

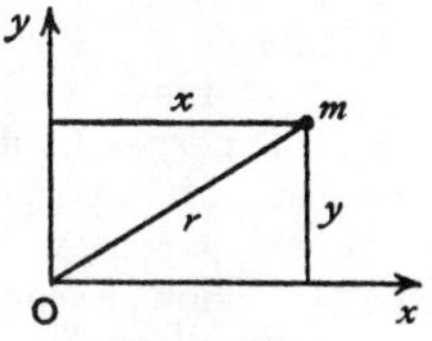

FIG. 165.

$$=\Sigma m(x^2+y^2)=\Sigma mx^2+\Sigma my^2=I_x+I_y.$$

The same result holds for a lamina in the plane xOy, since it may be regarded as made up of particles, each sum being replaced by an integral.

(ii) The M.I. of a body of mass M about any axis AB is equal to the M.I. about the parallel axis through the centre of gravity *plus* $\mathsf{M}h^2$, where h is the distance between the two parallel axes.

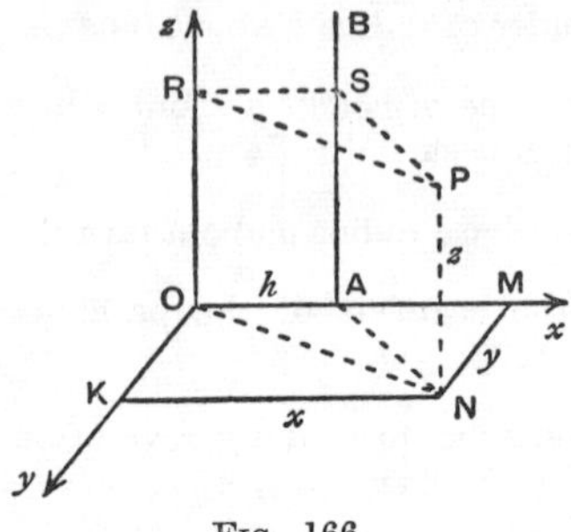

FIG. 166.

Consider first any system of particles in space. Take the origin O at the centre of gravity of the system, the line through O parallel to AB as axis of z, the perpendicular OA from O to AB as axis of x, and the line perpendicular to Ox, Oz as axis of y, see Fig. 166; PR, PS are the perpendiculars from any point P(x, y, z) to Oz, AB.

If m is the mass of a particle at P its M.I. about Oz is $m.\mathsf{PR}^2$, $=m(x^2+y^2)$, and its M.I. about AB is $m.\mathsf{PS}^2$, $=m\{(x-h)^2+y^2\}$. Hence the M.I. of the system about AB

$$=\Sigma m\{(x-h)^2+y^2\}$$

$$=\Sigma m(x^2+y^2)+\Sigma mh^2-2\Sigma mxh$$

$$=\text{M.I. about O}z+h^2\Sigma m-2h\Sigma mx.$$

But $\Sigma m=\mathsf{M}$ and $\Sigma mx=0$ since the centre of gravity of the system is at the origin; hence

$$\text{M.I. about AB}=\text{M.I. about O}z+\mathsf{M}h^2.$$

The same result holds for a solid body, since it may be regarded as made up of particles, each sum being replaced by an integral.

Example 4. Find the M.I. of a uniform rectangular lamina ABCD of mass M and sides $2a$, $2b$ about an axis Oz through its centre O perpendicular to its plane. Find also the M.I. about an axis through A perpendicular to its plane.

By the extension noted at the end of Example 1, pp. 438, 439, the M.I. about axes through O parallel to the sides are $\frac{1}{3}\mathsf{M}a^2$, $\frac{1}{3}\mathsf{M}b^2$. Hence by the general theorem (i),

$$\text{M.I. about O}z=\tfrac{1}{3}\mathsf{M}(a^2+b^2).$$

It follows by the general theorem (ii) that the M.I. about the parallel axis through A, see Fig. 162, p. 439, is

$$\tfrac{1}{3}\mathsf{M}(a^2+b^2)+\mathsf{M}\,.\,\mathsf{AO}^2,\text{ that is }\tfrac{4}{3}\mathsf{M}(a^2+b^2).$$

Also for a cuboid of mass M with edges $2a$, $2b$, $2c$, the M.I. about an axis through the centre parallel to the edges $2c$ is $\frac{1}{3}\mathsf{M}(a^2+b^2)$.

Example 5. Find the M.I. of a uniform circular lamina of mass M and radius a about a diameter.

Let O be the centre. Denote the M.I. about the radius Ox by I; then that about the perpendicular radius Oy is also I. Hence by the general theorem (i) the M.I. about the axis Oz perpendicular to the plane of the lamina is I + I, = 2I.

But by Example 2, p. 439, $2I = \frac{1}{2}Ma^2$,

$$\therefore I = \tfrac{1}{4}Ma^2.$$

Example 6. Find the M.I. of a uniform sphere of mass M and radius a about a diameter.

Let O be the centre and Ox, Oy, Oz three mutually perpendicular radii.

Then with the notation of p. 442, see Fig. 166, the M.I. about Oz is expressed by the limit of $\Sigma m(x^2 + y^2)$.

By symmetry, $\Sigma mx^2 = \Sigma my^2 = \Sigma mz^2$,

$$\therefore \Sigma m(x^2 + y^2) = \tfrac{2}{3}\Sigma m(x^2 + y^2 + z^2)$$

$$= \tfrac{2}{3}\Sigma m \,.\, OP^2,$$

since $x^2 + y^2 + z^2 = ON^2 + NP^2 = OP^2$.

Divide the sphere into thin shells bounded by spheres of radii $r, r + \delta r$, centre O; then if ρ is the mass per unit volume, the mass of a shell may be taken to be $\rho 4\pi r^2 \delta r$. Hence the M.I. of the sphere about Oz

$$= \tfrac{2}{3}\int_0^a r^2 \,.\, \rho 4\pi r^2\, dr = \tfrac{8}{3}\pi\rho \int_0^a r^4\, dr$$

$$= \tfrac{8}{15}\pi\rho a^5.$$

But $M = \frac{4}{3}\pi a^3 \rho$, $\therefore$ M.I. $= \frac{2}{5}Ma^2$.

Example 7. Find the M.I. of a thin uniform spherical shell of mass M and radius a about a diameter.

With the notation of Example 6 the M.I. about Oz

$$= \Sigma m(x^2 + y^2) = \tfrac{2}{3}\Sigma m(x^2 + y^2 + z^2)$$

$$= \tfrac{2}{3}\Sigma ma^2 = \tfrac{2}{3}Ma^2.$$

Routh's Rule. There is a mnemonic for the M.I. of a few simple bodies about axes of symmetry through the centre of gravity which some students find useful :

If GX, GY, GZ are perpendicular axes of symmetry through the centre of gravity G, cutting the body at X_1, Y_1, Z_1,

$$\text{M.I. about GX} = \text{mass} \times \frac{GY_1^2 + GZ_1^2}{3 \text{ or } 4 \text{ or } 5},$$

where the denominator is

3 for a rod, a rectangular lamina, a cuboid ;

4 for a circular or elliptic disc ;

5 for a sphere or ellipsoid.

Thus for a circular disc, centre G, radius a, if GX is a radius, $GY_1 = a$, $GZ_1 = 0$; M.I. about $GX = M(a^2 + 0)/4$; and if GX is perpendicular to the plane of the disc, $GY_1 = a$, $GZ_1 = a$; and M.I. about $GX = M(a^2 + a^2)/4$.

Example 8. Find the M.I. of a uniform circular cone of mass M, height h, and semivertical angle α about a line through the vertex parallel to the base.

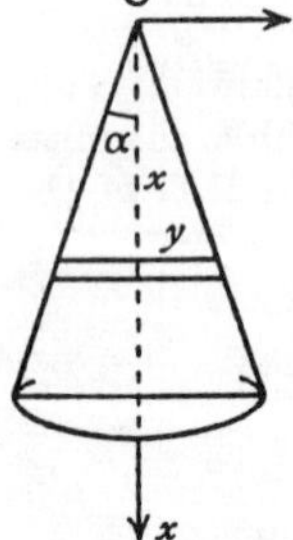

FIG. 167.

Take the axis of the cone as Ox and the given line through the vertex as Oy. Consider a slice of the cone bounded by parallel planes at distances x, $x + \delta x$ from O ; if ρ is the density per unit volume, the mass of the slice can be taken to be $\rho\pi y^2 \delta x$, and therefore, by Example 5, its M.I. about its diameter parallel to Oy is $\frac{1}{4}(\rho\pi y^2 \delta x)y^2$. Hence by the general theorem (ii), p. 442, its M.I. about Oy is $(\frac{1}{4}y^2 + x^2)\rho\pi y^2 \delta x$.

$\therefore$ the M.I. of the cone about Oy

$$= \tfrac{1}{4}\rho\pi \int_0^h (y^2 + 4x^2)y^2\,dx$$

$$= \tfrac{1}{4}\rho\pi \int_0^h (x^2 \tan^2 \alpha + 4x^2)x^2 \tan^2 \alpha\,dx$$

$$= \tfrac{1}{4}\rho\pi \tan^2 \alpha\,(\tan^2 \alpha + 4) \int_0^h x^4\,dx$$

$$= \tfrac{1}{20}\rho\pi h^5 \tan^2 \alpha\,(\tan^2 \alpha + 4).$$

But $M = \frac{1}{3}\pi(h \tan \alpha)^2 h\rho$, $\therefore$ M.I. $= \frac{3}{20} M h^2 (\tan^2 \alpha + 4)$.

EXERCISE XXII b

[Assume in this Exercise that each body is of mass M and, unless otherwise stated, of uniform density.]

1. Find the M.I. of a wire in the form of a rectangle with sides $2a$, $2b$ about an axis perpendicular to its plane at one corner.

2. Find the M.I. of a solid cuboid with edges of lengths l, m, n about one of the edges of length l.

3. Find the M.I. of a circular lamina of radius a about a tangent.

4. Find the M.I. of a solid sphere of radius a about a straight line which touches it.

5. Find the M.I. of a lamina bounded by concentric circles of radii a and b about a diameter.

6. Find the M.I. about a diameter of the solid bounded by concentric spheres of radii a and b.

7. A closed cylinder of base-radius a and height h is made of thin sheeting. Find its M.I. about a generating line.

8. In Example 8, p. 444, find the M.I. about a diameter of the base.

9. Find the M.I. of a rod AB of length l about an axis through A perpendicular to AB if the density varies as the cube of the distance from A.
What is its least M.I. about any line perpendicular to AB ?

10. The density at any point of a circular lamina of radius a varies as the square of its distance from the centre. Find its M.I. about (i) a line through the centre perpendicular to the lamina, (ii) a diameter, (iii) a tangent.

11. Find the M.I. of a semi-circular lamina of radius a about the tangent parallel to the bounding diameter.

12 Find the M.I. of a solid cylinder of base-radius a and height h about (i) a diameter of its base, (ii) a tangent to its base.

13 Find the M.I. of the surface in No. 7 about a diameter of the base.

14. Find the M.I. of the curved surface of a cone of base-radius r about the axis of the cone.

15. A lamina is in the form of an equilateral triangle of side $2a$. Find its M.I. about an axis through its centre of gravity perpendicular to its plane.

16. The density of a solid sphere of radius a varies as the square of the distance from the centre. Find its M.I. about a diameter.

Centre of fluid pressure. If one face of a plane lamina is wholly in contact with a fluid, each point of that face is subjected to a normal pressure, and, disregarding the pressure of the atmosphere, this normal pressure is proportional to the depth of the point below the surface of the fluid.

The total normal pressure on the face can therefore be found as the resultant of a system of parallel forces; the point of the lamina at which it acts is called the *centre of pressure.*

To determine the position of this point, take as x-axis the line in which the plane of the lamina meets the free surface of the fluid and as y-axis the line of greatest slope of the lamina through its centre of gravity G. Let the coordinates of G be $(0, \bar{y})$ and those of the centre of pressure (ξ, η); also let α be the angle between the lamina and the horizontal, and w the weight per unit volume of the fluid.

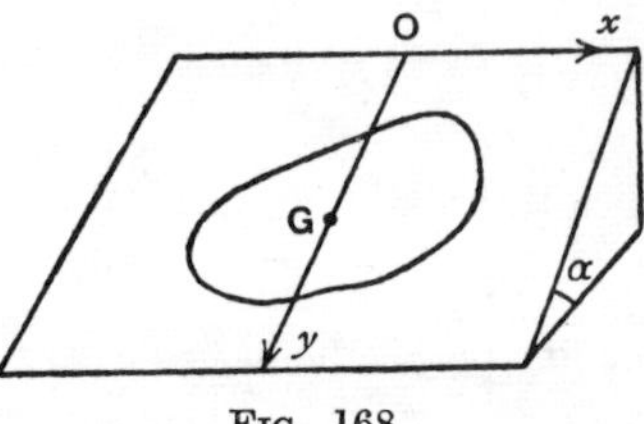

Fig. 168.

Consider an element of area δA of the lamina at the point (x, y). The depth at this point is $y \sin \alpha$, and therefore the pressure on the element is known to be $wy \sin \alpha \, \delta A$; taking moments about the axes for parallel forces of which this is typical,

$$\xi \int wy \sin \alpha \, dA = \int xwy \sin \alpha \, dA, \quad \eta \int wy \sin \alpha \, dA = \int ywy \sin \alpha \, dA;$$

$$\therefore \xi = \frac{\int xy \, dA}{\int y \, dA}, \quad \eta = \frac{\int y^2 \, dA}{\int y \, dA}.$$

These formulae determine the centre of pressure. A convenient form can be found for the depth of the centre of pressure in terms of the radius of gyration, k, of the lamina about Ox.

For if ρ is the mass per unit area, k is given by

$$A\rho k^2 = \int \rho y^2 \, dA \, ;$$

also $\bar{y}$ is given by $\quad A\rho\bar{y} = \int \rho y \, dA \, ;$

$\therefore \eta = k^2/\bar{y}$, and the depth of the centre of pressure is $(k^2 \sin \alpha)/\bar{y}$.

The integral $\int wy \sin \alpha \, dA$ represents the total normal pressure, and since it equals $w \sin \alpha \, A\bar{y}$,

total normal pressure = (area of lamina)(pressure at C.G.)

Example 9. A uniform rectangular lamina in a vertical plane has sides of lengths $2a$, $2b$ and is wholly immersed in a fluid with the sides of length $2a$ horizontal. The centre of gravity of the lamina is at a depth h below the surface. Find the depth of the centre of pressure and the total pressure on one face of the lamina.

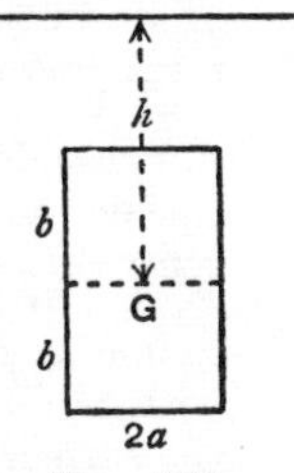

FIG. 169.

The M.I. of the rectangle about the line joining the mid-points of the edges $2b$ is $\frac{1}{3}Mb^2$; therefore the M.I. about the line in which the plane of the rectangle meets the surface is $\frac{1}{3}Mb^2 + Mh^2$. Also the depth of the centre of gravity is h.

Therefore, with the notation explained above,

$$\text{depth of centre of pressure} = (k^2 \sin \alpha)/\bar{y}$$
$$= (\tfrac{1}{3}b^2 + h^2)/h.$$

Also the pressure at the centre of gravity is wh and the area of the lamina is $4ab$,

$$\therefore \text{total normal pressure} = 4abwh.$$

EXERCISE XXII c

[Assume in this exercise that the plane of the lamina is vertical when not otherwise stated, and that the weight of the fluid is w per unit volume.]

1. Find the total pressures on the upper and lower halves of a rectangle with sides $2a$, $2b$ if the upper side ($2a$) is in the surface and the plane of the lamina is (i) vertical, (ii) at 60° to the vertical.

2. Find the total pressure on a circular lamina, radius a, if its highest point is (i) in the surface, (ii) at depth h below the surface. Also find the depth of the centre of pressure.

3. Find the total pressure on a lamina in the form of an equilateral triangle whose upper side, of length $2a$, is in the surface and whose plane is inclined at 30° to the vertical.

4. Find the centre of pressure of an isosceles triangle of height h whose base is horizontal and vertex is (i) in the surface, (ii) at depth c.

5. The segment of $y^2 = 4ax$ cut off by $x = c$ is placed with the origin in the surface and Ox vertically downwards. Find the total pressure and the depth of the centre of pressure.

6. A square lamina, side a, is placed with its upper corner in the surface and one diagonal vertical. Find the depth of the centre of pressure.

7. A lamina in the form of an ellipse $x^2/a^2 + y^2/b^2 = 1$ is placed with the y-axis horizontal at depth h, greater than a, below the surface. Find the depth of the centre of pressure.

8. Find the depth of the centre of pressure of a semicircular area of radius a with its base in the surface and vertex downwards.

9. A lamina OABC in the form of a trapezium is placed with its upper edge OA in the surface. OC is perpendicular to the parallel edges OA, CB. If OA $= a$, OC $= h$, CB $= b$, prove that the depth of the centre of pressure is $h(a + 3b)/(2a + 4b)$.

10. A lamina is in the form of a rectangle, sides $2a$, $2b$, together with two isosceles right-angled triangles having the sides $2a$ as hypotenuses. It is placed so that the sides $2b$ are vertical and the top point of one triangle is in the surface. Find the depth of the centre of pressure.

APPENDIX

TO VOLUME II

THE subject matter of each exercise is identical with that of the exercise which has the same heading in the book itself. To avoid confusion, the Appendix examples are numbered from 51 upwards. There are no test papers in this volume, but a collection of 125 Miscellaneous Examples is given at the end of the Appendix.

Selected numbers suitable for a first course are printed in brackets at the head of each exercise.

EXERCISE XI a

[51, 53, 58, 62, 64, 65, 66, 68, 70, 72]

Use the given substitutions to find the values of the integrals in Nos. 51-65 :

51. $\int x(3-2x)^4\,dx$; $3-2x=z$.

52. $\int x(x-1)^{-4}\,dx$; $x-1=z$.

53. $\int\{x^2/\sqrt{(1-x)}\}\,dx$; $1-x=z^2$.

54. $\int x(a+bx)^{10}\,dx$; $a+bx=z$.

55. $\int (x+a)\sqrt{(x+b)}\,dx$; $x+b=z^2$.

56. $\int x\sqrt{(1+x^2)}\,dx$; $x^2=z$.

57. $\int x(a+bx^2)^n\,dx,\ n\neq -1$; $x^2=z$.

58. $\int \cos x \operatorname{cosec}^4 x\,dx$; $\sin x=z$.

59. $\int \cot^2 x \operatorname{cosec}^2 x\,dx$; $\cot x=z$.

60. $\int\{1/(a^2+x^2)\}\,dx$; $x=a\tan\theta$.

61. $\int x^3\sqrt{(1-9x^2)}\,dx$; $x^2=u$, and then $1-9u=z$.

62. $\displaystyle\int\frac{x^2\,dx}{(1+x^2)^2}$; $x=\tan\theta$.

63. $\displaystyle\int\frac{dx}{\sqrt{(a^2-x^2)}}$; $x=a\sin\theta$.

64. $\displaystyle\int\sqrt{\frac{a+x}{a-x}}\,dx, a>0$; $x=a\cos 2\theta$.

65. $\displaystyle\int\frac{dx}{x\sqrt{(x^2-1)}}$; $x=\sec\theta$.

Find the values of the integrals in Nos. 66-74:

66. $\int \frac{x\,dx}{(a+bx)^n}$, $n \neq 1$, $n \neq 2$.

67. $\int \frac{x^2\,dx}{\sqrt{(1+x)}}$.

68. $\int x\sqrt{(3x+4)}\,dx$.

69. $\int x^2\sqrt{(1+x^3)}\,dx$.

70. $\int \sin x \sec^4 x\,dx$.

71. $\int \sec^2 x \tan^2 x\,dx$.

72. $\int \frac{dx}{(a^2+x^2)^{3/2}}$.

73. $\int \frac{x^3\,dx}{\sqrt{(1-9x^2)}}$.

74. $\int \frac{dx}{x^2\sqrt{(1-x^2)}}$.

EXERCISE XI b

[51, 53, 55, 57-60, 64, 66, 68]

Find the values of the integrals in Nos. 51-65:

51. $\int_0^1 \frac{dx}{\sqrt{(a+bx)}}$.

52. $\int_0^1 \frac{x\,dx}{\sqrt{(a+bx)}}$.

53. $\int_{\pi/6}^{\pi/2} \frac{\cos x}{\sin^5 x}\,dx$.

54. $\int_0^{2/\sqrt{3}} \sqrt{(x^6+x^{10})}\,dx$.

55. $\int_0^{\pi/2} \cos^5 x \sin x\,dx$.

56. $\int_{1/2}^{3/2} \frac{dx}{\sqrt{(2x-x^2)}}$.

57. $\int_0^{1/4} \frac{dx}{\sqrt{(1-4x^2)}}$.

58. $\int_0^1 \frac{x\,dx}{\sqrt{(1+4x^2)}}$.

59. $\int_0^{2/3} \frac{dx}{4+9x^2}$.

60. $\int_2^4 \frac{dx}{x^2\sqrt{(x^2-1)}}$, by the substitution $\frac{1}{x^2}=1-z$.

61. $\int_1^2 \frac{x^2-1}{(x^2+1)^2}\,dx$, by the substitution $x+\frac{1}{x}=z$.

62. $\int_1^{\sqrt{2}} \frac{x^2+1}{x^4+1}\,dx$, by the substitution $x-\frac{1}{x}=z$.

63. $\int_0^1 \sqrt{(4-x^2)}\,dx$, by the substitution $x=2\sin\theta$.

64. $\int_{3/2}^{5/2} \frac{dx}{\sqrt{\{(x-1)(3-x)\}}}$, by the substitution $x=\cos^2\theta+3\sin^2\theta$.

65. $\int_{\pi/4}^{3\pi/4} \frac{\sin\theta+\cos\theta}{3-\sin 2\theta}\,d\theta$, by the substitutions $\theta=\frac{1}{4}\pi+\phi$, $\sin\phi=z$.

66. Use the substitution $y=1/x^2$ to find the values of $\int_{+2}^{+1}\frac{dx}{x^2}$ and $\int_{-2}^{-1}\frac{dx}{x^2}$.

67. Calculate the value of $\int_{-1}^{+1} x^2\,dx$ by making the substitution $x^2=z^3$ and check the result by direct integration.

68. Use the substitution $x=\tan\theta$ to find the values of $\int_a^b \frac{x^2-1}{(x^2+1)^2}\,dx$, for (i) $a=1$, $b=\sqrt{3}$; (ii) $a=-1$, $b=-\sqrt{3}$; (iii) $a=-1$, $b=+1$.

EXERCISE XI c

[51, 54, 57, 59, 62, 63]

Evaluate the integrals in Nos. 51-56 :

51. $\int x\cos nx\,dx$.
52. $\int x^2\sin 2x\,dx$.
53. $\int x^3\cos x\,dx$.
54. $\int x\tan^{-1}x\,dx$.
55. $\int x\tan x\sec^2 x\,dx$.
56. $\int x\cot^{-1}x\,dx$.

57. By writing $\int\cos^6\theta\,d\theta$ in the form $\int\cos^5\theta\,d(\sin\theta)$ and integrating by parts, obtain a relation between $\int_0^{\pi/2}\cos^6\theta\,d\theta$ and $\int_0^{\pi/2}\cos^4\theta\,d\theta$. In the same way obtain a relation between $\int_0^{\pi/2}\cos^4\theta\,d\theta$ and $\int_0^{\pi/2}\cos^2\theta\,d\theta$. Hence evaluate $\int_0^{\pi/2}\cos^6\theta\,d\theta$.

58. Use the method of No. 57 to evaluate $\int_0^{\pi/2}\sin^6\theta\,d\theta$.

59. Prove that $\int x^n\cos x\,dx=x^n\sin x-n\int x^{n-1}\sin x\,dx$.

$$=x^n\sin x+nx^{n-1}\cos x-n(n-1)\int x^{n-2}\cos x\,dx.$$

Hence find the value of $\int x^4\cos x\,dx$.

60. Find a relation between $\int x^n\sin x\,dx$ and $\int x^{n-2}\sin x\,dx$. Hence find the value of $\int x^4\sin x\,dx$.

61. Find the value of $\int_0^{\pi} x\sin^2 x\cos^2 x\,dx$.

62. Find the values of $\int x\sin x\sin 4x\,dx$ and $\int x\cos x\cos 4x\,dx$.

63. Prove that $\int_0^1 x^m(1-x)^n\,dx = \dfrac{n}{m+1}\int_0^1 x^{m+1}(1-x)^{n-1}\,dx$ where m and n are positive integers. Use the result to find the value of $\int_0^1 x^5(1-x)^4\,dx$.

64. Verify that $\int\sqrt{(1-x^2)}\,dx = x\sqrt{(1-x^2)} + \int\dfrac{x^2}{\sqrt{(1-x^2)}}\,dx$ and that $\int\dfrac{x^2}{\sqrt{(1-x^2)}}\,dx = \sin^{-1}x - \int\sqrt{(1-x^2)}\,dx$.

Deduce that $\int_0^1 \sqrt{(1-x^2)}\,dx = \frac{1}{4}\pi$.

65. Prove that $\int\dfrac{x^2\,dx}{(1-x^2)^{3/2}} = \pm\left\{\dfrac{x}{\sqrt{(1-x^2)}} - \sin^{-1}x\right\}$.

66. Prove that

$$\int\sec^n x\,dx = \sec^{n-2}x\tan x - (n-2)\int(\sec^n x - \sec^{n-2}x)\,dx$$

and deduce a relation between $\int\sec^n x\,dx$ and $\int\sec^{n-2}x\,dx$.

EXERCISE XII a

[51, 56, 57, 58, 61-64, 66, 68, 69, 70]

Differentiate with respect to x the functions in Nos. 51-65:

51. $\log(x^n)$. **52.** $\log\sec x$. **53.** $\sin\log x$.

54. $(\log x)^3$. **55.** $(\log x)/\sqrt{x}$. **56.** $\log\log x$.

57. $\log(\sin x\cos x)$. **58.** $(\log\sin x)\cos x$.

59. $\log\{(1+x^2)/(1-x^2)\}$. **60.** $\log(a^2+x^2)$.

61. $\log\{x+\sqrt{(x^2-1)}\}$. **62.** $\log(\sec x+\tan x)$.

63. $\log\tan(\frac{1}{4}\pi+\frac{1}{2}x)$. **64.** $\log\{x+\sqrt{(a^2+x^2)}\}$.

65. $x^m(\log x)^n$.

66. If $y=(\log x)^2$, prove that $x^2\dfrac{d^3y}{dx^3}+3x\dfrac{d^2y}{dx^2}+\dfrac{dy}{dx}=0$.

67. What are the least possible values of $x\log x$ and $x^2\log x$?

68. Sketch the graph of $y=\log(x^2+x^{-2})$ and find its points of inflexion.

69. Prove for the curve $y = a \log \sec \frac{x}{a}$ that $x = a\psi$.

70. If $x > 0$, prove that the functions $\log(1+x) - (x - \frac{1}{2}x^2)$ and $(x - \frac{1}{2}x^2 + \frac{1}{3}x^3) - \log(1+x)$ increase as x increases.

71. If $x > 1$, prove that $\tan^{-1} x < \frac{1}{4}\pi + \frac{1}{2}\log x$.

72. What value of θ between 0 and $\frac{1}{2}\pi$ gives a minimum value of $\tan\theta + 3\log\cos\theta + \theta$?

EXERCISE XII b

[52, 53, 55, 57, 58, 62, 63, 64, 66, 70, 71, 73]

Integrate with respect to x the functions in Nos. 51-56:

51. $\dfrac{2}{1-x}$.   52. $\dfrac{1}{p-qx}$.   53. $\dfrac{x+a}{x^2+2ax+b}$.

54. $\dfrac{\sec^2 x}{\tan x}$.   55. $\dfrac{\log x}{x}$.   56. $\dfrac{1}{x\log x}$.

Evaluate the integrals in Nos. 57-68:

57. $\displaystyle\int_a^b \frac{dx}{x}$.   58. $\displaystyle\int_{ac}^{bc} \frac{dx}{x}$.   59. $\displaystyle\int_e^{e^2} \frac{dx}{x}$.

60. $\displaystyle\int_{-9}^{-3} \frac{dx}{x}$.   61. $\displaystyle\int_{\pi/2}^{3\pi/4} \cot x\, dx$.   62. $\displaystyle\int_{\pi/3}^{\pi/2} \tan\tfrac{1}{2}x\, dx$.

63. $\displaystyle\int_0^1 \frac{2x+1}{x+1}\, dx$.   64. $\displaystyle\int_1^3 \frac{x^3\, dx}{1+x^2}$.   65. $\displaystyle\int_e^{e^e} \frac{dx}{x\log x}$.

66. $\displaystyle\int_1^e \frac{\log x}{x^2}\, dx$.   67. $\displaystyle\int_0^1 x\log(1+\tfrac{1}{2}x)\, dx$.   68. $\displaystyle\int_0^1 x^2 \tan^{-1} x\, dx$.

69. Find the area bounded by the x-axis, the ordinates $x = pa$, $x = qa$, $(1 < p < q)$, and the curve $y(x-a) = c^2$, $(a > 0)$.

70. Find the area bounded by the x-axis, the ordinates $x = 0$, $x = \frac{1}{4}\pi$, and the curve $y = \tan^3 x$.

71. Find the volume generated by rotating about Ox the segment of $xy^2 = c^2(a-x)$, $(a > 0)$, cut off by $x = \lambda a$, $(0 < \lambda < 1)$.

72. Prove directly from the definition (cf. pp. 251, 254) that

$$\log a^b = b \log a.$$

73. The acceleration of a body falling vertically from rest in a resisting medium is $g - v/k$ where g, k are constants and v is the velocity. Prove that its velocity is $\frac{1}{2}gk$ after time $k \log 2$ and that the distance fallen in this time is

$$\tfrac{1}{2}k^2 g(\log 4 - 1).$$

74. The retardation of a body projected vertically upwards in a resisting medium is $g + \dfrac{v^2}{k}$ where g, k are constants and v is the velocity. If the initial velocity is u, prove that at height s the velocity is given by the equation $v^2 = (u^2 + gk)e^{-2s/k} - gk$ and find the greatest height attained.

EXERCISE XII c

[51, 52, 56, 60, 63, 65, 66, 69, 73, 74, 75]

Differentiate with respect to x the functions in Nos. 51-59:

51. $\log(e^x)$.
52. $e^{\log x}$.
53. $x^n e^x$.
54. $e^{\sqrt{x}}$.
55. xe^{x^2}.
56. $e^{ax} \cos bx$.
57. $\dfrac{\sin ax}{e^{bx}}$.
58. $\dfrac{e^x - 1}{e^x + 1}$.
59. $\dfrac{e^x - e^{-x}}{e^x + e^{-x}}$.

Integrate with respect to x the functions in Nos. 60-65:

60. (i) $\log(e^{2x})$; (ii) $e^{x \log 2}$; (iii) $e^{2 \log x}$.

61. $e^{\log(2x)}$.
62. $e^x e^{e^x}$.
63. $e^x(x^{-1} + \log x)$.
64. $e^{x^2}(1 + 2x^2)$.
65. $e^{2x} \log(1 + e^{-x})$.

66. Prove that $\displaystyle\int_0^1 \frac{e^x - e^{-x}}{e^x + e^{-x}} dx = \log \frac{e^2 + 1}{2e}$.

67. Prove that $\displaystyle\int_0^{\pi/3} e^{\sec x} \sec x \tan x \, dx = e(e - 1)$.

68. Prove that $\displaystyle\int_0^{\log 2} \frac{e^x \, dx}{1 + e^x} = \log \tfrac{3}{2}$.

69. Find n such that $y = e^{nx}$ satisfies the equation

$$\frac{d^2y}{dx^2} - \frac{dy}{dx} - 6y = 0.$$

70. Prove that $y = \mathsf{A}e^{2x} + \mathsf{B}e^{3x}$ satisfies the equation

$$\frac{d^2y}{dx^2} - 5\frac{dy}{dx} + 6y = 0$$

for all constant values of A and B.

71. Find the values of x which give turning points on the curve $y = e^{-x} \sin (x\sqrt{3})$.

72. Find the length of the subtangent at any point of the curve $y = ae^{bx}$, proving that it is constant.

73. Find the values of x which give points of inflexion on the curve $y = e^{-2x^2}$ and sketch the curve.

74. PN is the ordinate at any point P of the catenary $y = \frac{1}{2}c(e^{x/c} + e^{-x/c})$; prove that the length of the perpendicular from N to the tangent at P is constant.

75. The space-time relation for a particle moving along the x-axis is $x = e^{-at} \sin (at)$, $(a > 0)$; find the times when the particle is momentarily at rest. What is its greatest distance from the origin ?

76. The tension T at any point of a rope which is slipping round a rough cylindrical post, coefficient of friction μ, is given by $\frac{d\mathsf{T}}{d\psi} = \mu\mathsf{T}$, where ψ is the angle the tangent at the point makes with a fixed line in the plane of the rope. Find the relation between the tensions at the ends of the rope if it makes one complete turn round the post.

EXERCISE XII d

[51, 52, 55, 57, 58, 60, 63, 66, 67]

Differentiate with respect to x the functions in Nos. 51-56 :

51. x^2a^x, $(a > 0)$. 52. $\log_p \sin x$. 53. $(ex)^x$.

54. $(1 + x)^m \log (1 + x)$. 55. $x^n e^x \tan x$. 56. $e^{-x}(x + 1)^{x+(1/2)}$.

Integrate with respect to x the functions in Nos. 57-59 :

57. 10^{2x}. 58. $\log_p x$. 59. $e^x (\tan x + \sec^2 x)$.

60. If $\log y = \tan x$, prove that $\frac{d^2y}{dx^2} \Big/ \frac{dy}{dx} = (1 + \tan x)^2$.

61. If $y = \log(1 + \sin x)$, prove that $\dfrac{d^3y}{dx^3} + \dfrac{d^2y}{dx^2}\dfrac{dy}{dx} = 0.$

62. If $\log x + \log y = x/y$, prove that $x(x+y)dy = y(x-y)dx$.

63. If $y = xe^{-x}\cos x$, prove that

$$x^2\frac{d^2y}{dx^2} + 2x(x-1)\frac{dy}{dx} + 2(x^2 - x + 1)y = 0.$$

64. If $\mathsf{A}e^{-x^2} - \mathsf{B}e^{-(x-c)^2}$ has a stationary value at $x = x_1$, prove that that value is $\mathsf{A}c(c - x_1)^{-1}e^{-x_1^2}$.

65. If $x = a\log\tan\frac{1}{2}t$, $y = a\sin t$, find $\dfrac{dy}{dx}$ in terms of t.

66. If $y = a + x\log y$, $(a > 0)$, prove that when $x = 0$, $\dfrac{dy}{dx} = \log a$ and $a\dfrac{d^2y}{dx^2} = 2\log a$.

67. If $y = e^{\theta}$ and $\sin\theta = 1 - 2x^2$, $(0 < \theta < \frac{1}{2}\pi)$, prove that

$$(1 - x^2)\frac{d^2y}{dx^2} - x\frac{dy}{dx} - 4y = 0.$$

68. If $x = e^y$ and u is a function of x, prove that

$$x^2\frac{d^2u}{dx^2} = \frac{d^2u}{dy^2} - \frac{du}{dy}.$$

EXERCISE XII e

[52, 54, 56, 58, 60, 62-65, 70, 72]

Integrate with respect to x the functions in Nos. 51-62 :

51. $(\sqrt{x})\log x$. **52.** $x^n\log x$. **53.** $e^{2x}\sin 3x$.

54. $e^{2x}\cos 3x$. **55.** $\dfrac{\log(ax)}{\sqrt{x}}$. **56.** $x(\log x)^2$.

57. $e^x\sin 3x\sin 2x$. **58.** $e^x\cos 3x\cos 2x$. **59.** $(x\log x)^2$.

60. $\log\{x + \sqrt{(x^2 - 1)}\}$. **61.** $\log\{x + \sqrt{(a^2 + x^2)}\}$.

62. $(x\sin^{-1}x)/\sqrt{(1 - x^2)}$.

Find the values of the integrals in Nos. 63-68 :

63. $\int_1^e \frac{\log x}{x^3}\,dx.$

64. $\int_4^5 \log\{x+\sqrt{(x^2-16)}\}\,dx.$

65. $\int_0^1 x^2 \log(1+x)\,dx.$

66. $\int_0^1 x^4 \tan^{-1} x\,dx.$

67. $\int_0^2 x\log(1+\frac{1}{2}x^2)\,dx.$

68. $\int_0^1 x\sin^{-1} x\,dx.$

69. Prove that $\int \log\left(1+\frac{1}{x^2}\right)dx = x\log\left(1+\frac{1}{x^2}\right)+2\tan^{-1} x.$

70. Prove that $\int_0^\pi x(\pi-x)\sin x\,dx = 4$ by using the substitution $x=\frac{1}{2}\pi+y.$

71. Prove that $\int_2^e\left\{\frac{1}{\log x}-\frac{1}{(\log x)^2}\right\}dx = e-\frac{2}{\log 2}.$

72. Obtain a relation between $\int x^n e^x\,dx$ and $\int x^{n-1}e^x\,dx.$

73. Prove that $(n^2+1)\int e^x \sin^n x\,dx$

$$= n(n-1)\int e^x \sin^{n-2} x\,dx + (\sin x - n\cos x)e^x \sin^{n-1} x.$$

EXERCISE XIII b

[51, 52 (i), 53, 54, 55, 62-65, 68, 69, 71]

51. Prove that $\text{th}\,(\theta-\phi)=\frac{\text{th}\,\theta-\text{th}\,\phi}{1-\text{th}\,\theta\,\text{th}\,\phi}.$

52. Prove that

(i) $\text{sh}\,3\theta = 3\,\text{sh}\,\theta + 4\,\text{sh}^3\theta$; (ii) $\text{ch}\,3\theta = 4\,\text{ch}^3\theta - 3\,\text{ch}\,\theta.$

Write down the formula for $\text{th}\,3\theta$ in terms of $\text{th}\,\theta$.

Write down simple alternative forms of the expressions in Nos. 53-61 :

53. $1-\text{ch}\,2x.$

54. $\text{th}\,x+\coth x.$

55. $2\,\text{sh}\,x\,\text{sh}\,y.$

56. $\text{sh}\,\theta\,\text{ch}\,\phi.$

57. $\text{ch}\,\theta\,\text{ch}\,\phi.$

58. $\text{ch}\,2A-\text{ch}\,4A.$

59. $(\text{ch}\,x-\text{sh}\,x)^{-1}.$

60. $(\text{ch}\,x+\text{sh}\,x)^n.$

61. $\text{sh}\,x/(1-\text{ch}\,x).$

62. If $\operatorname{sh} x = 2$, find $\operatorname{th} x$.

63. If $\operatorname{th} x = \frac{1}{2}$, find $\operatorname{ch} x$.

64. If $\operatorname{ch} 2x = k$, find $\operatorname{sh} x$.

65. If $\operatorname{th} \frac{1}{2}x = t$, find $\operatorname{ch} x$.

66. If $\operatorname{th} x = \frac{1}{3}$, find e^{2x} and prove that $x = \frac{1}{2}\log 2$.

67. If $x = \sin a \operatorname{ch} b$ and $y = \cos a \operatorname{sh} b$, prove that

$$x^2 \operatorname{cosec}^2 a - y^2 \sec^2 a = 1$$

and find a relation between x, y, b.

68. Simplify $\operatorname{sh}(\log x)$ and $\operatorname{ch}(\log x)$.

69. If $\operatorname{sh} x = \tan\theta$ and $0 < \theta < \frac{1}{2}\pi$, express $\operatorname{ch} x$ and $\operatorname{th} x$ in terms of θ and prove that $x = \log(\sec\theta + \tan\theta)$. How are the results modified if $\frac{1}{2}\pi < \theta < \pi$? [Remember that $\operatorname{ch} x$ is always positive.]

70. If $\operatorname{sh} x \cos y = \sin a$ and $\operatorname{ch} x \sin y = \cos a$, prove that

$$\operatorname{sh}^2 x = \cos^2 y = |\sin a|.$$

71. Simplify $(1 + \operatorname{sh} x + \operatorname{ch} x)/(1 - \operatorname{sh} x - \operatorname{ch} x)$.

72. Sketch the graphs of $\operatorname{sech} x$ and $\operatorname{cosech} x$.

EXERCISE XIII c

[51-59, 63-67, 75, 78, 81]

Differentiate with respect to x the functions in Nos. 51-62:

51. $\operatorname{ch}^2 x$.　**52.** $\operatorname{sh} x \operatorname{ch} x$.　**53.** $\operatorname{cosech} x$.

54. $\operatorname{coth} 2x$.　**55.** $\operatorname{ch}^2 x + \operatorname{sh}^2 x$.　**56.** $2x + \operatorname{sh} 2x$.

57. $\log \operatorname{ch} x$.　**58.** $\tan^{-1}(\operatorname{coth} x)$.　**59.** $\log(\operatorname{ch} x + \operatorname{sh} x)$.

60. $\operatorname{sh}^n (ax)$.　**61.** $3 \operatorname{th} x - \operatorname{th}^3 x$.　**62.** $\sin x \operatorname{sh} x + \cos x \operatorname{ch} x$.

Integrate with respect to x the functions in Nos. 63-71:

63. $\operatorname{sh} 3x$.　**64.** $\operatorname{th} \frac{1}{2}x$.　**65.** $\operatorname{ch}^2 3x$.

66. $\operatorname{th}^2 2x$.　**67.** $\operatorname{cosech} 2x$.　**68.** $\log(\operatorname{ch} x - \operatorname{sh} x)$.

69. $\operatorname{ch} 3x \operatorname{ch} 2x$.　**70.** $\operatorname{sh} 3x \operatorname{sh} 5x$.　**71.** $\operatorname{ch} 3x \operatorname{sh} 2x$.

72. Find the values of

$$\text{(i)}\ \frac{d}{dx}(\operatorname{ch} x \cos x + \operatorname{sh} x \sin x); \quad \text{(ii)}\ \int \operatorname{sh} x \cos x\, dx.$$

73. Find the value of $\int_0^{\pi/2} 2 \operatorname{ch} x \sin x\, dx$.

74. If $y = \cos x \operatorname{ch} x$, prove that $\dfrac{d^2y}{dx^2} = -2\sin x \operatorname{sh} x$.

75. If $\tan\theta = e^{kt}$, prove that $\dfrac{d\theta}{dt} = \frac{1}{2}k \operatorname{sech}(kt)$.

76. Use the method of integration by parts to find the values of $\int x^2 \operatorname{sh} x\, dx$ and $\int x^3 \operatorname{ch} x\, dx$.

77. Differentiate $x^{\operatorname{sh} x}$ with respect to x.

78. Integrate $e^{ax} \operatorname{sh} bx$ with respect to x.

79. Integrate $e^{-x}(\operatorname{sech}^2 x - \operatorname{th} x)$ with respect to x.

80. Prove that $\displaystyle\int_0^{\log\sqrt{2}} \operatorname{sh} x \operatorname{sh} 2x \operatorname{sh} 3x\, dx = \tfrac{5}{192}$.

81. Prove that $\displaystyle\int_0^{\log 2} \frac{\operatorname{sh} x\, dx}{(1+\operatorname{ch} x)(2+\operatorname{ch} x)} = \log\tfrac{27}{26}$.

EXERCISE XIII d

[53, 54, 55, 56, 60]

(Miscellaneous Applications)

51. Find the condition that $a \operatorname{ch} x + b \operatorname{sh} x$ steadily increases as x increases.

52. Prove that the length of the normal at any point (x, y) of the curve $y = c \operatorname{ch}(x/c)$ is y^2/c.

53. Prove that $e^{-x} \operatorname{sh}(\frac{1}{2}x)$ has a maximum value when $x = \log 3$.

54. Prove that the length of the tangent at any point of the curve $x = a(t - \operatorname{th} t)$, $y = a \operatorname{sech} t$ is constant.

55. If the tangent at any point (x, y) of the curve $y = c \operatorname{ch}(x/c)$ makes an angle ψ with Ox, find $\tan\psi$ in terms of x and prove that, if ψ is acute, $y = c\sec\psi$ and $x = c\log(\sec\psi + \tan\psi)$.

56. Find the area bounded by the curve $y = \operatorname{sech} x$, the x-axis, and the ordinates $x = 0$ and $x = a$.

Does the area tend to a limit when a increases indefinitely?

57. Find the angle of intersection of the curves $y = 1 + \operatorname{ch} x$ and $y = e^{-x}$.

58. Prove that the parametric equations $x = -a\,\text{th}^2 u$, $y = a\,\text{sech}\,u$ represent part of a parabola and find the area bounded by the curve and the axes.

59. Sketch on the same diagram the curves represented parametrically by:

(i) $x = \cos 3t$, $y = \cos 4t$; (ii) $x = \text{ch}\,3t$, $y = \text{ch}\,4t$.

60. The area bounded by the straight lines $x = 0$, $x = k$, $(k > 0)$, $y = 1$, and the part of the graph of $y = \text{th}\,x$ for which x is positive rotates about the line $y = 1$. Find the volume of the solid so formed; also find its approximate value when k is large.

EXERCISE XIII e

[51, 52, 53, 56, 57, 58, 63, 64, 65]

Find from first principles the values in the form of logarithms of the expressions in Nos. 51-53:

51. $\text{ch}^{-1}\frac{5}{3}$. 52. $\text{coth}^{-1} 3$. 53. $\text{cosech}^{-1} 1$.

54. Discuss the behaviour of the function $\text{cosech}^{-1} x$ as x increases from $-\infty$ to $+\infty$.

55. Sketch the graphs of $\text{Sech}^{-1} x$ and $\text{coth}^{-1} x$.

56. Prove that $d(\text{sech}^{-1} x) = -dx/\{x\sqrt{(1 - x^2)}\}$, $(0 < x < 1)$.

Differentiate with respect to x the functions in Nos. 57-63:

57. $\text{ch}^{-1}(2x + 1)$. 58. $\text{th}^{-1}(\tan\frac{1}{2}x)$. 59. $\text{ch}^{-1}\sqrt{(1 + x^2)}$.

60. $\text{sh}^{-1}\dfrac{1 - x}{1 + x}$. 61. $\text{th}^{-1}\sqrt{\{\frac{1}{2}(1 + x)\}}$. 62. $\text{coth}^{-1}\dfrac{1 + x^2}{2x}$.

63. (i) $\text{sh}^{-1}\tan x$; (ii) $\text{ch}^{-1}\sec x$; (iii) $\text{th}^{-1}\sin x$.

64. Find the values of

$$\text{(i) } \frac{d}{dx}\{x\sqrt{(1 + x^2)} + \text{sh}^{-1} x\}; \quad \text{(ii) } \int_0^1 \sqrt{(1 + x^2)}\, dx.$$

65. Find the value of x for which $2\sec^{-1} x - \text{ch}^{-1} x$ is a maximum.

66. If $\sin\theta$ is positive, prove that

$$\text{sh}^{-1}\cot\theta = \log(\cot\theta + \text{cosec}\,\theta).$$

Express $\text{sh}^{-1}\cot\theta$ as a logarithm if $\sin\theta$ is negative.

67. Express $\text{th}^{-1} x + \text{th}^{-1} y$ in the form $\text{th}^{-1} z$.

EXERCISE XIII f

[51, 53, 55, 59, 60, 63, 64, 72, 74, 75, 80]

(Assume in this exercise that x is positive except in Nos. 54, 55.)

Find the values of the integrals in Nos. 51-56 :

51. $\int \frac{dx}{\sqrt{(4x^2+9)}}$. **52.** $\int \frac{dx}{\sqrt{(25x^2-1)}}$. **53.** $\int \frac{(x+1)\,dx}{\sqrt{(9-4x^2)}}$.

54. $\int_{-1}^{0} \frac{dx}{\sqrt{(x^2+4)}}$. **55.** $\int_{-3}^{-4} \frac{dx}{\sqrt{(x^2-4)}}$. **56.** $\int_{0}^{1} \frac{x\,dx}{\sqrt{(x^4+1)}}$.

Use hyperbolic substitutions to integrate with respect to x the functions in Nos. 57-62 :

57. $\frac{\sqrt{(x^2+1)}}{x^2}$. **58.** $\frac{1}{x\sqrt{(1+x^2)}}$. **59.** $\frac{1}{x\sqrt{(1-x^2)}}$.

60. $\frac{1}{\sqrt{(x^2-4x)}}$. **61.** $\sqrt{(x^2+6x)}$. **62.** $\sqrt{\{(x+a)^2+b^2\}}$.

Use integration by parts to find the values of the integrals in Nos. 63-68 :

63. $\int \operatorname{ch}^{-1}(\tfrac{1}{2}x)\,dx$. **64.** $\int x \operatorname{sh}^{-1} x\,dx$. **65.** $\int x^3 \operatorname{th}^{-1} x\,dx$.

66. $\int \coth^{-1}(2x)\,dx$. **67.** $\int x^2 \operatorname{ch}^{-1} x\,dx$. **68.** $\int \frac{x \operatorname{sh}^{-1} x}{\sqrt{(1+x^2)}}\,dx$.

69. Find the value of $\frac{d}{dx}\operatorname{sh}^{-1}\frac{x}{a}$ for (i) $a>0$, (ii) $a<0$.

70. Differentiate $\operatorname{ch}^{-1}(x/a)$ with respect to x, when $a>0$.

71. Verify that $\int \frac{dx}{a^2-x^2} = c + \frac{1}{a}\operatorname{th}^{-1}\frac{x}{a} = c + \frac{1}{2a}\log\frac{a+x}{a-x}$.

Is the result altered when a is changed to $-a$?

Differentiate with respect to x the functions in Nos. 72-76 :

72. $\operatorname{sech}^{-1}(x/a)$. **73.** $\operatorname{cosech}^{-1}(x/a)$. **74.** $\coth^{-1}(x/a)$.

75. $x\sqrt{(x^2-a^2)} - a^2 \operatorname{ch}^{-1}(x/a)$. **76.** $x\sqrt{(a^2+x^2)} + a^2 \operatorname{sh}^{-1}(x/a)$.

77. Prove that the curves $y=\operatorname{sh}^{-1} x$, $y=\operatorname{ch}^{-1}(2x)$ intersect where $x=\frac{1}{3}\sqrt{3}$ and find the area bounded by the x-axis and these curves.

78. Sketch the graphs of $\operatorname{ch}^{-1}(\sec x)$ and $\operatorname{sh}^{-1}(\tan x)$.

79. If $y = \text{th}^{-1}(\sin x)$, prove that $\frac{d^2y}{dx^2} = \frac{1}{2}\,\text{sh}\,2y$.

80. A particle starts from the point $(a, 0)$ and moves in the x-axis so that its velocity, v, at the point $(x, 0)$ is given by $v = k\sqrt{(x^2 - a^2)}$ where k is a positive constant. Find the time it takes to reach the point $(b, 0)$.

EXERCISE XIV a

[51, 54-59, 64, 65, 68-71, 74, 75, 76, 78-81]

Write down or obtain the integrals with respect to x of the following functions:

51. $\sqrt{x^3}$. 52. $\frac{1}{\sqrt{x^3}}$. 53. $\frac{1}{1+e^{-x}}$. 54. $\frac{1}{1+e^x}$.

55. $\sin nx$. 56. $\cot nx$. 57. $\text{sech}^2\, 2x$. 58. $\text{cosech}^2\, \frac{1}{2}x$.

59. $\frac{\sin x}{1 - \cos x}$. 60. $\frac{\text{sh}\, x}{1 - \text{ch}\, x}$. 61. $\frac{x}{4 + x^2}$. 62. $\frac{(\log x)^2}{x}$.

63. $\frac{\cot x}{\log \sin x}$. 64. $\text{cosec}\, \frac{1}{2}x$. 65. $\sec x \tan x$.

66. $\text{cosec}^n\, x \cot x$. 67. $\frac{1}{\sqrt{(16x^2 + 25)}}$. 68. $\frac{x}{\sqrt{(25 - 16x^2)}}$.

69. $\frac{1}{\sqrt{(25 - 16x^2)}}$. 70. $\frac{x + 1}{\sqrt{(16x^2 + 25)}}$. 71. $\coth \frac{1}{2}x$.

72. $\text{th}\,(2x + 3)$. 73. $\text{cosech}\, 2x$. 74. $\text{sech}\, \frac{1}{2}x$.

75. $\frac{1}{(2x+1)^2 + 1}$. 76. $\frac{1}{(2x+1)^2 - 1}$. 77. $\frac{1}{1 - (2x+3)^2}$.

78. $\frac{1}{\sqrt{\{(2x+1)^2 + 1\}}}$. 79. $\frac{1}{\sqrt{\{(2x+3)^2 - 1\}}}$. 80. $\frac{1}{\sqrt{\{1 - (2x-1)^2\}}}$.

81. $\frac{1}{x^2 + 2x + 2}$. 82. $\frac{1}{\sqrt{\{x^2 + 4x + 3\}}}$. 83. $\frac{1}{\sqrt{\{4x - x^2 - 3\}}}$.

84. $\int \frac{dx}{\sqrt{\{(x+k)^2 + a^2\}}}$. 85. $\int \frac{dx}{\sqrt{\{(x+k)^2 - a^2\}}}$. 86. $\int \frac{dx}{\sqrt{\{a^2 - (x+k)^2\}}}$.

87. $\int \frac{dx}{a^2 - (x+k)^2}$; $a > (x + k) > 0$. 88. $\int \frac{dx}{\sqrt{(a^2 - x^2)}}$; $a < 0$.

EXERCISE XIV b

[51, 52, 54, 56, 60-63, 66, 71, 72, 75, 76]

Integrate with respect to x the following functions :

51. $\dfrac{7}{8-9x}$; $(9x<8)$.
52. $\dfrac{7}{8-9x}$; $(9x>8)$.
53. $\dfrac{5}{6x-7}$.

54. $\dfrac{3x}{1-2x}$.
55. $\dfrac{x-1}{x+1}$.
56. $\dfrac{x^2-1}{2-x}$.

57. $\dfrac{ax^2+bx+c}{x+p}$.
58. $\dfrac{x^4}{x-1}$.
59. $\dfrac{x}{3x^2-5}$.

60. $\dfrac{2x}{x^4+1}$.
61. $\dfrac{x-1}{(x+1)^2}$.
62. $\dfrac{6}{x^2-9}$.

63. $\dfrac{7x-8}{(x+1)(4x-1)}$.
64. $\dfrac{x^2+1}{x^2-1}$.
65. $\dfrac{8x+9}{14x^2-41x+15}$.

66. $\dfrac{x^2-4x+2}{x^2-5x+6}$.
67. $\dfrac{(x-1)(x-4)}{(x-2)(x-3)}$.
68. $\dfrac{2x^2+1}{x(x-1)^2}$.

69. $\dfrac{x^4+x^2}{x^2-1}$.
70. $\dfrac{x^3-1}{x^2(x+1)}$.
71. $\dfrac{2x+1}{(x^2-1)(x+2)}$.

72. $\dfrac{x}{(x-1)^3}$.
73. $\dfrac{2x^2}{x^3-1}$.
74. $\dfrac{x}{(x-1)(x+3)(x+5)}$.

75. $\dfrac{3x^2-2x-3}{(x-1)(x-2)(x-3)}$.
76. $\dfrac{3x+1}{(x-1)(x^2-1)}$.

77. $\dfrac{x+1}{(x-3)(x-1)^3}$.
78. $\dfrac{x^2}{(x-a)(x-b)}$.

79. $\dfrac{px+q}{(x-a)(x-b)}$.
80. $\dfrac{x^4}{(x+a)(x+b)}$.

EXERCISE XIV c

[51, 54, 56, 58, 60, 63, 64, 68, 70, 73, 76, 78]

Integrate with respect to x the following functions :

51. $\dfrac{1}{4x^2-25}$.
52. $\dfrac{x-1}{4x^2-3}$.
53. $\dfrac{2x-2}{x^2+2x-1}$.

54. $\frac{x}{x^2-7}$. 55. $\frac{x}{x^2+7}$. 56. $\frac{2x+3}{x^2+7}$.

57. $\frac{x^2+2x}{x^2+2x+2}$. 58. $\frac{x+1}{(x-1)^2+5}$. 59. $\frac{x^4}{x^2+3}$.

60. $\frac{x}{x^4-1}$. 61. $\frac{1-2x}{1+x^3}$. 62. $\frac{1}{(x+3)^2-2}$.

63. $\frac{1}{x^2+8x+11}$. 64. $\frac{1}{x^2+8x+21}$. 65. $\frac{x+1}{x^2+8x+11}$.

66. $\frac{2x+3}{x^2+8x+21}$. 67. $\frac{1}{1-2x-x^2}$. 68. $\frac{x+1}{1-2x-x^2}$.

69. $\frac{3x+5}{1-2x-x^2}$. 70. $\frac{3+2x+x^2}{(1+x)(1+x^2)}$. 71. $\frac{2x-1-x^2}{(1+x)(1+x^2)}$.

72. $\frac{1}{x^3+x^2+x}$. 73. $\frac{1}{x^5+x^3}$. 74. $\frac{x^3}{(x-2)(x^2+1)}$.

75. $\frac{x^3+2x^2+x+1}{(x^2+1)(x^2+x+1)}$. 76. $\frac{x^2}{(x^2+1)(x^2+2)}$. 77. $\frac{x^4-2x+1}{x(x^2+1)^2}$.

78. $\frac{1}{(x^2+a^2)^2}$. 79. $\frac{x+b}{(x^2+a^2)^2}$. 80. $\frac{x+3}{(x^2+2x+3)^2}$.

EXERCISE XIV d

[51, 52, 53, 55-59, 61, 63, 65, 66, 68, 70, 72, 77, 84]

Integrate with respect to x the functions in Nos. 51-83 :

51. $\frac{1+\sqrt{x}}{x}$. 52. $\frac{1}{(2-x)\sqrt{x}}$. 53. $\frac{1}{x\sqrt{(1-x)}}$.

54. $\frac{1-\sqrt[3]{x}}{x}$. 55. $\frac{1}{x(1+\sqrt[3]{x})}$. 56. $\frac{1}{\sqrt{x}+\sqrt{(x+1)}}$.

57. $\frac{1}{\sqrt{(x^2-2x)}}$. 58. $\frac{1}{\sqrt{(2x-x^2)}}$. 59. $\frac{1}{\sqrt{(x^2-2x+2)}}$.

60. $\frac{x+2}{\sqrt{(x^2-1)}}$. 61. $\frac{x-1}{\sqrt{(x^2+2x)}}$. 62. $\frac{2x+3}{\sqrt{(x^2+3x+2)}}$.

63. $\frac{2x+3}{\sqrt{(x^2+2x+2)}}$. 64. $\frac{x+1}{\sqrt{(x^2+x+1)}}$. 65. $\frac{x+a}{\sqrt{(2ax-x^2)}}$.

66. $\sqrt{\dfrac{x-1}{x+1}}$.　67. $\dfrac{2x+1}{\sqrt{(1-x^2)}}$.　68. $\dfrac{x+2}{\sqrt{(x^2+1)}}$.

69. $\dfrac{1}{x\sqrt{(2x^2-x)}}$.　70. $\dfrac{1}{x\sqrt{(x^2+4x)}}$.　71. $\dfrac{1}{x\sqrt{(3x^2+4x+1)}}$.

72. $\dfrac{1}{(x+1)\sqrt{(x^2+1)}}$.　73. $\dfrac{3x+2}{(x-1)\sqrt{(x^2+4)}}$.　74. $\dfrac{1}{x\sqrt{(x^4-1)}}$.

75. $\dfrac{x^2}{\sqrt{(1-x^2)}}$.　76. $\dfrac{\sqrt{(x^2-1)}}{x^3}$.　77. $\dfrac{1}{x^3\sqrt{(x^2-1)}}$.

78. $\dfrac{(2x-1)(x-2)}{\sqrt{(1-x^2)}}$.　79. $\dfrac{1}{\sqrt{(x+1)}-\sqrt{(x-1)}}$.

80. $\dfrac{1}{x-\sqrt{(x^2-1)}}$.　81. $\dfrac{x^2+2a^2}{\sqrt{(x^2+a^2)}}$.

82. $\dfrac{2x+a+b}{\sqrt{(x^2+xa+a^2)}}$.　83. $\dfrac{x^2-1}{x\sqrt{(x^4+1)}}$.

84. Find the value of $\displaystyle\int\frac{dx}{\sqrt{\{(x-a)(b-x)\}}}$, where $a<x<b$, by the substitution $x=a\cos^2\theta+b\sin^2\theta$.

85. Find the value of $\int\sqrt{\{(x-a)(x-b)\}}dx$, where $x>a>b$, by the substitution $x=a\,\mathrm{ch}^2\,\theta-b\,\mathrm{sh}^2\,\theta$.

86. Find the value of $\displaystyle\int\sqrt{\frac{x-a}{b-x}}\,dx$, where $a<x<b$.

87. Prove that

$$\int\frac{dx}{(x-a)\sqrt{\{(x-a)(b-x)\}}}=\frac{2}{a-b}\sqrt{\frac{b-x}{x-a}},\text{ where } a<x<b.$$

EXERCISE XIV e

[51-54, 57, 60-64, 67, 77, 78]

Integrate with respect to x the functions in Nos. 51-74:

51. $\operatorname{cosec}^2 x\sec^2 x$.　52. $(1-\cos x)^{-1}$.　53. $\sec^4 x$.

54. $\sin x\tan x$.　55. $\cos x\operatorname{cosec}^2 x$.　56. $\operatorname{cosec}^6 x$.

57. $\dfrac{\sin x+\cos x}{\sin x-\cos x}$.　58. $\dfrac{1+\tan^2 x}{\sqrt{(1+\tan x)}}$.　59. $\dfrac{\cos x}{a+b\sin x}$.

60. $\dfrac{1}{8-17\cos x}$. 61. $\dfrac{2}{17-15\cos x}$. 62. $\dfrac{1}{25+24\sin x}$.

63. $\dfrac{\sin x}{3\cos x+2\sin x}$. 64. $\dfrac{1}{5\cos^2 x-1}$. 65. $\dfrac{1}{1-\cos x+\sin x}$.

66. $\dfrac{6+3\sin x+14\cos x}{3+4\sin x+5\cos x}$. 67. $\dfrac{\sin x\cos x}{a\cos^2 x-b\sin^2 x}$.

68. $\dfrac{\cos^2 x}{(1+\cos x)^2}$. 69. $\dfrac{(2\cos x+1)(2-\cos x)}{\cos^2 x}$.

70. $\dfrac{\sin x}{\cos 2x}$. 71. $\dfrac{\sin 2x}{(a+b\cos x)^2}$. 72. $\dfrac{\sin x\cos x}{\sin^4 x+\cos^4 x}$.

73. $\dfrac{1+\sin x}{5-4\cos^2 x}$. 74. $\dfrac{\sqrt{(a^2+b^2\cos^2 x)}}{\cos x}$.

In Nos. 75-80 find the values of the integrals by the substitutions indicated :

75. $\displaystyle\int\frac{dx}{\sqrt{(a^2+x^2)}}$; $x=a\cot\theta$. 76. $\displaystyle\int\frac{dx}{\sqrt{(x^2-a^2)}}$; $2x=z+\dfrac{a^2}{z}$.

77. $\displaystyle\int\frac{dx}{x\sqrt{(1-x^2)}}$; $x=\sin\theta$.

78. $\int\sqrt{(1+\sec x)}\,dx$; $\sin\frac{1}{2}x=z$.

79. $\displaystyle\int\frac{dx}{\sqrt{\{(1+\sin x)(2+\sin x)\}}}$; $1+\sin x=\dfrac{1}{z}$.

80. $\int\{\sqrt{(\tan x)}+\sqrt{(\cot x)}\}\,dx$; $\sin x-\cos x=z$.

EXERCISE XIV f

[51-54, 58, 60, 64]

Integrate with respect to x the functions in Nos. 51-59 :

51. $\sin^5 x$. 52. $\cos^2 3x$. 53. $\sin^8 x\cos^5 x$.

54. $\sin^4 x\cos^2 x$. 55. $\cos^6 x$. 56. $\sin^3 x\cos^n x$.

57. $\sin^3 x\sec^4 x$. 58. $\operatorname{cosec} x\sec^4 x$. 59. $\cos^2 x\sec 2x$.

Prove the results in Nos. 60-64 :

60. $\int \operatorname{sh}^3 x\operatorname{ch}^4 x\,dx=\frac{1}{7}\operatorname{ch}^7 x-\frac{1}{5}\operatorname{ch}^5 x$.

61. $\int \operatorname{sh}^2 x\operatorname{ch}^4 x\,dx=\frac{1}{48}\operatorname{sh}^3 2x+\frac{1}{64}\operatorname{sh} 4x-\frac{1}{16}x$.

62. $\int \frac{2\,dx}{x^3\sqrt{(1-x^2)}} = \log\{1-\sqrt{(1-x^2)}\} - \log x - x^{-2}\sqrt{(1-x^2)}$.

63. $\int \frac{2x^4\,dx}{(1+x^2)^2} = 3x - \frac{x^3}{1+x^2} - 3\tan^{-1} x$.

64. $\int \frac{2+x+x^3}{(1+x^2)^2}\,dx = \frac{x}{1+x^2} + \tan^{-1}x + \frac{1}{2}\log(1+x^2)$.

EXERCISE XIV g

[51, 52, 53, 55, 56, 57, 62, 64, 66]

51. If $u_n = \int x^n e^{ax}\,dx$, prove that $au_n + nu_{n-1} = x^n e^{ax}$.

52. Find a reduction formula for $\int (x^2+a^2)^{-n}\,dx$.

53. If $u_n = \int \cot^n x\,dx$, find a relation between u_n and u_{n-2}.

54. If $u_n = \int \sec^n x\,dx$, prove that

$$(n-1)u_n = \sin x \sec^{n-1} x + (n-2)u_{n-2}.$$

Use reduction formulae to find the integrals in Nos. 55-60:

55. $\int \sin^7\theta\,d\theta$. 56. $\int \cos^6\theta\,d\theta$. 57. $\int \cos^3\theta\sin^6\theta\,d\theta$.

58. $\int \cos^2\theta\sin^8\theta\,d\theta$. 59. $\int \operatorname{cosec}^7\theta\,d\theta$. 60. $\int \sec^8\theta\,d\theta$.

61. If $u_n = \int x^m(\log x)^n\,dx$, prove that

$$(m+1)u_n = x^{m+1}(\log x)^n - nu_{n-1}.$$

Hence find $\int x^{m-1}(\log x)^2\,dx$.

62. If $u_n = \int x^n\cos(ax)\,dx$ and $v_n = \int x^n\sin(ax)\,dx$, prove that

(i) $au_n = x^n\sin(ax) - nv_{n-1}$

(ii) $av_n = -x^n\cos(ax) + nu_{n-1}$.

Hence find a relation between u_n and u_{n-2}. Deduce the value of $\int x^4\cos 2x\,dx$.

Find reduction formulae for the integrals in Nos. 63-67:

63. $\int x^{2n}(1+x^2)^{-1}\,dx$. 64. $\int \operatorname{ch}^n x\,dx$. 65. $\int e^x\sin^n x\,dx$.

66. $\int x^n \operatorname{sh} x\,dx$. 67. $\int x\operatorname{cosec}^n x\,dx$.

68. Find a reduction formula for $\int \frac{dx}{(1-x^2)^n}$ and use it to evaluate $\int \frac{dx}{(1-x^2)^3}$. Verify by the substitution $x=(1+z)/(1-z)$.

69. If $u_n = \int \frac{x^n\,dx}{\sqrt{(x+a)}}$, prove that

$$(2n+1)u_n = 2x^n\sqrt{(x+a)} - 2nau_{n-1}.$$

70. If $u_n = \int \frac{x^n\,dx}{\sqrt{(ax^2+2bx+c)}}$, prove that

$$nau_n = x^{n-1}\sqrt{(ax^2+2bx+c)} - (2n-1)bu_{n-1} - (n-1)cu_{n-2}.$$

EXERCISE XV a

[51-59, 61, 62, 66, 70, 71, 73, 75, 81, 84, 86]

Find the values of the integrals in Nos. 51-74 :

51. $\int_0^{\pi/2} \sin^9\theta\,d\theta$. 52. $\int_0^{\pi/2} \cos^8\theta\,d\theta$. 53. $\int_0^{\pi/2} \sin^2\theta\cos^4\theta\,d\theta$.

54. $\int_0^{\pi} \sin^n\theta\,d\theta$. 55. $\int_0^{2\pi} \cos^n\theta\,d\theta$. 56. $\int_0^{\pi/2} \sin^4\tfrac{1}{2}\theta\,d\theta$.

57. $\int_0^{\pi/2} \sin^8\theta\cos^5\theta\,d\theta$. 58. $\int_0^{\pi/2} \cos^6\theta\sin^6\theta\,d\theta$.

59. $\int_0^{\pi} (1-\cos\theta)^5\,d\theta$. 60. $\int_0^{\pi/2} \sin^3 2\theta\cos^4\theta\,d\theta$.

61. $\int_0^1 x^2\sqrt{(1-x)}\,dx$. 62. $\int_0^a x^3\sqrt{(a^2-x^2)}\,dx$.

63. $\int_0^a x^4\sqrt{(a-x)}\,dx$. 64. $\int_0^a x(a-x)^n\,dx$.

65. $\int_0^{\pi/4} \sec^6\theta\,d\theta$. 66. $\int_{-\pi/2}^{+\pi/2} x\sin 2x\,dx$.

67. $\int_0^{\pi/2} \cos^3 x\sin 5x\,dx$. 68. $\int_0^{\pi/2} x\sin x\cos^3 x\,dx$.

69. $\int_0^{\pi/2} x(\tfrac{1}{2}\pi - x)\cos x\,dx$.

70. $\int_{-1}^{+1} (1+x)^{3/2}(1-x)^2\,dx$; put $x=\cos\theta$.

71. $\displaystyle\int_{-1}^{+1} \sqrt{\{(1+x)(1-x)^5\}}\,dx.$ 72. $\displaystyle\int_{1}^{2} (x-1)^{3/2}(2-x)^{5/2}\,dx.$

73. $\displaystyle\int_{0}^{\pi/4} \tan^7\theta\,d\theta.$ 74. $\displaystyle\int_{\pi/4}^{\pi/2} \cot^8\theta\,d\theta.$

75. Prove that $\displaystyle\int_0^1 x^m(1-x)^n\,dx = \frac{m!\,n!}{(m+n+1)!}.$

76. Prove that $\displaystyle\int_{-1}^{+1} (1+x)^m(1-x)^n\,dx = \frac{m!\,n!\,2^{m+n+1}}{(m+n+1)!}.$

77. Use the substitution $x=(1-y)/(1+y)$ to find the value of $\displaystyle\int_0^1 (1-x)^m x^n (1+x)^{-m-n-2}\,dx.$

78. Use the substitution $x = 2\,\mathrm{ch}^2\theta - \mathrm{sh}^2\theta$ to find the value of $\displaystyle\int_2^3 \sqrt{\{(x-1)(x-2)\}}\,dx.$

79. If $v_n = \displaystyle\int_0^{\log 2} \mathrm{th}^n x\,dx$, find the relation between v_n and v_{n-2}. Hence find the value of v_5.

80. Prove that $\displaystyle\int_0^{\log k} \frac{dx}{\sin^2\alpha + \mathrm{sh}^2 x} = 2\,\mathrm{cosec}\,2\alpha \tan^{-1}\frac{(k^2-1)\cos\alpha}{(k^2+1)\sin\alpha}.$

81. The curve $y^2=(b-x)(x-a)$, $b>a>0$, is rotated about $\mathsf{O}x$; prove that the volume generated is $\pi(b-a)^3/6$.

82. Prove that the area of the loop of $y^2(a+x)=x^2(a-x)$ is $a^2(2-\frac{1}{2}\pi)$ and that the volume generated by rotating the upper half of the loop about the x-axis is $2\pi a^3(\log 2 - \frac{2}{3})$.

83. **AB** and **CD** are perpendicular diameters of a circle, radius a. Prove that the mean value of the reciprocal of the distance of **A** from points on the semicircle **CBD** is $\dfrac{2}{a\pi}\log(1+\sqrt{2})$.

84. Prove that the centroid of the area contained by the loop of the curve $4y^2=(x-1)(x-3)^2$ is at a distance $1\frac{6}{7}$ from the origin.

85. Evaluate $\displaystyle\int_0^{\pi/2} \cos^3 x \sqrt[4]{\sin x}\,dx.$

86. Prove that $\displaystyle\int_0^{\pi} \frac{dx}{a^2\cos^2 x + b^2\sin^2 x} = \frac{\pi}{ab}$, where ab is positive.

87. Prove that $\displaystyle\int_0^{\pi/2} \sin\theta \tan^{-1}\sin\theta\,d\theta = \tfrac{1}{2}\pi(\sqrt{2}-1).$

EXERCISE XV b

[51, 52, 53, 55, 58, 59, 63, 65]

Discuss the infinite integrals in Nos. 51-62 and when possible find their values:

51. $\int_1^\infty \frac{dx}{x^k}$; $k>1$. 52. $\int_0^\infty \frac{x\,dx}{x^2+4}$. 53. $\int_{-\infty}^{-2} \frac{dx}{x^2-1}$.

54. $\int_2^\infty \frac{dx}{(x-1)\sqrt{x}}$. 55. $\int_0^\infty e^{-x}\,dx$. 56. $\int_0^\infty xe^{-x^2}\,dx$.

57. $\int_0^\infty \frac{x^2\,dx}{(x^2+1)(x^2+4)}$. 58. $\int_0^\infty \frac{x\,dx}{(x^2+1)(x^2+4)}$.

59. $\int_2^\infty \frac{dx}{x\log x}$. 60. $\int_1^\infty \frac{dx}{x(1+x^2)}$.

61. $\int_{-\infty}^{+\infty} \frac{dx}{(a^2+x^2)^2}$. 62. $\int_{\pi/2}^\infty \frac{\sin^2 x}{e^{2x}}\,dx$.

63. If $a>0$, prove that $\int_0^\infty xe^{-ax}\,dx = \frac{1}{a^2}$.

64. If $a>0$, prove that $\int_0^\infty e^{-ax}\cos bx\,dx = \frac{a}{a^2+b^2}$.

65. If $-\pi<\alpha<\pi$ and $\alpha\neq 0$, prove that $\int_0^\infty \frac{dx}{x^2+2x\cos\alpha+1} = \frac{\alpha}{\sin\alpha}$.

66. Prove that $\int_0^\infty \frac{dx}{a^2e^x+b^2e^{-x}} = \frac{1}{ab}\tan^{-1}\frac{b}{a}$, where $ab\neq 0$.

EXERCISE XV c

[51-54, 57, 58, 62, 64, 65, 67]

Discuss the infinite integrals in Nos. 51-62 and when possible find their values:

51. $\int_0^1 \frac{dx}{x^k}$; $0<k<1$. 52. $\int_0^a \frac{dx}{a-x}$.

53. $\displaystyle\int_0^a \frac{dx}{\sqrt{(a-x)}}$. 54. $\displaystyle\int_0^a \frac{dx}{\sqrt{(a^2-x^2)}}$. 55. $\displaystyle\int_{-1}^{+1} \frac{x^2\,dx}{\sqrt{(1-x)}}$.

56. $\displaystyle\int_0^1 \frac{\sqrt{x^3}\,dx}{\sqrt{(1-x)}}$. 57. $\displaystyle\int_{-\infty}^{+\infty} \frac{dx}{x^3}$. 58. $\displaystyle\int_0^1 \frac{dx}{\sqrt{(x-x^2)}}$.

59. $\displaystyle\int_0^1 \frac{dx}{(1+x)\sqrt{(1-x^2)}}$. 60. $\displaystyle\int_0^\pi \frac{\sin x\,dx}{\cos^2 x}$.

61. $\displaystyle\int_0^{\pi/2} \frac{\cos^3 x\,dx}{\sqrt[4]{\sin x}}$. 62. $\displaystyle\int_0^a \log x\,dx$.

63. If $-a<0<b$, prove that $\displaystyle\int_{-a}^b \frac{dx}{\sqrt[3]{x}} = \tfrac{3}{2}(b^{2/3}-a^{2/3})$.

64. If $n>-1$, prove that $\displaystyle\int_0^1 x^n \log x\,dx = -1/(n+1)^2$.

65. If $0<a<b$, prove that $\displaystyle\int_a^b \frac{dx}{x\sqrt{\{(b-x)(x-a)\}}} = \frac{\pi}{\sqrt{(ab)}}$.

66. If $0<\alpha<\pi$, prove that $\displaystyle\int_1^\infty \frac{dx}{(x+\cos\alpha)\sqrt{(x^2-1)}} = \frac{\alpha}{\sin\alpha}$.

67. If $u_n = \displaystyle\int_0^\infty \operatorname{sech}^n u\,du$, and $n>2$, prove that

$$(n-1)u_n = (n-2)u_{n-2}.$$

Hence find the values of u_8 and u_7.

68. If $-1<n<0$, prove that $\displaystyle\int_0^1 \frac{x^n+x^{-n-1}}{1+x}\,dx = \int_0^\infty \frac{x^n}{1+x}\,dx$.

69. If $u_n = \displaystyle\int_0^1 \frac{x^n e^x\,dx}{\sqrt{(1-x)}}$ and $n \geqslant 1$, prove that

$$2u_{n+1} + (2n-1)u_n - 2nu_{n-1} = 0.$$

70. Prove that if $0<\theta<\tfrac{1}{2}\pi$, $\theta \operatorname{cosec}\theta < 2$, and that the integrals $\displaystyle\int_0^1 \log\frac{1}{\theta}\,d\theta$ and $\displaystyle\int_0^{\pi/2} \log\operatorname{cosec}\theta\,d\theta$ exist.

71. Prove that $\displaystyle\int_0^1 \frac{1-x^2}{(1+x^2)^2}\,dx = \int_2^\infty \frac{1}{z^2}\,dz$. $\left(\text{Put } x+\dfrac{1}{x}=z.\right)$

Find the value of $\displaystyle\int_{-1}^{+1} \frac{1-x^2}{(1+x^2)^2}\,dx$. Is it equal to $\displaystyle\int_{-2}^{+2} \frac{1}{z^2}\,dz$?

EXERCISE XV d

[51, 53, 54, 55, 58, 60]

51. Sketch the curve $y^2(a-x)=x^3$ where $a>0$ and show that the area between the curve and the line $x=a$ is $\frac{3}{4}\pi a^2$.

52. Sketch the curve $y^2=c^2(x-a)/(b-x)$ where $0<a<b$ and prove that the area between the curve and the line $x=b$ is $\pi c(b-a)$.

53. Sketch the curve $y^2(a-x)=x^2(a+x)$ where $a>0$ and prove that the area between the line $x=a$ and the part of the curve for which x is positive is $\frac{1}{2}(\pi+4)a^2$. Also find the area of the loop.

54. Find the area between the y-axis, the line $y=b$, $(b>0)$, and the part of the graph of $y=b \operatorname{th} x$ for which x is positive.

55. Discuss the existence of a volume for the solid generated by revolving about $\mathsf{O}x$ the area between $y=0$, $x=0$, $x=1$, and the part of the curve $y=x^{-k}$ for which x is positive.

56. Answer the same question as in No. 55 for the curve $y=\sqrt{\{-\log(1-x)\}}$.

57. Prove that the volume of the solid generated by revolution about $\mathsf{O}y$ of the area between the x-axis, the negative y-axis, and the part of the curve $y=\log x$ for which $x<1$ is $\frac{1}{2}\pi$.

58. Prove that the volume of the solid generated by revolution about the line $y=1$ of the area between the y-axis, $y=1$, and the part of $y=\operatorname{th} x$ for which x is positive is $\pi(\log 4-1)$.

59. The centre of a variable circle of constant radius a is on $x'\mathsf{O}x$; P is the point of contact of the tangent from O to the circle. Prove that the area between the part of the locus of P for which $y>0$ and the line $y=a$ is $\frac{1}{2}\pi a^2$. Sketch the locus.

60. Find the centre of gravity of the area between the x-axis and the curve $y=1/(1+x^2)$.

61. In No. 60 discuss the existence of a centre of gravity for the half of the area for which x is positive.

62. Find the area between the *witch* $y^2x=a^2(a-x)$, $a>0$, and its asymptote $x=0$ and show that the centre of gravity of the area is $(\frac{1}{4}a, 0)$.

63. The area between the positive axes and the curve $y=a\sqrt[4]{\{(a-x)/x\}}$ where $a>0$ is rotated about Ox. Show that the volume of the solid generated is $\frac{1}{2}\pi^2a^3$ and that the centre of gravity of the volume is $(\frac{1}{4}a, 0)$.

EXERCISE XVe

[51-56, 60, 63, 64, 65, 67, 71]

Evaluate the integrals in Nos. 51-56:

51. $\int_0^{\pi} \cos^7 x\,dx.$ 52. $\int_0^{\pi} \cos^8 x\,dx.$ 53. $\int_{-\pi/2}^{+\pi/2} \sin^9 x\,dx.$

54. $\int_{\pi/2}^{3\pi/2} \sin^{10} x\,dx.$ 55. $\int_0^{\pi} (a+b\cos x)^3\,dx.$ 56. $\int_{-\pi}^{+\pi} x^2 \sin x\,dx.$

57. If m, n are positive integers, under what condition is $\int_0^{2\pi} \sin^m \frac{1}{2}x \cos^n \frac{1}{2}x\,dx$ zero ?

58. If m, n are positive integers, prove that

$$\int_0^1 x^n(1-x)^n\{x^{m-n}-(1-x)^{m-n}\}\,dx=0.$$

59. If $ab \neq 0$, find the value of $\int_0^{\pi} \frac{\sin x\,dx}{\sqrt{(a^2-2ab\cos x+b^2)}}.$

60. Prove that $u, \equiv \int_0^{\pi/4} \log(1+\tan\theta)\,d\theta, = \int_0^{\pi/4} \log\frac{2}{1+\tan\phi}\,d\phi$ by putting $\theta=\frac{1}{4}\pi-\phi$. Deduce that $u=\frac{1}{8}\pi\log 2$.

61. If $0<e<1$, prove that $\int_0^{\pi} \frac{d\theta}{(1+e\cos\theta)^2}=\frac{\pi}{(1-e^2)^{3/2}}.$

62. If $0<b<a$, prove that $\int_0^{\pi} \frac{\cos\theta}{(a+b\cos\theta)^2}\,d\theta=-\frac{\pi b}{(a^2-b^2)^{3/2}}.$

63. Illustrate geometrically the inequalities in (iii), p. 320. Hence use the integral $\int_n^{n+1} \frac{1}{x}\,dx$ where $n>0$ to show that $1/(n+1)<\log(1+1/n)<1/n$; also prove that

$$1+\log n>\left(1+\tfrac{1}{2}+\tfrac{1}{3}+\,.\,.\,.\,.\,.\,+\frac{1}{n}\right)>\log(n+1).$$

64. Prove that $\int_0^\pi x \cos^4 x\, dx = \int_0^\pi (\pi - x) \cos^4 x\, dx$, and hence find the value of each integral.

65. Prove that

$$\int_0^{\pi/2} (a \cos^2 x + b \sin^2 x)\, dx = \int_0^{\pi/2} (a \sin^2 x + b \cos^2 x)\, dx$$

and deduce from the sum of these integrals the value of each.

66. If m and n are positive integers, evaluate $\int_a^b (x-a)^m (b-x)^n\, dx$ by the substitution $x = a + (b-a)z$.

67. Prove that $\int_0^\infty \frac{\log x}{1+x^2}\, dx = 0$ by dividing the range of integration $(0, \infty)$ into the two parts $(0, 1)$ and $(1, \infty)$.

68. Find ξ such that $0 < \xi < t$ and $\int_0^t e^x\, dx = te^\xi$.

69. Prove that $0{\cdot}5 < \int_0^{1/2} \frac{dx}{\sqrt{(1-x^4)}} < \sin^{-1} \frac{1}{4} + \frac{1}{\sqrt{15}} < 0{\cdot}511.$

70. Prove that $\int_0^{1/2} \frac{dx}{\sqrt{(1-x^3)}}$ lies between $0{\cdot}5$ and $\frac{1}{6}\pi$.

71. Prove that $\int_0^{\pi/2} \sqrt{(\sin x)}\, dx$ lies between 1 and $\frac{1}{6}\pi\sqrt{(2\pi)}$.

72. Prove that $\int_0^{\pi/2} \sin 2x \log \tan x\, dx = 0$.

73. Prove that $\int_0^\infty \left(\frac{\log x}{1-x}\right)^2 dx = 2\int_1^\infty \left(\frac{\log x}{1-x}\right)^2 dx.$

EXERCISE XVI a

[51, 52, 53, 56, 59]

51. If the length s of an arc of the *tractrix*

$$x = a(u - \operatorname{th} u),\ y = a \operatorname{sech} u$$

is measured from the point given by $u = 0$, prove that $ds = a \operatorname{th} u\, du$ and that $ds/dy = -a/y$. Also find y in terms of s.

52. If the tangent at any point P on the curve $x = a(t - \frac{1}{3}t^3)$, $y = at^2$ cuts Ox at T, prove that the arc OP equals twice OT.

53. Prove that the length of the arc of the curve $x = a\cos t$, $y = \frac{1}{4}a\cos 2t$ from $t = 0$ to $t = \frac{1}{2}\pi$ is $\frac{1}{2}a\{\sqrt{2} + \log(1 + \sqrt{2})\}$.

54. Prove that, for the curve $y = \log(1 - x^2)$, $s = \log\left(\frac{1+x}{1-x}\right) - x$. From what point is s then measured?

55. Prove that the length of the arc of the curve $ay^2 = x^3$ from $x = 0$ to $x = c$ is $8a\{\sqrt{(1 + 9c/4a)^3} - 1\}/27$.

56. Prove that the length of the loop of the curve $3y^2 = x(1 - x)^2$ is $\frac{4}{3}\sqrt{3}$.

57. Prove that the whole length of the curve $8a^2y^2 = x^2(a^2 - x^2)$ is $\pi a\sqrt{2}$.

58. Prove that the arc of the curve

$$x = a\cos t - \tfrac{1}{3}(a - b)\cos^3 t,\ y = b\sin t + \tfrac{1}{3}(a - b)\sin^3 t$$

is given by $s = \frac{1}{2}(a + b)t - \frac{1}{4}(a - b)\sin 2t$.

59. Prove that the arc of the curve

$$x = (a + b)\cos\frac{b}{a}\theta + b\cos\frac{a+b}{a}\theta,\ y = (a + b)\sin\frac{b}{a}\theta + b\sin\frac{a+b}{a}\theta,$$

for $-\pi < \theta < +\pi$, is given by $as = 4b(a + b)\sin\frac{1}{2}\theta$, and find the intrinsic equation of this part of the curve.

60. Find a in terms of b, c if the length of the perimeter of the ellipse $x^2/a^2 + y^2/b^2 = 1$ is equal to that of one complete period of $y = c\sin(x/b)$.

EXERCISE XVI b

[51, 52, 53, 55, 56, 58]

51. For the curve $y = \log\coth\frac{1}{2}x$, find $\frac{ds}{dx}$ and express s in terms of x.

52. For the semi-cubical parabola $x = at^2$, $y = at^3$, prove that

when $y > 0$, $$\frac{ds}{dx} = \sqrt{\left(1 + \frac{9x}{4a}\right)}.$$

53. For the curve given by $ax = c^2 \cos^3 t$, $by = c^2 \sin^3 t$, $(c^2 = a^2 - b^2)$, prove that for $ab > 0$

$$\frac{ds}{dt} = \frac{3c^2 |\sin 2t|}{2ab\sqrt{2}} \sqrt{\{(a^2 + b^2) - (a^2 - b^2)\cos 2t\}}.$$

54. For the curve $y = e^x$, prove that $\dfrac{ds}{dy} = \dfrac{1}{y}\sqrt{(1 + y^2)}$.

55. If the intrinsic equation of a curve is $s = c\psi$, prove that the cartesian equation is $(x - \alpha)^2 + (y - \beta)^2 = c^2$.

56. Express dx and dy in terms of s, ds for the curve $s = c \sec \psi$. Also, if $x = y = 0$ when $s = c$, express x and y in terms of s and prove that $\{\pm y + c \sec^{-1}(e^{x/c})\}^2 + c^2 = c^2 e^{2x/c}$.

With the usual notation for a curve prove the results in Nos. 57-60:

57. If $y = \log \tan \frac{1}{2}\psi$, then $x + \operatorname{cosec} \psi$ is constant.

58. If $\sec \psi + \tan \psi = e^{s/a}$, then $dx = a\, d\psi$.

59. If $\sec x = e^y$, then $\operatorname{cosec}^2 x = (ds/dy)^2$.

60. If $x^2 = a^2 + s^2$, then $x\, dy = \pm a\, ds$.

61. If $y = c \operatorname{ch}(x/c)$, prove that $\dfrac{d^2x}{ds^2} = -\dfrac{1}{c} \operatorname{sh}(x/c) \operatorname{sech}^3(x/c)$ and $\dfrac{d^2y}{ds^2} = \dfrac{1}{c} \operatorname{sech}^3(x/c)$.

EXERCISE XVI c

[52, 53, 56, 57, 59, 64]

51. Express the equation $x^3 + y^3 = 3axy$ in polar coordinates.

52. Express the equation $l/r = \pm 1 + e \cos \theta$ in cartesian coordinates.

53. Prove that the parameter θ of the equations

$$x = a(2\cos\theta - \cos 2\theta - 1), \quad y = a(2\sin\theta - \sin 2\theta)$$

may be regarded as the polar θ, and find the polar equation of the curve.

54. Find what curve is represented by the polar equation $1/r = 1 - 2\cos\theta$, (i) when r is restricted to be positive, and (ii) when there is no such restriction.

55. Sketch the limaçon $r = 3 + 2\cos\theta$ and give an integral formula for its total length.

56. Sketch the curve $r = 2a\cos^3 \frac{1}{3}\theta$. Prove that, for this curve, $ds = 2a\cos^2 \frac{1}{3}\theta\, d\theta$ and show that the total length is $3\pi a$.

57. Prove that the length of the arc of the spiral $r\theta = a$ from $\theta = \frac{1}{2}$ to $\theta = 1$ is $a\{\sqrt{5} - \sqrt{2} + \log(2 + \sqrt{8}) - \log(1 + \sqrt{5})\}$.

58. For the curve $r^n \sin n\theta = a^n$, prove that $ds/d\theta = |r^{n+1}/a^n|$.

59. For the curve $r^n = a^n \cos n\theta$, prove that $a^{2n}\dfrac{d^2r}{ds^2} + nr^{2n-1} = 0$.

60. If b/a is small, prove that the perimeter of the limaçon $r = a + b\cos\theta$ is approximately $2\pi a(1 + b^2/4a^2)$.

61. Prove that the perimeter of the lemniscate $r^2 = a^2 \cos 2\theta$ is $4a\displaystyle\int_0^{\pi/4} \sqrt{(\sec 2\theta)}\, d\theta$ and use the substitution $\sqrt{2}\sin\theta = \sin\phi$ to reduce it to the form $2a\sqrt{2}\displaystyle\int_0^{\pi/2} \frac{d\phi}{\sqrt{(1 - \frac{1}{2}\sin^2\phi)}}$.

62. Show that $x = \dfrac{b\lambda^3}{1 + \lambda^4}$, $y = \dfrac{b\lambda}{1 + \lambda^4}$ are parametric equations of a lemniscate whose perimeter is $\displaystyle\int_0^1 \frac{4b\, d\lambda}{\sqrt{(1 + \lambda^4)}}$.

63. Prove that the length of the arc of $r = na\sin n\theta$ from $\theta = \beta$ to $\theta = \gamma$ is equal to that of the arc of the ellipse $x = a\cos t$, $y = an\sin t$ from $t = n\beta$ to $t = n\gamma$.

64. Prove that the cartesian coordinates of a point in the first quadrant on the lemniscate $r^2 = 2a^2 \cos 2\theta$ are given by $x = a\sqrt{(t + t^2)}$, $y = a\sqrt{(t - t^2)}$ where $t = \cos 2\theta$, and that the arc from $t = t_1$ to $t = t_2$, where $0 < t_1 < t_2 < 1$, is $\frac{1}{2}a\sqrt{2}\displaystyle\int_{t_1}^{t_2} \frac{dt}{\sqrt{(t - t^3)}}$.

Prove also that if $(t_1 + 1)(t_2 + 1) = 2$, the arc from $t = 0$ to $t = t_1$ equals that from $t = t_2$ to $t = 1$, and hence find the mid-point of the arc of the lemniscate in the first quadrant.

EXERCISE XVI d

[52, 53, 54, 58, 61]

Find the p, r equations of the curves in Nos. 51-56 :

51. $r = a\cos\theta$. **52.** $xy = c^2$. **53.** $r = ae^{\theta\cot\alpha}$.

54. $r^n = a^n \cos n\theta$. **55.** $r = a\,\text{cosec}\, n\theta$. **56.** $a = r\,\text{ch}\, m\theta$.

57. Express $\tan\phi$ in terms of θ for the curve $r^m\theta^n = k^m$.

58. Prove that the curves $r = a^2e^{\theta}$, $r = b^2e^{-\theta}$ intersect at the point $(ab, \log(b/a))$ where $ab > 0$ and find their angle of intersection.

59. Obtain the formula $\tan\phi = rd\theta/dr$ directly from $\tan(\psi - \theta)$.

60. Prove that the curves
$r^n\cos(n\theta - \alpha) = c^n \sin 2\alpha$, $r^n = 2c^n \sin(n\theta + \alpha)$ cut orthogonally.

61. Prove that three points exist on the curve $r^2 = a^2\cos^2\theta + b^2\sin^2\theta$ for which $\theta + \phi = \pi$ and $r\sin\theta > 0$ provided that $a^2 > 2b^2$ and interpret this geometrically. Sketch the curve when (i) $a = 2b$, (ii) $a = \frac{4}{3}b$.

EXERCISE XVI e

[51, 52, 54, 57, 59, 61, 62, 66, 67, 69, 70]

51. Find the polar equation of the pedal of $r = ae^{\theta\cot\alpha}$ with respect to the pole.

52. For the curve given by

$$x = a(2\cos t + \cos 2t),\ y = a(2\sin t - \sin 2t),$$

prove that $p = a\sin 3\psi$.

53. Prove that the p, ψ equation of the curve $r^n = a^n \sin n\theta$ is $p = a\sin^{m}(\psi/m)$ where $m = 1 + 1/n$ and $a > 0$.

54. Prove that $pds = r^2 d\theta$.

Find the p, r equations of the pedals with respect to the pole of the curves in Nos. 55-57 :

55. $r(1 + \cos\theta) = 2a$. **56.** $r = a\theta$. **57.** $r^n = a^n\cos n\theta$.

Find the polar equations of the pedals with respect to the pole of the curves in Nos. 58-60 :

58. $r^n \sin n\theta = a^n$, where n is an integer and a is positive.

59. $r^3 = a^2 p$. 60. $p = f(\psi)$.

61. Find a cartesian equation of the curve $p = a \sin^2 \frac{1}{2}\psi$.

62. Find the polar equation of the curve $pr = a^2$.

63. Express the cartesian coordinates of a point of the curve $s = \text{ch}^{-1} \sec \psi$ in terms of s.

64. Use the relation $s = \int \frac{r\, dr}{\sqrt{(r^2 - p^2)}}$ to find the length of the arc of the parabola $p^2 = ar$ from $r = a$ to $r = 2a$. (Cf. No. 72.)

65. Show that the length of the arc of the pedal curve of $r = f(\psi)$ with respect to the pole is given by $\int r\, d\psi$.

66. The curve $s = c \sin^n \psi$ touches Ox at O, and the tangent at any point P cuts Ox at T. Prove that the arc OP equals $(1 + 1/n)$TP.

67. If $\text{ch}\,(s/c) = e^{-y/c}$, prove that $\text{th}\,(s/c) = -\sin \psi$.

68. Show that the pedal of the curve $r = 2a \operatorname{cosec} 2\theta$ with respect to the pole has a cartesian equation

$$(x^2 + y^2)(x^2 + y^2 - a^2)^3 = 27a^4x^2y^2.$$

69. If a body moves in a known orbit under the action of a force towards the pole it can be proved that the force varies as $\frac{1}{p^3}\frac{dp}{dr}$. Find for what law of force the following orbits are described :

(i) the ellipse $b^2/p^2 = 2a/r - 1$ (focus as pole),
(ii) the ellipse $r^2 = a^2 + b^2 - a^2b^2/p^2$ (centre as pole),
(iii) the curve $r \cos n\theta = a$.

70. A body is attracted towards a fixed point O with a force which varies inversely as the square of the distance from O. Use the result stated in No. 69 to prove that its path is a conic.

71. Find the polar equations of the curves

(i) $p = a \operatorname{cosec} \psi$; (ii) $p = -a \operatorname{cosec} \psi$

where a is positive.

72. Prove that, subject to a certain restriction, the length of the arc of the curve $p=f(r)$ from $r=r_1$ to $r=r_2$ is equal to $\int_{r_1}^{r_2} \frac{r\,dr}{\sqrt{(r^2-p^2)}}$.

It can be proved that the p, r equation of the astroid $x=a\cos^3 t$, $y=a\sin^3 t$ is $3p^2+r^2=a^2$. Point out the mistake in the following argument :

For the points $t=0$ and $t=\frac{1}{2}\pi$, $r=a$;

$\therefore$ the arc from $t=0$ to $t=\frac{1}{2}\pi$ is $\int_a^a \frac{r\,dr}{\sqrt{\{r^2-\frac{1}{3}(a^2-r^2)\}}}$, that is, zero.

Use the formula to find the true length of this arc.

EXERCISE XVII a

[51, 53, 54, 55, 56, 57, 60, 62, 64]

51. For the *catenary of equal strength* given by

$$\sec\psi+\tan\psi=e^{s/c},$$

prove that the curvature varies as $\cos\psi$.

52. Find the intrinsic equation of the curve for which $\rho^2+s^2=a^2$.

53. If the radius of curvature of a curve at P varies as the length of the arc AP measured from a fixed point A on the curve, find the intrinsic equation.

54. For the parabola $cy=x^2$, prove that $2c^2\rho=(c^2+4x^2)^{3/2}$ and determine the sign of the root.

55. Find the curvature at the origin of the curve $y=x+2x^2+cx^3$.

56. Find ρ in terms of t for the curve $x=at$, $y=at^3$.

57. Find $|\kappa|$ for the ellipse $\frac{(x-a)^2}{a^2}+\frac{y^2}{b^2}=1$ at the origin.

58. Find $|\kappa|$ for the ellipse in No. 57 at the point (a, b) by transferring the origin to that point.

59. Find ρ in terms of t for the curve $x=3at^2$, $y=at(3-t^2)$.

60. Show that the curvature at the origin of $y=x^p$ is zero when $p>2$ and also when $0<p<\frac{1}{2}$.

61. Prove that $|\kappa|$ for $y = \frac{1}{3}x^3 - x$ is greatest at $x \simeq \pm 1{\cdot}07$.

62. Find $|\kappa|$ at the origin for each branch of the curve $x^2 + xy + y^3 = 0$.

63. For the tractrix $y = ae^{-s/a}$ prove that

$$a \sin \psi = -y \quad \text{and} \quad a \cot \psi = -\rho.$$

Also show that the length of the normal varies as the curvature.

64. Prove that the length of the chord of curvature parallel to **O**x of the curve $\sin\left(\frac{y}{a} - c\right) = be^{x/a}$ where a, b, c are constants is $2a$.

65. Find $|\rho|$ at $(0, -4a)$ for $x(x^2 + y^2) = a(x + y)(4a - x + y)$.

66. Prove that for any curve

$$\frac{d\kappa}{ds} = \frac{dx}{ds}\frac{d^3y}{ds^3} - \frac{dy}{ds}\frac{d^3x}{ds^3}.$$

EXERCISE XVII b

[51, 52, 53, 55, 60, 63, 65]

51. Find κ in terms of p, r for the curve $r^m p^n = k^{m+n}$.

52. Use the p, r equation $r^2 = a^2 + kp^2$, where $k > 0$, of an epicycloid to find the radius of curvature at any point in terms of r.

53. Prove that the curvature at any point of the reciprocal spiral $r = a/\theta$ is $p^3 r^{-4}$.

54. Find the curvature at any point of the spiral of Archimedes given by $r = a\theta$.

55. Find the curvature at the pole of $r = a \sin n\theta$.

56. Prove that the chord of curvature through the pole at any point of the cardioid $r = a(1 + \cos\theta)$ is $\frac{4}{3}r$.

57. Find the radii of curvature of the conic $r = l/(1 + e\cos\theta)$ at the points $\theta = 0$, $\theta = \frac{1}{2}\pi$, $\theta = \pi$.

58. What is the geometrical significance of the fact that $p + \dfrac{d^2p}{d\psi^2}$ is zero when $p = a\sin\psi$ or $p = a\cos\psi$?

59. Express in terms of ψ the radius of curvature at any point of the curve for which $\dfrac{dp}{d\psi} = e^{\psi} \cos \psi$.

60. Prove that for the curve $r = a(\sin n\theta)^{1/n}$, $(n+1)\rho = r \operatorname{cosec} n\theta$ and deduce the value of ρ in terms of θ for the curve $r = a \sin^3 \frac{1}{3}\theta$.

61. Find ρ in terms of θ for the curve $r = a \sec^2 \theta$.

62. If for the cardioid $r = a(1 + \cos \theta)$ the arc s is measured from the point $\theta = 0$, prove that $9\rho^2 + s^2 = 16a^2$ for $0 < \theta < \pi$.

63. If $r = a \sec t$ and $\theta = \tan t - t$ are polar parametric equations of a curve, prove that for this curve $\phi = t$ and $\rho = p$.

64. If C is the centre of curvature at any point P on the curve $p = a \sin n\psi$ and the line joining the pole O to C meets the tangent at P in S, prove that $OC = n^2 OS$.

65. If a curve is given by the equation $r = f(\theta)$ and if $u = 1/r$, $u_1 = du/d\theta$, and $u_2 = du_1/d\theta$, show that the formula for the curvature is $\kappa\sqrt{(u^2 + u_1{}^2)^3} = u^3(u + u_2)$ where s is measured so that it increases with θ.

66. Prove that, for any curve, $r = f(\theta)$,

$$\rho\left\{\frac{1}{r} - \frac{1}{r}\left(\frac{dr}{ds}\right)^2 - \frac{d^2r}{ds^2}\right\} = \left\{1 - \left(\frac{dr}{ds}\right)^2\right\}^{1/2},$$

where the sign of the root is that of r.

EXERCISE XVII c

[51, 53, 57, 58, 59, 61, 63, 65]

Find the envelopes of the systems in Nos. 51-60 where t, s are parameters :

51. $x \cos t + y \sin t = a \sec t$.

52. $x \cos^3 t + y \sin^3 t = a$.

53. $x \sec t + y \tan t = a$.

54. $ax \sec t - by \operatorname{cosec} t = c^2$.

55. $x \log t - yt^2 = c$.

56. $x\sqrt{(\cos t)} + y\sqrt{(\sin t)} = a$.

57. $y = t^2(x - 2a) - 2at^3$.

58. $(x - a \cos t)^2 + (y - a \sin t)^2 = b^2$.

59. $x/s + y/t = 1$; $s^2 + t^2 = a^2$.

60. $x^2/s^2 + y^2/t^2 = 1$; $\sqrt{s} + \sqrt{t} = \sqrt{a}$.

61. Find the envelope of normals of $y^2 = 4ax$.

62. Find the envelope of normals of $x = at^2$, $y = at^3$.

63. What are the envelopes, when t varies, of
(i) $(a_1t + a_2)x + (b_1t + b_2)y + (c_1t + c_2) = 0$,
(ii) $(a_1t^2 + 2a_2t + a_3)x + (b_1t^2 + 2b_2t + b_3)y + (c_1t^2 + 2c_2t + c_3) = 0$?

64. P is a variable point on a given curve and O is a fixed point. Prove that the circle on OP as diameter touches the pedal of the curve with respect to O.

65. Show that the envelope of a circle having its centre on $y^2 = 4ax$ and passing through the origin is $x(x^2 + y^2) + 2ay^2 = 0$.

EXERCISE XVII d

[51, 53, 54, 56, 57, 61, 62]

51. Find the coordinates of the centre of curvature at any point (x_1, y_1) of the curve $y = \log \operatorname{cosec} x$.

52. Find the coordinates of the centre of curvature of the rectangular hyperbola $x = ct$, $y = c/t$ at the point t_1.

53. Find the coordinates of the centre of curvature at any point of the cycloid $x = a(\theta + \sin\theta)$, $y = a(1 - \cos\theta)$.

54. Find the length of the portion of the evolute of the parabola $y^2 = 4ax$ which corresponds to the arc from $(0, 0)$ to $(a, 2a)$.

55. C is the centre of curvature at any point P of the equiangular spiral $r = ae^{\theta \cot \alpha}$; O is the pole. Prove that $\angle COP = \frac{1}{2}\pi$ and that the evolute is $r = be^{\theta \cot \alpha}$ where $b \tan \alpha = ae^{-\frac{1}{2}\pi \cot \alpha}$.

56. Find the radius of curvature of the envelope of $x \sec t + y \operatorname{cosec} t = a$ at an arbitrary point.

57. Find the p, ψ equation of the evolute of the epicycloid $p = c \cos k\psi$.

58. If C is the centre of curvature at any point P of the curve $p = a \operatorname{sh} n\psi$ and the line joining the pole O to C meets the tangent at P in S, prove that $OC = n^2 SO$.

59. If C is the centre of curvature corresponding to a point P, (p, ψ), of the curve $p = f(\psi)$ and O is the pole, prove that $OC^2 = \{f'(\psi)\}^2 + \{f''(\psi)\}^2$.

60. An epicycloid is represented by the equation $\rho^2/a^2 + s^2/b^2 = 1$, where s and ρ are the arc and radius of curvature. If s_1 and ρ_1 are the corresponding quantities for the evolute, prove that $\rho_1 = \rho\, d\rho/ds$ and deduce that the ρ, s equation of the evolute can be written $\rho_1{}^2/a^2 + s_1{}^2/b^2 = a^2/b^2$.

61. Prove that the p, r equation of the evolute of the curve $p = f(\psi)$ may be found by eliminating ψ from the equations

$$p_1 = f'(\psi) \quad \text{and} \quad r_1{}^2 - p_1{}^2 = \{f''(\psi)\}^2.$$

62. Prove that the p, r equation of the evolute of the curve $p = 4a \sin \frac{1}{2}\psi$ is $4(r^2 - a^2) = 3p^2$.

63. Given the p, r equation of a curve, show how to find the p, r equation of its evolute. Apply the method to the parabola $p^2 = ar$.

64. Prove that the curvature of the pedal curve of $r = f(p)$ with respect to the pole is $(2r^2 - p\rho)/r^3$.

EXERCISE XVIII a

[52, 54, 55, 58, 61, 62]

51. Find the area of the sector of $r = a + b\theta$ bounded by the radii $\theta = -1$, $\theta = +1$, $(a > b)$.

52. Sketch the conchoid $r = a \sec \theta + b$, $(ab > 0)$, and find the area bounded by $\theta = \frac{1}{4}\pi$, $\theta = 0$, $r = a \sec \theta$, and the curve.

Find the area of a loop in each of Nos. 53-56 :

53. $r = a \sin n\theta$.

54. $r = a \sin \theta\,(1 + \cos \theta)$.

55. $r^2 = a^2 \sin \theta$.

56. $r(\cos \theta + \sin \theta) = a \sin 2\theta$.

57. Prove that the area of the sector of $l = r(1 + \cos \theta)$ bounded by $\theta = 0$, $\theta = 2\alpha$ is $\frac{1}{4} l^2 \tan \alpha\,(1 + \frac{1}{3} \tan^2 \alpha)$.

58. Find the area bounded by the lines $\theta=0$, $\theta=\frac{1}{2}\pi$ and the parts of the spiral $r=a\theta$ given by $0<\theta<\frac{1}{2}\pi$ and $2\pi<\theta<\frac{5}{2}\pi$.

59. Answer the same questions as in No. 58 for the equiangular spiral $r=ae^{b\theta}$.

60. Sketch the curve $2\theta=r+r^{-1}$ and find the area bounded by the line $2\theta=\pi$ and the part of the curve given by $r_1<r<r_2$ where r_1, r_2 are the roots of the equation $r^2-\pi r+1=0$.

61. Prove that the area common to the ellipses

$$r^2(a^2\cos^2\theta+b^2\sin^2\theta)=1,\ r^2(b^2\cos^2\theta+a^2\sin^2\theta)=1$$

where $0<a<b$ is $4(ab)^{-1}\tan^{-1}(a/b)$.

62. Show that for the curve $r=a(2\cos\theta+\cos 3\theta)$ there are maximum values of r equal to $3a$ and $\frac{1}{9}a\sqrt{3}$ and that the areas of the loops are $\frac{1}{12}a^2(10\pi+9\sqrt{3})$ and $\frac{1}{24}a^2(5\pi-9\sqrt{3})$.

63. An area is formed of the circle $r=a$ and the portions of the four loops of $r=2a\sin 2\theta$ exterior to the circle. Show that the whole area of the figure is $\frac{1}{3}(5\pi+3\sqrt{3})a^2$.

EXERCISE XVIII b

[51, 53, 54, 57, 59, 61]

51. Sketch the curve $x=\cos^2 t$, $y=2\sin t$ and find the area enclosed by the curve and the y-axis.

52. Sketch the curve $x=a\cos^3 t$, $y=b\sin^3 t$ and find the area enclosed by it.

53. Find the area of a loop of the curve $x=at^2$, $y=bt(t^2-1)$.

54. Find the area of a loop of the curve

$$x=a\cos t,\ y=a\cos t\,(1+\sin t).$$

55. If $n-1$ is a positive integer, prove that the area enclosed by $x=n\cos t-\cos nt$, $y=n\sin t-\sin nt$ is $n(n+1)\pi$.

56. Prove that the area enclosed by

$$x=a\cos\theta+b\sin\theta+c,\ y=a_1\cos\theta+b_1\sin\theta+c_1$$

is $\pi\,|ab_1-a_1b|$.

57. Prove that the area of the sector of the curve

$$x = a(\cos t + t \sin t),\ y = a(\sin t - t \cos t)$$

bounded by the lines joining the origin to $t = 0$, $t = t_1$ is $\frac{1}{6}a^2 t_1^3$.

58. Prove that the area of the loop of

$$x = a \sin 3t \operatorname{cosec} t,\ y = a \sin 3t \sec t$$

is $3a^2\sqrt{3}$ and equals the area between the curve and its asymptote, $x = -a$.

59. Prove that the area of each loop of $x^6 + y^6 = a^2 x^2 y^2$ is $\frac{1}{12}\pi a^2$.

60. Prove that the area of the curve

$$x = a/(1 + t + t^2),\ y = bt/(1 + t + t^2) \text{ is } \tfrac{2}{9}\pi ab\sqrt{3}.$$

61. Sketch the curve $(x - y)^2 = y^2(1 - y^2)$ and prove that the area of a loop is $\frac{2}{3}$.

EXERCISE XVIII c

[51, 53, 54, 60, 61, 62]

Find the areas of the surfaces generated by revolution about Ox of the arcs in Nos. 51-60 :

51. $x = \cos t,\ y = 2 + \sin t$; $t = 0$ to $t = 2\pi$.

52. $x = a(\log \tan t + \cos 2t),\ y = a \sin 2t$; $t = 0$ to $t = \frac{1}{4}\pi$.

53. $y = 2 \operatorname{cosec} x$; $x = \frac{1}{3}\pi$ to $x = \frac{1}{2}\pi$.

54. $3ay^2 = x(a - x)^2$; $x = 0$ to $x = a$.

55. $y = \cos x$; $x = 0$ to $x = \frac{1}{2}\pi$.

56. $y = e^x$; $x = 0$ to $x = \log c$.

57. $3y = 2\sqrt{x^3}$; $x = 0$ to $x = 1$.

58. $y = \sqrt{(x^2 - c^2)}$; $x = c$ to $x = \frac{3}{2}c\sqrt{2}$.

59. $r(1 + \cos\theta) = l$; $\theta = 0$ to $\theta = \alpha$, $(0 < \alpha < \pi)$.

60. $r = 2a \sin\theta$; $\theta = 0$ to $\theta = \pi$.

61. Find the area of the surface generated by revolution about Oy of the arc of the curve $y = \frac{1}{8}x^2 - \log(\frac{1}{2}x)$ from $x = 1$ to $x = 2$.

62. Find the area of the surface generated by revolution about Oy of the arc of the ellipse $x = a\sqrt{(1 - y^2/b^2)}$ from $x = 0$ to $x = a$.

EXERCISE XVIII d

[51, 52, 54, 55, 58]

51. The ellipse $x^2/a^2 + y^2/b^2 = 1$ rotates about the tangent at $(a, 0)$. Find the volume of the solid generated.

52. Assuming the approximate formula $\pi\{a + b + \frac{1}{2}(\sqrt{a} - \sqrt{b})^2\}$ for the perimeter of the ellipse in No. 51, find the approximate area of the surface of the solid generated by the revolution of $y^2 = 1 - \frac{1}{16}x^2$ about the tangent $x = 4$.

53. If the perimeter of the ellipse in No. 52 is 17·1587, find the percentage error in the approximation given, and also in the approximation $\pi\{3(a + b) - \sqrt{(3a^2 + 10ab + 3b^2)}\}$.

54. Give the approximate volume of the solid generated by revolution about the initial line of the small sector bounded by an arc of the curve $r = f(\theta)$ and the radii $\theta = \alpha$, $\theta = \alpha + \delta\alpha$.

55. Apply the result of No. 54 to find the volume of the solid generated by revolution about the initial line of the upper half of a loop of the lemniscate $r^2 = a^2 \cos 2\theta$.

56. Find the volume of the solid generated by revolving a loop of the curve $r^2 = a^2 \sin 2\theta$ about the initial line.

57. The area bounded by the lines $y = 0$, $x = a$, and the part of the parabola $x = at^2$, $y = 2at$ given by values of t from -1 to 0, revolves about the line $x = a$. Find the volume of the cup-shaped solid so formed and the area of its curved surface.

58. A solid is formed by the revolution of the ellipse $x^2/p^2 + y^2/q^2 = 1$ about the line $x = c$; p and q vary so that when the plane has turned through an angle θ, $p = a(1 + \sin\theta)$, $q = b(1 + \cos\theta)$, where a, b, c are constant and $c > 2a$. Find the volume of the solid.

EXERCISE XVIII e

[51, 52, 54, 55, 57, 59, 61, 63]

Evaluate the integrals in Nos. 51-58 :

51. $\int_0^1 \int_0^2 x^2\, dx\, dy.$ **52.** $\int_0^a dx \int_0^b (x + y)\, dy.$

53. $\int_0^c \int_0^a r^2 \sin\theta \, dr \, d\theta.$ 54. $\int_0^a \int_0^{c \sin\theta} r \cos\theta \, d\theta \, dr.$

55. $\int_0^a dx \int_0^b dy \int_0^c xy \, dz.$ 56. $\int_0^1 dx \int_0^1 dy \int_0^1 xy^2z^3 \, dz.$

57. (i) $\int_0^1 dx \int_0^x dy \int_0^{xy} yz \, dz$; (ii) $\int_0^a dr \int_0^\pi d\theta \int_0^{2\pi} r^3 \sin\theta \, d\phi.$

58. $\int_0^a \int_0^{\pi/2} \int_0^{\pi/2} \int_0^{\pi/2} r^2 \sin^2\theta \sin\phi \, dr \, d\theta \, d\phi \, d\psi.$

59. O is the centre of a rectangular plate and A is a fixed point in the plate. The thickness at any point P is λAP^2 where λ is constant. If the lengths of the sides are $2a$, $2b$, prove that the volume is $4\lambda ab(a^2 + b^2 + 3OA^2)/3$.

60. The density of a lamina in the shape of a cardioid $r = a(1 + \cos\theta)$ at any point is λr^2 where λ is constant. Find the mass of the lamina.

61. The density of a semicircular lamina, radius a, varies as the distance from the bounding diameter. Prove that the centre of gravity is at a distance $\frac{3}{16}\pi a$ from the centre.

62. The base of a right cylinder is one loop of the lemniscate $r^2 = a^2 \cos 2\theta$. Prove that the volume of the cylinder which lies inside the sphere, centre the pole and radius a, is $(3\pi + 20 - 16\sqrt{2})a^3/18$.

63. Find the volume which is interior to each of two cylinders of radius a whose axes intersect at right angles.

64. Find the volume which is interior to each of three cylinders of radius a whose axes are three mutually perpendicular intersecting lines.

EXERCISE XIX a

[52, 56, 58, 63, 65, 67, 68, 70]

Find f_x, f_y for the functions $f(x, y)$ in Nos. 51-56 :

51. x^py^q. 52. $xy/(x + y)$. 53. $e^{ax} \sin by$.

54. $\dfrac{x^2-y^2}{x^2+y^2}$. 55. $\operatorname{sh}^{-1}\dfrac{x}{y}$. 56. $\log\dfrac{x+\sqrt{(x^2+y^2)}}{y}$.

Find $f_{xx}, f_{xy}, f_{yx}, f_{yy}$ for the functions $f(x, y)$ in Nos. 57-62 :

57. $x^n e^y$. 58. $\log(x^2y^2+xy^2)$. 59. $a\sin(bx+cy)$.

60. $\log(x^y)$. 61. $e^{x^2+xy+y^2}$. 62. $\operatorname{sh} x \operatorname{ch} y$.

63. Find $\dfrac{\partial^2 u}{\partial x \partial y}$ if $u=x^2\tan^{-1}(y/x)-y^2\tan^{-1}(x/y)$.

64. If $z^2=(x+b)^2/(a^2-y^2)$, prove that $yz\left(\dfrac{\partial z}{\partial x}\right)^2=\dfrac{\partial z}{\partial y}$.

65. If $V^2=x^2+y^2+z^2$, prove that $\dfrac{\partial^2 V}{\partial x^2}+\dfrac{\partial^2 V}{\partial y^2}+\dfrac{\partial^2 V}{\partial z^2}=\dfrac{2}{V}$.

66. If $z=f(x^2+y^2)$, prove that $yz_x=xz_y$.

67. If $uz=e^{-(x^2+y^2)/4z}$, prove that $\dfrac{\partial^2 u}{\partial x^2}+\dfrac{\partial^2 u}{\partial y^2}=\dfrac{\partial u}{\partial z}$.

68. Prove that, for a suitable numerical value of n, $\theta=t^n e^{-r^2/4t}$ is a solution of $\dfrac{\partial}{\partial r}\left(r^2\dfrac{\partial \theta}{\partial r}\right)=r^2\dfrac{\partial \theta}{\partial t}$.

69. Verify that $1/z=\sqrt{(1-2ax+a^2)}$ satisfies the equation

$$\frac{\partial}{\partial x}\left\{(1-x^2)\frac{\partial z}{\partial x}\right\}+\frac{\partial}{\partial a}\left(a^2\frac{\partial z}{\partial a}\right)=0.$$

70. If $z=y^2f(x/y)$, prove that

$$x^2\frac{\partial^2 z}{\partial x^2}+2xy\frac{\partial^2 z}{\partial x \partial y}+y^2\frac{\partial^2 z}{\partial y^2}=2z.$$

71. If $z^2=(y^2-nx)^3$, prove that

$$\frac{\partial^2 z}{\partial x^2}\frac{\partial^2 z}{\partial y^2}-\left(\frac{\partial^2 z}{\partial x \partial y}\right)^2=\tfrac{9}{4}n^2.$$

72. If $z=y(3x-y^2)+(y^2-2x)^{3/2}$, prove that

$$\frac{\partial^2 z}{\partial x^2}\frac{\partial^2 z}{\partial y^2}=\left(\frac{\partial^2 z}{\partial x \partial y}\right)^2.$$

EXERCISE XIX b

[52, 54, 58, 60, 63, 66, 68]

51. If $z = x^p y^q$ where $x = a \operatorname{sh} u$, $y = a \operatorname{ch} u$, prove that

$$\frac{dz}{du} = \frac{z}{xy}(qx^2 + py^2).$$

52. If $\mathsf{V} = x^p y^q$ where $x = a(t - \sin t)$, $y = a(1 - \cos t)$, prove that

$$\frac{d\mathsf{V}}{dt} = \frac{\mathsf{V}}{xy}(py^2 - qx^2 + aqxt).$$

53. If $\mathsf{V} = \sin^p x \cos^q y$, prove that if $0 < x, y < \frac{1}{2}\pi$,

$$\frac{\delta \mathsf{V}}{\mathsf{V}} \simeq p \cot x \, \delta x - q \tan y \, \delta y.$$

Find $\dfrac{dy}{dx}$ at any point (x, y) of the curves in Nos. 54-56 :

54. $x^3 + y^3 = 3axy$. **55.** $\log y = y \log x$. **56.** $(x + y)^{p+q} = x^p y^q$.

Find the equation of the tangent at any point (x_1, y_1) to the curves in Nos. 57-59 :

57. $ax^2 + 2bxy + cy^2 = 1$. **58.** $\sqrt{(ax)} + \sqrt{(by)} = 1$. **59.** $x^p y^q = 1$.

60. Find the equation of the normal at (x_1, y_1) to the curve $f(x, y) = 0$.

61. Find $\dfrac{d^2y}{dx^2}$ at the point (x, y) of the curve

(i) $x^3 + y^3 = 3axy$; (ii) $x \log y = c$.

62. Prove that the curves $\sin x \operatorname{ch} y = a$, $\cos x \operatorname{sh} y = b$ cut at right angles.

63. If the angle A of a triangle is expressed in terms of the sides, prove that $bc \sin \mathsf{A} \, \delta \mathsf{A} \simeq a(\delta a - \delta b \cos \mathsf{C} - \delta c \cos \mathsf{B})$, if $\mathsf{B}, \mathsf{C} \neq \frac{1}{2}\pi$.

64. If the side c of a triangle is expressed in terms of a, b, and the circumradius R, prove that $\dfrac{\partial c}{\partial a} = -\cos \mathsf{C} \sec \mathsf{A}$ and find $\dfrac{\partial c}{\partial \mathsf{R}}$.

65. The radius R of the circumcircle of the triangle ABC is expressed in terms of a, b, C; prove that $\dfrac{\partial R}{\partial C} = \dfrac{R \cos A \cos B}{\sin C}$ and $\dfrac{\partial R}{\partial a} = \dfrac{R \cos B}{c}$. Write down an approximation for $\delta R/R$ due to small errors δa, δb, δC in a, b, C, assuming A, $B \neq \frac{1}{2}\pi$.

66. If $V = x^y$ where x and y are functions of t, find a relation between dV, dx, and dy.

67. Find the gradient at the point $(r\cos\theta, r\sin\theta)$ of the curve $x^2 \tan^{-1}(y/x) - y^2 \tan^{-1}(x/y) = c$.

68. Prove that the angle of intersection of the curves $f(x, y) = 0$, $g(x, y) = 0$ is $\tan^{-1}\{(f_x g_y - f_y g_x)/(f_x g_x + f_y g_y)\}$. What is the condition for the curves to cut at right angles ?

69. Use the condition obtained in No. 68 to prove that the curves $x^2 + y^2 = e^{2x}$ and $y = x \tan y$ cut at right angles.

70. Prove that at any point of the curve $f(x, y) = 0$,

$$\frac{d^2y}{dx^2} = -\{f_y^2 f_{xx} - 2f_x f_y f_{xy} + f_x^2 f_{yy}\}/f_y^3$$

and that the curvature is given by

$$\{f_y^2 f_{xx} - 2f_x f_y f_{xy} + f_x^2 f_{yy}\}/(f_x^2 + f_y^2)^{3/2}.$$

EXERCISE XIX c

[51, 53, 54, 55, 56, 59, 62, 63, 65, 66]

51. If $V = f(x, y)$ and $x = r\cos\theta$, $y = r\sin\theta$, express $\left(\dfrac{\partial V}{\partial x}\right)^2 + \left(\dfrac{\partial V}{\partial y}\right)^2$ in terms of r, θ.

52. If $x = r\cos\theta$, $y = r\sin\theta$, express $\left\{1 + \left(\dfrac{dy}{dx}\right)^2\right\}^{3/2} \div \dfrac{d^2y}{dx^2}$ in terms of r, θ.

53. Express $\dfrac{d^3y}{dx^3}$ in terms of $\dfrac{dx}{dy}, \dfrac{d^2x}{dy^2}, \dfrac{d^3x}{dy^3}$.

54. If $V=f(x, y)$ and $x=\xi \sin \alpha+\eta \cos \alpha$, $y=\xi \cos \alpha-\eta \sin \alpha$ where α is constant, express $\left(\frac{\partial V}{\partial x}\right)^2+\left(\frac{\partial V}{\partial y}\right)^2$ in terms of ξ, η.

55. With the data of No. 54 express $\frac{\partial^2 V}{\partial x^2}+\frac{\partial^2 V}{\partial y^2}$ in terms of ξ, η.

56. If $u=f_1(x, y)$, $v=f_2(x, y)$ and $\phi(u, v)=0$ for all values of x and y, prove that $\frac{\partial u}{\partial x} \frac{\partial v}{\partial y}=\frac{\partial u}{\partial y} \frac{\partial v}{\partial x}$.

57. If u, v are functions of r, θ such that $r\frac{\partial u}{\partial r}=-\frac{\partial v}{\partial \theta}$ and $\frac{\partial u}{\partial \theta}=r\frac{\partial v}{\partial r}$ prove that $\frac{\partial^2 u}{\partial r^2}+\frac{1}{r} \frac{\partial u}{\partial r}+\frac{1}{r^2} \frac{\partial^2 u}{\partial \theta^2}=0$.

58. If $z=f(x, y)$ is transformed into $Z=F(X, Y)$ by the substitutions $x=aX+bY+Z$, $y=a_1X+b_1Y+Z$, $z=X+Y-Z$, prove that

$$\frac{\partial Z}{\partial X}=\frac{1-ap-a_1q}{1+p+q}, \quad \frac{\partial Z}{\partial Y}=\frac{1-bp-b_1q}{1+p+q}$$

where $p=\frac{\partial z}{\partial x}$, $q=\frac{\partial z}{\partial y}$.

59. If $V=f(x, y)$ and $x=r(\sec \theta+\tan \theta)$, $y=r(\sec \theta-\tan \theta)$, prove that

$$4\frac{\partial V}{\partial x} \frac{\partial V}{\partial y}=\left(\frac{\partial V}{\partial r}\right)^2-\frac{1}{r^2} \cos^2 \theta\left(\frac{\partial V}{\partial \theta}\right)^2.$$

60. If $z=f(x, y)$ and if the functions u, v are defined by $ux+vy=1$, $vx+uy=uv$, prove that

$$(u^2-v^2)^2\frac{\partial z}{\partial u}=(v^2-1)\left\{(u^2+v^2)\frac{\partial z}{\partial x}-2uv\frac{\partial z}{\partial y}\right\}.$$

61. In No. 60 write down an expression for $\frac{\partial z}{\partial v}$ and prove that

$$\frac{\partial z}{\partial x}=\frac{u^2+v^2}{v^2-1} \frac{\partial z}{\partial u}+\frac{2uv}{u^2-1} \frac{\partial z}{\partial r}.$$

62. If $z=f(x, y)$ and $x+y=\log (u+v)$, $x-y=\log (u-v)$, prove that

$$\frac{\partial^2 z}{\partial x^2}-\frac{\partial^2 z}{\partial y^2}=(u^2-v^2)\left(\frac{\partial^2 z}{\partial u^2}-\frac{\partial^2 z}{\partial v^2}\right).$$

63. If $V = f(x, y)$ and if $x \cos\theta - y \sin\theta = x \sin\theta + y \cos\theta = r$, prove that

$$\frac{\partial^2 V}{\partial x^2} + \frac{\partial^2 V}{\partial y^2} = \frac{1}{2}\left(\frac{\partial^2 V}{\partial r^2} + \frac{1}{r}\frac{\partial V}{\partial r} + \frac{1}{r^2}\frac{\partial^2 V}{\partial \theta^2}\right).$$

64. If $z = f(x, y)$ and $u = x + y$, $v = y(x + y)$, express $\frac{\partial^2 z}{\partial x^2} + \frac{\partial^2 z}{\partial y^2}$ in terms of u, v.

65. Find the greatest value of xy^2z^3 if $x + y + z = 6$.

66. Find the greatest value of $\sin A \sin B \sin C$ if $A + B + C = \pi$.

EXERCISE XX a

[52, 53, 55, 56, 59]

51. If $y = a \log x$, prove that $y\frac{dx}{dy} = x \log x$.

52. If $x^2 + y^2 = c^2$, prove that $\frac{dy}{dx} = -\frac{x}{y}$ and interpret the result geometrically.

53. If $x = ae^{nt} + be^{-nt}$, prove that $\frac{d^2x}{dt^2} = n^2t$ and interpret the result kinematically.

54. Write down the differential equation of all circles of unit radius.

55. Eliminate a, b from the relation

(i) $y = ax^2 + bx$; (ii) $y = a \log x + b$.

56. If $y = (a + bx)e^{nx}$, prove that

$$\frac{d^2y}{dx^2} - 2n\frac{dy}{dx} + n^2y = 0.$$

57. Obtain the differential equation of the third order satisfied by $y = (a + bx + cx^2)e^{nx}$.

58. If $xy = ae^{nx} + be^{-nx}$, prove that

$$\frac{d^2y}{dx^2} + \frac{2}{x}\frac{dy}{dx} - n^2y = 0.$$

59. If $y=a\cos(\log x)+b\sin(\log x)$, prove that

$$x^2\frac{d^2y}{dx^2}+x\frac{dy}{dx}+y=0.$$

60. Prove that the differential equation of all circles is

$$\left\{1+\left(\frac{dy}{dx}\right)^2\right\}\frac{d^3y}{dx^3}=3\frac{dy}{dx}\left(\frac{d^2y}{dx^2}\right)^2.$$

EXERCISE XX b

[51-54, 57, 59, 60, 62, 65, 66, 67, 74, 75, 79]

Solve the differential equations in Nos. 51-65 :

51. $\frac{dy}{dx}=ax^n$. **52.** $\frac{dy}{dx}=ay^n$. **53.** $\frac{d\theta}{dr}=ce^{-n\theta}$.

54. $x^2y\frac{dy}{dx}=k$. **55.** $x\frac{dy}{dx}=ny$. **56.** $\left(\frac{dy}{dx}\right)^2=y$.

57. $\frac{dy}{dx}=\tan x\tan y$. **58.** $\frac{dy}{dx}=\sec^2 y$. **59.** $xy^2\frac{dy}{dx}=1+y^3$.

60. $(a^2-x^2)\frac{dy}{dx}=y$. **61.** $\frac{dy}{dx}=(x+a)(y+b)$.

62. $\frac{dy}{dx}+y\log x=0$. **63.** $x^2\frac{dy}{dx}=y(x+y)$.

64. $\frac{dy}{dx}-\frac{y}{x}=\tan\frac{y}{x}$. **65.** $y^2\frac{dx}{dy}=x^2+xy$.

66. Find the curves for which the subtangent is constant.

67. Find the curves $r=f(\theta)$ for which $\phi=2\theta$.

68. Find the curves $r=f(\theta)$ for which OT is constant where T is the point where the tangent at P meets the perpendicular at O to OP.

69. Find the curves for which $(1+y^2)^{3/2}=\frac{dy}{dx}$.

70. Find the curves for which $(1+p^2)^{3/2}=\frac{dp}{dx}$ where $p=\frac{dy}{dx}$. Explain the connection between this result and the formula on p. 346.

71. What are the curves for which $\left\{1+\left(\frac{dy}{dx}\right)^2\right\}^{3/2}=c\frac{d^2y}{dx^2}$?

72. Use the substitutions $x-y+1=u$, $x+y-3=v$ to solve $(x+y-3)dy=(x-y+1)dx$.

73. Solve the equation $\frac{dy}{dx}\left(\frac{dy}{dx}+x-y\right)=xy$.

74. Solve the equation $x\,dy+y\,dx=x^3\,dx$.

75. Express $d(y/x)$ in terms of x, y. Hence solve the equation

$$x\frac{dy}{dx}-y=x^2y\frac{dy}{dx}.$$

76. Solve the equation $x(1-x)dy=ydx$.

77. Solve the equation $(kx-y)dy=(x+ky)dx$.

78. Find the curves for which the length of the normal is constant.

79. Find the curves $r=f(\theta)$ for which r^2 varies as p.

EXERCISE XX c

[51, 53, 54, 56, 59, 61]

Solve the differential equations in Nos. 51-60 :

51. $x\frac{dy}{dx}+y=x^n$.

52. $x\frac{dy}{dx}+ny=x^m$.

53. $(1-x^2)\frac{dy}{dx}-2xy=1$.

54. $\operatorname{cosec} x\frac{dy}{dx}-y\sec x=\cos 2x$.

55. $\frac{dy}{dx}+2xy=2x(x^2+1)$.

56. $\frac{dy}{dx}+ay=b\sin cx$.

57. $\frac{dy}{dx}+y\operatorname{th} x=x$.

58. $\frac{dy}{dx}+y=x^2y^2$.

59. $\frac{dx}{dy}+x=e^{-y}$.

60. $\frac{d^2y}{dx^2}=x+\frac{dy}{dx}$.

61. Find a complete primitive for

$$\left(y - x\frac{dy}{dx}\right)\left(1 + \frac{dy}{dx}\right) + c^2\frac{dy}{dx} = 0.$$

Is there any singular solution?

62. If the tangent to $y = f(x)$ makes intercepts l, m on $\mathbf{O}x$, $\mathbf{O}y$, and if these are connected by a relation $m = \phi(-m/l)$, prove that the coordinates of the point of contact satisfy

$$y = x\frac{dy}{dx} + \phi\left(\frac{dy}{dx}\right).$$

Hence find the curve for which $lm = a^2$ and interpret the result. [Use No. 11.]

EXERCISE XX d

[52, 53, 55, 56, 57]

Find the orthogonal trajectories of the systems in Nos. 51-59:

51. $3x + 4y = c$. **52.** $x^2 + y^2 = 2c(x - y)$. **53.** $cy = \text{ch}\, x$.

54. $2cy = x^2 - c^2$. **55.** $y = cx^n$. **56.** $r\theta = a$.

57. $r = c\sin^2\theta$. **58.** $r^n = a^n \sin n\theta$. **59.** $r^2\cos^2 2\theta = c^2 \sin 2\theta$.

60. Determine the curves which intersect the lemniscates $r^2 = c^2\cos 2\theta$ at a constant angle α.

EXERCISE XX e

[51, 52, 53, 55, 59, 65, 66]

Solve the differential equations in Nos. 51-65:

51. $xy_2 = a$. **52.** $y_2 \sin^2 x = a$. **53.** $y_2 = ay_1^2$.

54. $yy_2 = ny_1^2$. **55.** $y_2^2 = 1 - y_1^2$. **56.** $y_1^n y_2 = a$.

57. $y_3 = ky_2$. **58.** $xy_2 + 2y_1 = 1$. **59.** $y_2 = (1 + y)y_1$.

60. $(1 - x^2)y_2 = xy_1$. **61.** $y_2 = y_1(1 - y_1)$. **62.** $(1 - x^2)y_2 = 2xy_1$.

63. $yy_2 = y_1(1 - y_1)$. **64.** $y\dfrac{d^2x}{dy^2} = \dfrac{dx}{dy} - 1$. **65.** $\dfrac{d^3x}{dy^3} = m^2\dfrac{dx}{dy}$.

66. The motion of a particle along Ox is given by the relation

$$x^2\frac{d^2x}{dt^2} = -2k^2$$

and it starts from rest at a distance a from O. Prove that it passes through O after a time $\frac{1}{4}\pi k^{-1}a^{3/2}$.

67. Find the curves $y=f(x)$ for which the radius of curvature is equal to the normal.

EXERCISE XXf

[51, 52, 53, 56, 57, 59, 60, 62, 63, 66, 67, 72, 73, 76, 78]

Find complete primitives of the equations in Nos. 51-56 :

51. $y_2 - y_1 = 12y$.

52. $y_2 + 10y_1 + 25y = 0$.

53. $y_2 + 4y = 0$.

54. $y_2 = 2(y_1 + y)$.

55. $y_3 - y_2 = 2y_1$.

56. $y_2 + 8y_1 + 17y = 0$.

Find values of A, B, C such that the equations in Nos. 57-62 have the particular integrals indicated :

57. $y_2 + 4y_1 + 3y = 5x$; $y = \mathsf{A}x + \mathsf{B}$.

58. $2y_2 + 3y_1 + 4y = 4x^2 + 2x + 5$; $y = \mathsf{A}x^2 + \mathsf{B}x + \mathsf{C}$.

59. $y_2 - 3y_1 - 4y = 12e^{2x}$; $y = \mathsf{A}e^{2x}$.

60. $y_2 - 3y_1 - 4y = 10e^{4x}$; $y = \mathsf{A}xe^{4x}$.

61. $y_2 + y_1 - 20y = \sin 2x$; $y = \mathsf{A}\sin 2x + \mathsf{B}\cos 2x$.

62. $y_2 + 16y = \sin 4x$; $y = x(\mathsf{A}\sin 4x + \mathsf{B}\cos 4x)$.

Find particular integrals of the equations in Nos. 63-70 :

63. $y_2 + y_1 + y = x^3$.

64. $y_2 - 3y_1 - 18y = e^{5x}$.

65. $y_2 - 5y_1 - 14y = \sin x$.

66. $y_2 - 3y_1 - 18y = e^{6x}$.

67. $y_2 - 2y_1 - 15y = \sin 5x$.

68. $y_2 - 6y_1 + 9y = e^{2x}$.

69. $y_2 + y_1 = 2x + 3$.

70. $y_2 - 6y_1 + 9y = e^{3x}$.

Solve the differential equations in Nos. 71-80 :

71. $y_2 - 6y_1 + 8y = e^x + e^{-x} + x^2$.

72. $y_2 - 6y_1 + 8y = e^{2x} + \sin 2x$.

73. $y_2 + 6y_1 + 9y = e^{-3x}$.

74. $y_2 + 6y_1 + 9y = \cos 3x + \sin 3x$.

75. $y_2 - 4y = e^x \sin x$.

76. $y_2 + 4y = e^x \cos x$.

77. $y_2 + 2by_1 + (b^2 + c^2)y = \cos nx$.

78. $x^2y_2 + 3xy_1 + y = 0$. [Put $x = e^z$.]

79. $x^2y_2 + 3xy_1 + y = 9x^2$.

80. $xy_2 + 2y_1 = xy$. [Put $y = vz$, where $v = f(x)$, and choose v so that the coefficient of z_1 is zero.]

EXERCISE XXI a

[51, 52, 56, 57, 59, 61, 63, 66, 67, 69]

Use Maclaurin's series to expand the functions in Nos. 51-56 in powers of x and write down expressions for the *remainders after n terms*:

51. e^{ax}. 52. $\sin ax$. 53. $\cos bx$.

54. $(1 + x)^p$. 55. $\operatorname{ch} ax$. 56. $\log\{(1 + x)/(1 - x)\}$.

Find polynomial approximations of degree n for the functions in Nos. 57-62:

57. $\tan x$; $n = 5$. 58. $\sec^2 x$; $n = 4$.

59. $\operatorname{th} x$; $n = 3$. 60. $x \operatorname{cosec} x$; $n = 4$.

61. $e^x \sin x$; $n = 4$. 62. $\tan^{-1} x$; $n = 5$.

63. Verify that $\log(1 + \cos x) \simeq \log 2 - \frac{1}{4}x^2 - \frac{1}{96}x^4$.

64. Verify that $\log(\sec x + \tan x) \simeq x + \frac{1}{6}x^3 + \frac{1}{24}x^5$.

65. Use Taylor's series to expand $\operatorname{ch}(x + h)$ in powers of h.

66. If $y = a + x \log(y/b)$ and x is small, prove that

$$y \simeq a + x \log(a/b) + x^2a^{-1} \log(a/b).$$

Evaluate the limits when $x \to 0$ of the functions in Nos. 67-70:

67. $(x^2 - \operatorname{sh} x \sin x)/x^6$. 68. $(x \operatorname{cosec} x - 1 - \frac{1}{6}x^2)/x^4$.

69. $\dfrac{a \sin x - \sin ax}{x(\cos x - \cos ax)}$. 70. $\dfrac{\tan ax - a \tan x}{\sin ax - a \sin x}$.

EXERCISE XXI b

[51, 52, 54, 59, 62, 64, 65, 67, 69, 71, 74, 75]

Find the nth derivatives of the functions in Nos. 51-68 :

51. $(ax+b)^{2n}$. **52.** $1/\sqrt{(ax+b)}$. **53.** $\log(ax+b)$.

54. $\sin mx$. **55.** $\cos 2nx$. **56.** $e^{ax}\operatorname{ch} bx$.

57. $x^3\cos 2x$. **58.** $x^4\sin x$. **59.** $x^2\sin^2 x$.

60. $x^n\log x$. **61.** $x^2(1-x)^n$. **62.** $e^x\cos x$.

63. $(1+x+x^2)\operatorname{ch} x$. **64.** $\sin^3 x$. [See No. 10.]

65. $x\sin^3 x$. **66.** $\cos^4 x$.

67. $1/(x^2-x-2)$. **68.** $e^{3x}\sin 4x$.

69. If $f(x)=e^x\cos x$, show that $f^n(0)=(\sqrt{2^n})\cos\frac{1}{4}n\pi$ and hence write down the expansion of $f(x)$ in powers of x.

70. If $y=a\cos\log x+b\sin\log x$, prove that

$$x^2y_{n+2}+(2n+1)xy_{n+1}+(n^2+1)y_n=0.$$

71. If $y=f(x)=\log\{x+\sqrt{(x^2+1)}\}$, prove that

(i) $(1+x^2)y_2+xy_1=0$; (ii) $f^{n+2}(0)=-n^2f^n(0)$.

Hence expand y in a series of powers of x.

72. Expand $\sin(m\sin^{-1}x)$ in a series of powers of x.

73. If $\log y=\tan^{-1}x$, prove that

$$(1+x^2)y_{n+1}+(2nx-1)y_n+n(n-1)y_{n-1}=0,\ n>1.$$

74. Find a cubic polynomial $a+by+cy^2+dy^3$ which is an approximation to $x\sin x$ for values of x near to $\frac{1}{2}\pi$ where $y=\frac{1}{2}\pi-x$.

75. Expand $(x-\frac{1}{4}\pi)\cot x$ in powers of $x-\frac{1}{4}\pi$ as far as the cubic term.

76. If $y=\sqrt{\{x+\sqrt{(1+x^2)}\}}$, prove that $a_{n+2}=(\frac{1}{4}-n^2)a_n$, where a_r is the value for $x=0$ of $\dfrac{d^ry}{dx^r}$, and hence obtain the expansion of y in powers of x.

EXERCISE XXI c

[51-54, 58-62, 66, 70, 72, 74, 76]

Sketch the forms near the origin of the curves in Nos. 51-59 :

51. $y^3 = x$. **52.** $y^3 = x^2$. **53.** $y = x - x^3$.

54. $y^4 = x^5$. **55.** $y^4 = x^3$. **56.** $y^3 = x^5$.

57. $y = x^2 - x^4$. **58.** $y = x + \sqrt{x^3}$. **59.** $y = x - x^{3/2}$.

Find approximations for y in terms of x (or for x in terms of y) when x and y are small for the relations in Nos. 60-71 and sketch the portions of the graphs near the origin :

60. $y^3 = (x - y)^2$. **61.** $x^5 + y^5 = x^2y^2$. **62.** $y(1 - x)^2 = x^3$.

63. $x^5 = (y - x^2)^2$. **64.** $2x^3 + y^3 = x^2 - y^2$. **65.** $x^3 + y^3 = x^2 + y^2$.

66. $y^2(1 - x) = x^3 + x^4$. **67.** $(y - x)^2(y + x) = 2x$.

68. $(y - x)^2(y + x) = 2x^2$. **69.** $y = \log \cos (x + y)$.

70. $1 + x = \cos (x + y)$. **71.** $\sin x = \log (1 + y)$.

Find the value of $|\kappa|$ at the origin for the curves in Nos. 72-77 :

72. $y = ax^2 + 2hxy + by^2$. **73.** $x^3 + y^2 + 2xy = y + 2x$.

74. $2x^3 + y^3 = x^2 - y^2$. **75.** $(x^2 - y^2)(x + y) = 4axy$.

76. $y(y - x)(y - 3x) = x^4$. **77.** $e^{2y} = x + y + 1$.

EXERCISE XXI d

[52, 53, 55, 56, 61, 62, 63, 66, 67, 70, 71, 74, 77, 78]

Find the asymptotes of the curves in Nos. 51-62 :

51. $x^2y^2 = a^2x^2 + b^2y^2$. **52.** $x^3(y - 1) = y^2(x + 1)(x - 2)$.

53. $y(y^2 - x^2) = 3x + 4y$. **54.** $x(x^2 - y^2) = x^2 + y^2$.

55. $xy(y - x) = x^2 + 2y^2$. **56.** $x^4 - y^4 = 2xy^2$.

57. $x(x^6 - y^6) = y$. **58.** $(y^2 - 4x^2)(y + 2x) = 4x^2$.

59. $y^3(y^2 - x^2) = x^4$. **60.** $x^2(y - x) + x(y + x) + x + 2 = 0$.

61. $(x - y)^2(2x + y) = x + y$.

62. $(x - y)^2(x^2 + y^2) + (y - x)(2x^2 + y^2) + x^2 = 0$.

63. Prove that approximations for the branches of the curve $y(y+x)(y-2x)=6x^2$ are $y=-3-9/(2x)$, $y=-x+2+16/(3x)$, and $y=2x+1-5/(6x)$.

64. Prove that the curve $x^5+y^5=3x^2y^2$ approaches its asymptote from the same side at both ends by obtaining the approximation $y=-x+\frac{3}{5}-27/(125x^2)$.

65. Obtain the approximation $y=-2x\pm 2\sqrt{x}-5\pm 29/(4\sqrt{x})$ for the parabolic branch of the curve $xy(y+2x)^2=8x^3+2y^3$.

66. Find an approximation of the form $y=2x\pm \mathsf{A}\sqrt{x}+\mathsf{B}\pm \mathsf{C}/\sqrt{x}$ for the parabolic branch of the curve $y(y-2x)^2=2x^2$.

67. Prove that the three points where the curve

$$x(x^2-y^2)+x^2+y^2+x+y=0$$

meets its asymptotes lies on the line $2x+y+1=0$.

68. Prove that the asymptotes of the curve $x(x^2-y^2)=x^2+y$ meet the curve on the line $x+4y=0$.

69. Prove that the eight points where the curve

$$xy(y^2-x^2)-8xy^2=x^2-16xy+6y^2-25$$

meets its asymptote lie on an ellipse.

70. Find the equation of the cubic curve which has asymptotes $y=1$, $y=x+4$, $y=4-x$, meets them on the line $y=2x+3$ and passes through the point (2, 3).

Sketch the curves in Nos. 71-80 and find their asymptotes:

71. $xy(x-y)+y(x+y)+(x-y)=0$.

72. $xy(x-y)+(x-y)=1$.

73. $(y-x)^2(y+x)-(y-x)(4y-x)+(2y-x)=0$.

74. $x^2y=(2y+x)(y-2x)$.

75. $y^2(y-x)^2+(y-x)(y^2+x^2)=(x+y)^2$.

76. $(x-2)y^3=(y-3)x^2$.

77. $y^2(1+x^2)=4y-1$.

78. $x^4=ay^2(y-2x)$.

79. $x^2(x+y)(x+2y)+2xy+3x+y=0$.

80. $x^2y+x^4=(x^2+y^2)(x-y)^2y$.

EXERCISE XXII a

[51, 52, 53, 55, 56, 59, 60]

51. Find the M.I. of a rod AB about a line AC where AB $=2a$ and $\angle$BAC $=\alpha$.

52. Find the M.I. of a rod AB of length $2a$ about a line perpendicular to AB and meeting it at a distance c from the middle point.

53. A lamina is bounded by concentric circles of radii a, b. Find its M.I. about a line through the centre perpendicular to the plane of the lamina.

54. A cylinder of lead ($6\frac{1}{2}$ oz. per cu. in.) is 3 ft. long and of base-radius 1 in.; a cylinder of base-radius $\frac{1}{4}$ in. on the same axis is bored away. Find the M.I. about its axis of the solid that remains.

55. A triangular lamina ABC has AB $=3c$, BC $=4c$, CA $=5c$. Find its M.I. (i) about AB, (ii) about CA.

56. Find the M.I. about Ox of a lamina in the form of the segment of $ay^2=x^3$ cut off by the line $x=b$.

57. Find the M.I. of a segment of height h of a sphere of radius a about the axis of the segment.

58. Find the M.I. about the line $x=b$ of a lamina in the form of a segment of a parabola $y^2=4ax$ cut off by the line $x=b$.

59. Find the M.I. about Ox of the solid formed by revolution about Ox of the area bounded by $y=0$, $x=a$, and an arc of the curve $y=x^3$.

60. A solid anchor-ring is formed by the revolution of a circle of radius a about a line in its plane at distance b, greater than a, from the centre. Find the M.I. of the solid about its axis of revolution.

EXERCISE XXII b

[51, 53, 54, 56, 59, 60, 61, 64, 65]

51. Find the M.I. of a circular wire of radius a about a diameter.

52. Find the M.I. of the elliptic lamina $x^2/a^2+y^2/b^2=1$ about a line through the centre perpendicular to its plane.

53. Find the M.I. of the lamina $4x^2+25y^2=100$ about the line $x=5$.

54. ABC is a uniform rod; $AB=BC=a$. PN is a straight line which meets ABC at right angles at N, and PQ is a line perpendicular to the plane of PN and ABC at P. Find the M.I. of the rod about PQ (i) when N is at B, (ii) when N is at A.

55. Find the M.I. of a thin tin sheet in the form of a closed cube with edges of length l about an edge.

56. Find the M.I. of a solid cylinder of base-radius a about a generating line.

57. Find the M.I. of a solid hemisphere of radius a about a diameter of its base.

58. Find the M.I. of a hemisphere of radius a about a tangent line at its vertex.

59. Find the M.I. of an isosceles triangular lamina of height h about

    (i) the line through the vertex parallel to the base;
    (ii) the line through the centre of gravity parallel to the base;
    (iii) the base.

60. A solid is formed by revolution of the curve $y^4=a^2x(a-x)$ about Ox. Prove that its M.I. about Ox is $2\mathrm{M}a^2/(3\pi)$.

61. Find the M.I. of a thin hemispherical shell of radius a about a tangent line drawn at the vertex.

62. Find the M.I. of a thin shell in the form of the anchor-ring of Exercise XXII a, No. 60, about its axis.

63. A solid is generated by the revolution about Oy of the curve $c^2y^2=(x^2-a^2)(b^2-x^2)$. Prove that its volume is $\pi^2(b^2-a^2)^2/4c$ and that its M.I. about Oy is $\frac{1}{2}\mathrm{M}(a^2+b^2)$.

64. The density of a solid sphere of radius a varies as the distance from the centre. Find the M.I. about a diameter and about a tangent line.

65. A circular lamina of radius a has a density which varies as the distance from a point P of its circumference. Find the M.I. (i) about the perpendicular at P to the plane of the lamina, and (ii) about the tangent at P.

66. Find the M.I. about the directrix $y=0$ of a uniform wire in the form of the arc of the catenary $y=c\,\mathrm{ch}\,(x/c)$ for which $y\leqslant k$.

EXERCISE XXII c

[51, 52, 54, 55]

(Assume, except in No. 55, that the plane of the lamina is vertical.)

51. Find the centre of pressure of a triangle ABC with BC horizontal and A in the surface of the fluid.

52. Find the centre of pressure of a triangle ABC with BC in the surface of the fluid.

53. A trapezium has parallel sides of lengths a, $2a$ at distance h apart and is immersed in a fluid vertically with the side a in the surface. Prove that the depth of the centre of pressure is $\frac{7}{10}h$.

54. Find the depth of the centre of pressure of a rectangle with sides $2a$, $2b$ when the upper side $(2a)$ is in the surface.

55. Find the depth of the centre of pressure of an isosceles triangle of height h if its base is in the surface, its vertex downwards, and its plane inclined to the vertical at an angle α.

56. Find the depth of the centre of pressure of a semicircular area of radius a if its base is horizontal and its vertex is in the surface.

MISCELLANEOUS EXAMPLES

These examples are arranged by subjects and therefore a selection of the odd or even numbers will secure satisfactory revision. Selections of easy and harder examples are given:

[Easy: 2, 6, 10, 12, 15, 17, 22, 27, 30, 34, 40, 41, 42, 51, 52, 57, 60, 61, 65, 67, 70, 75, 77, 80, 85, 86, 88, 90, 92, 94, 99, 101, 103, 106, 108, 113, 116, 119, 121, 124.]

[Harder: 5, 7, 11, 13, 16, 18, 23, 29, 35, 44, 45, 47, 49, 55, 56, 58, 63, 69, 74, 76, 78, 83, 87, 89, 96, 105, 107, 109, 112, 118, 120, 122, 125.]

Differentiate with respect to x the functions in Nos. 1-8 and sketch the graphs of those in 6, 7.

1. $\dfrac{1+2\cos x}{2+\cos x}$. **2.** $\left(\dfrac{1-x}{1+x}\right)^n$. **3.** $\dfrac{x^n}{\tan x+\cot x}$.

4. $\sec^{-1}\sqrt{\dfrac{2}{1+x}}$. **5.** $\cot^{-1}\sqrt{\left(\dfrac{1}{x^2}-1\right)}$. **6.** $\tan^{-1}\dfrac{\cos x-\sin x}{\cos x+\sin x}$.

7. $\cot^{-1}(1-x)-\cot^{-1}(1+x)-\sin^{-1}\dfrac{2x}{\sqrt{(4+x^4)}}$.

8. $\{\sqrt{(x^2+a^2)}-\sqrt{(x^2-a^2)}\}/\{\sqrt{(x^2+a^2)}+\sqrt{(x^2-a^2)}\}$.

Find $\dfrac{d^2y}{dx^2}$ in terms of t for the relations Nos. 9-11 :

9. $x=a\cos t,\ y=b\sin t$. **10.** $x=t+\sin t,\ y=1+\cos t$.

11. $x=a\sin t-b\sin(at/b),\ y=a\cos t-b\cos(at/b)$.

12. Two sides of a triangle have fixed lengths a, b and include a variable angle θ. If the area is y when the third side is x, prove that $dy=\frac{1}{2}x\cot\theta\,dx$.

13. A particle moves a distance s along a straight line in time t. If the acceleration varies as the cube of the velocity, prove that $\dfrac{d^2t}{ds^2}$ is constant.

14. If $y=x(1-x)(1+ax+bx^2)$ and if $\dfrac{d^2y}{dx^2}$ is zero for $x=0$ and $x=1$, find a, b and prove $\dfrac{dy}{dx}$ is zero for $x=\frac{1}{2}$, $x=\frac{1}{2}(1\pm\sqrt{3})$.

15. $f(x)$ is defined by the relations
$f(x)\equiv 1+x,\ x<0$; $f(x)\equiv 1,\ 0\leqslant x\leqslant 1$; $f(x)\equiv 2x^2-4x+3,\ x>1$;
find $f'(x)$ for all values of x for which it exists.

16. If $f(x)\equiv -\frac{1}{2}x^2,\ x\leqslant 0$ and $f(x)\equiv x^n\sin\dfrac{1}{x},\ x>0$, find whether $f'(0)$ exists if (i) $n=2$, (ii) $n=1$.

17. Prove that $(\tan x)/x$ increases as x increases from 0 to $\frac{1}{2}\pi$.

18. Prove that $3\sin x/(2+\cos x)<x$ if $x>0$.

Find the values of x for which the functions in Nos. 19-23 have maximum or minimum values ; distinguish between them.

19. $\dfrac{x-c^2}{x^2+1}$. **20.** $\dfrac{x^4}{(x-1)(x-3)^3}$. **21.** $\sqrt{(1+x)}-\sqrt{(1-x)}$.

22. $(x-a)^2(x-b)^3$ if (i) $a>b$, (ii) $a=b$, (iii) $a<b$.

23. $\sin x+\frac{1}{3}\sin 3x+\frac{1}{5}\sin 5x$.

24. If $\tan y = 2 \tan x$, prove that the maximum value of $y - x$ is $\cot^{-1}\sqrt{8}$, assuming $0 < x < y < \frac{1}{2}\pi$.

25. $f(x) \equiv x^2 + x$, $0 \leqslant x \leqslant 1$; $f(x) \equiv 6 - x - 4/x$, $1 < x \leqslant 3$; for what values of x between 0 and 3 is $f(x)$ greatest?

26. $f(x) \equiv 1 - 2x + 3x^2$, $0 \leqslant x < 1$; $f(x) \equiv \frac{1}{2}(x + 3/x)$, $1 \leqslant x \leqslant 3$; find the greatest and least values of $f(x)$ in (0, 3) and find whether these are the same as the maximum and minimum values.

27. AB is a diameter of a given circle; PQ is a chord parallel to AB. Prove that the maximum perimeter of the quadrilateral APQB is $\frac{5}{2}$AB.

28. Prove that the area of the greatest rectangle that can be inscribed symmetrically in a sector of a circle, radius a, angle 2β, is $a^2 \tan \frac{1}{2}\beta$.

29. A rectangular sheet of paper OACB is folded over so that the corner O just reaches a point P on one of the sides AC, CB; OA $= a$, OB $= b$, $a < b$. Show that as P moves from A through C to B the end of the crease is either at A or B when the length of the crease is a maximum, and that it is greatest when the end is at A or B according as $a^2 >$ or $< \frac{1}{2}(\sqrt{5} - 1)b^2$.

30. Prove that the tangent and normal to $x^3 = ay(a - x)$ at $(\frac{1}{2}a, \frac{1}{4}a)$ form with the x-axis a triangle of area $5a^2/64$.

31. If the tangent at P to the curve $x = a\cos^3 t$, $y = a \sin^3 t$ cuts Ox at T, and if PN is the ordinate, prove that PT . OT $= a$. NT.

32. If the normal at P(x, y) to the curve $a^2/x^2 + b^2/y^2 = 1$ cuts Ox, Oy at G, H, prove that PG : PH $= a^2y^4 : b^2x^4$.

33. Find the length of the subnormal at (x, y) for the curve

$$by^2 = (a + x)^3.$$

34. Find the point where the tangent to $y(1 + x^2) = 2x$ at $x = t$ meets the curve again. What happens when $t^2 = 3$?

35. Prove that the tangent to $y^2 = x^3$ at $x = \frac{8}{9}$ is also a normal to the curve.

36. Prove that the tangents to $y^2(x - 1) = x^2(x + 1)$ at the points where $x = 2$ intersect at an angle $\frac{1}{3}\pi$.

37. Find the length of the tangent and find the subtangent at (x, y) for the curve $y = a \log (x^2 - a^2)$.

Find $\frac{dy}{dx}$ in terms of x for the relations in Nos. 38, 39:

38. $y=(1+x)^m \log(1+x)$. 39. $x=e^{xy}$.

Integrate with respect to x the functions in Nos. 40-45:

40. $\frac{1+3x}{x^2(x+1)}$. 41. $\frac{2x}{(x^2+1)(x+1)}$. 42. $\frac{1}{x(x^2+a^2)}$.

43. $\frac{\log x}{\sqrt{x}}$. 44. $\frac{\sqrt{(a^2-x^2)}}{x^2}$. 45. $\frac{1}{\sin x+\cos x}$.

46. Prove that $\int e^x \frac{1+\sin x}{1+\cos x}\,dx = e^x \tan \tfrac{1}{2}x + c$.

47. Evaluate (i) $\int_{-1}^{+1} \sqrt{(x^2)}\,dx$; (ii) $\int_0^1 \frac{x^3\,dx}{\sqrt{(1-x^2)}}$.

48. Prove that $\int_1^3 x \log\left(1+\frac{1}{x}\right) dx \simeq 1{\cdot}60$.

49. Find the function $f(x) \equiv ax+b$, for which $f(1)=1$ and $\int_0^1 |f(x)|\,dx$ has its minimum value. Prove that $f(x)$ is zero when $x=1-\tfrac{1}{2}\sqrt{2}$.

50. Prove that a cone of base-radius 2 in. and slant height 10 in. can just be passed through a loop of string of length 1 ft.

Prove the results in Nos. 51-55:

51. (i) $\int_0^1 \tan^{-1} x\,dx = \tfrac{1}{4}\pi - \tfrac{1}{2}\log 2$; (ii) $\int_{-1}^{\infty} e^{-x}(x^2-1)\,dx = 0$.

52. $\int_0^{\alpha} \frac{dx}{1-\cos x \cos \alpha} = \tfrac{1}{2}\pi \operatorname{cosec} \alpha,\ 0<\alpha<\pi$.

53. $\int_1^{\infty} \frac{dx}{(1+x)\sqrt{x}} = \tfrac{1}{2}\pi$.

54. $\int_0^1 \frac{\log x}{x^n}\,dx = -\frac{1}{(n-1)^2},\ n<1$.

55. $\int_0^{\pi} \frac{dx}{(a^2\cos^2 x+b^2\sin^2 x)^2} = \pi(a^2+b^2)/(2a^3b^3),\ a>0, b>0$.

56. (i) If $u_n = \int_0^{\pi/2} \cos^n \theta \cos n\theta \, d\theta$, where n is a positive integer, prove that $u_n = \pi/2^{n+1}$ by showing that $u_{n-1} = 2u_n$ or otherwise.

(ii) If $v_n = \int_0^{\pi/2} \sin n\theta \cos^{n+1} \theta \operatorname{cosec} \theta \, d\theta$, prove that $v_n - v_{n-1} = \pi/2^{n+1}$, and hence find v_n.

57. Find the area of the loop of $a^3y^2 = x^4(2x + a)$.

58. Find the area between the curve $x(x^2 + y^2) = y^2$ and the asymptote $x = 1$.

59. Prove that the area between the curve $xy^2 = 8(1 - x)$ and the asymptote $x = 0$ is $2\pi\sqrt{2}$.

60. v, s, t are connected by the relations $v = n\sqrt{(a^2 - s^2)}$, $s = a \sin nt$; v_1 is the mean value of v with respect to t from $t = 0$ to $t = \pi/(2n)$; v_2 is the mean value of v with respect to s from $s = 0$ to $s = a$. Prove that $\pi^2 v_1 = 8v_2$.

61. If $y = a \log(1 - x^2/a^2)$, prove that $s + x = a \log\{(a + x)/(a - x)\}$ if $s = 0$ when $x = 0$.

62. Prove that the "length of the tangent" to $s = c \log(c/y)$ is constant.

63. A uniform flexible chain of length l hangs between two points whose horizontal and vertical distances apart are a, b respectively, in the curve $y = c \operatorname{ch}(x/c)$. Prove that

$$2c \operatorname{sh}(a/2c) = \sqrt{(l^2 - b^2)}.$$

64. A point P moves along $r^n = a^n \sin n\theta$ so that $\frac{d\theta}{dt} = \omega$ where ω is constant; prove that the tangent at P rotates with angular velocity $(n + 1)\omega$.

65. Find p in terms of a, θ for the curve $r \cos^2 \theta = a \sin \theta$.

66. If the p, ψ equation of a curve is $p^n = a^n \cos n\psi$, where $a > 0$, prove that the p, ψ equation of the pedal with respect to the origin can be written $p^m = a^m \cos m\psi$ where $m = n/(n + 1)$.

67. For $r^n = a^n \sin n\theta$, prove that $|(n + 1)p\rho| = r^2$.

68. For $y = c \operatorname{sh}(x/c)$, prove that $cy\rho = \sqrt{(y^2 + 2c^2)^3}$, if $c > 0$.

69. For $r = ae^{m\theta}$, prove that the chord of curvature through the pole is $2r$.

70. For $x = a\cos^3 t$, $y = a\sin^3 t$, prove $\rho = -3p$.

71. For $y^2 = c^2 + s^2$, prove $|c\rho| = y^2$.

72. Find ρ in terms of λ for the curve given by $x = a(1 - 2\lambda^2)$, $y = a(3\lambda - 4\lambda^3)$.

73. If the normal at P to the parabola $4ay = x^2 + 4a^2$ cuts the x-axis at G, prove that the radius of curvature at P equals 2GP.

74. Prove that $\dfrac{d^2r}{ds^2} = \dfrac{\sin^2\phi}{r} - \dfrac{\sin\phi}{\rho}$.

Find the value of $|\rho|$ at the origin for the curves in Nos. 75-76:

75. $2x(x+1) = 5y^2(y+1)$.

76. $2x^2y + xy^2 - 4y^3 + y^2 - 4x^2 = 0$.

77. Prove that the envelope of

$x^3/(a^3\cos\lambda) + y^3/(b^3\sin\lambda) = 1$ where λ varies, is $x^2/a^2 + y^2/b^2 = 1$.

78. Find the envelope of the line $\lambda x(3 + \lambda^2) - 2y = 2a\lambda^3$ where λ varies. Find also the coordinates of the centre of curvature at the point of the envelope corresponding to the point λ.

79. Prove that the evolute of the curve given by

$x = 3\sin t - 2\sin^3 t$, $y = 3\cos t - 2\cos^3 t$ is $x^{2/3} + y^{2/3} = 4^{2/3}$.

80. Find the evolute of the curve given by

$$x = \cos t + t\sin t,\ y = \sin t - t\cos t.$$

81. Prove that the evolute of $r^2 = a^2\cos 2\theta$ is

$$9(x^{2/3} + y^{2/3})(x^{4/3} - y^{4/3}) = 4a^2.$$

82. Prove that the area of a loop of $r = a\sin n\theta + b\cos n\theta$ where n is a positive integer is $\pi(a^2 + b^2)/(4n)$.

83. Sketch the curve $x = a\cos t$, $y = a\sin 2t$, and prove that the area of a loop is $4a^2/3$.

84. Prove that the area of the curve $x = \sin t$, $y = \sin t\,(\cos t)^{1/2}$ is 1·6.

85. The thickness of a circular plate of radius a at distance r from the centre is $c\tan^{-1}(r/a)$. Find the volume of the plate.

Evaluate the integrals in Nos. 86-89 :

86. $\int_0^1 dx \int_x^{x^3} \sqrt{(xy)}\, dy.$ 87. $\int_0^{\pi/2} d\theta \int_0^{a\cos\theta} r^3 \cos 2\theta\, dr.$

88. $\int_0^1 \int_0^2 \int_0^3 z(x+y)\, dx\, dy\, dz.$ 89. $\int_0^c \int_0^\pi \int_0^{2\pi} r^4 \cos^2\theta \sin^2\phi\, dr\, d\theta\, d\phi.$

90. If $z = f(x^2/y)$, prove $x\dfrac{\partial z}{\partial x} + 2y\dfrac{\partial z}{\partial y} = 0.$

91. If $u = \log(x^3 + y^3 + z^3 - 3xyz)$, prove that

$$(x+y+z)\left(\frac{\partial u}{\partial x} + \frac{\partial u}{\partial y} + \frac{\partial u}{\partial z}\right) = 3.$$

92. If $u = \log r$ and $r^2 = x^2 + y^2 + z^2$, prove that

$$r^2\left(\frac{\partial^2 u}{\partial x^2} + \frac{\partial^2 u}{\partial y^2} + \frac{\partial^2 u}{\partial z^2}\right) = 1.$$

93. If $z = x^y$, prove that

$$\frac{\partial^2 z}{\partial x \partial y}\frac{\partial z}{\partial y} - \frac{\partial^2 z}{\partial y^2}\frac{\partial z}{\partial x} = \frac{1}{y}\frac{\partial z}{\partial x}\frac{\partial z}{\partial y}.$$

94. If $t = 2\pi\sqrt{(l/g)}$, express dg/g in terms of t, dt, l, dl.

95. A triangle with unequal sides a, b, c is inscribed in a circle of known radius R. Prove that the error in the area Δ due to small errors in b, c is given by

$$\frac{\delta\Delta}{\Delta} = \frac{2(a^2 - b^2)}{a^2 - b^2 + c^2}\frac{\delta b}{b} + \frac{2(a^2 - c^2)}{a^2 + b^2 - c^2}\frac{\delta c}{c}.$$

96. If $w = f(x, y)$, $x = u^3 - 3uv^2$, $y = 3u^2 v - v^3$, prove that

$$\text{(i)}\ \ 3(u^2+v^2)\left(u\frac{\partial w}{\partial x} + v\frac{\partial w}{\partial y}\right) = u\frac{\partial w}{\partial u} - v\frac{\partial w}{\partial v};$$

$$\text{(ii)}\ \ 9(u^2+v^2)^2\left(\frac{\partial^2 w}{\partial x^2} + \frac{\partial^2 w}{\partial y^2}\right) = \frac{\partial^2 w}{\partial u^2} + \frac{\partial^2 w}{\partial v^2}.$$

97. If $x = u + e^{-v}\sin u$, $y = v + e^{-v}\cos u$, prove that $\dfrac{\partial u}{\partial y} = \dfrac{\partial v}{\partial x}.$

98. If $y = ax\cos(b + n/x)$ where a, b, n are constants, prove that

$$x^4\frac{d^2 y}{dx^2} + n^2 y = 0.$$

99. Find the curves $r=f(\theta)$ for which $ap^2=r^3$.

100. Find the curves $r=f(\theta)$ for which $\phi=n\theta$.

Solve the differential equations in Nos. 101-110:

101. $y_1=(x+1)(y-1)$.

102. $yy_1+x=\sin x$.

103. $y_1+y\tan x=\cos^2 x$.

104. $y_2\cos^2 x=1$.

105. $y_2\tan x+y_1=0$.

106. $y_2-2y_1-8y=0$.

107. $y_2+12y_1+36y=0$.

108. $y_2+3y_1-4y=e^{2x}$.

109. $y_2+3y_1-4y=e^x$.

110. $y_2-12y_1+36y=e^{6x}$.

111. Evaluate (i) $\lim\limits_{t\to 1}\dfrac{1+\cos \pi t}{\tan^2 \pi t}$; (ii) $\lim\limits_{t\to \frac{1}{2}}(\sec \pi t-\tan \pi t)$.

112. Evaluate $\lim\limits_{t\to 1}\dfrac{\sin tx-t\sin x}{t(1-t^2)}$.

113. Determine a, b, c so that

$$\lim_{\theta\to 0}\frac{\theta(a+b\cos\theta)-c\sin\theta}{\theta^5}=1.$$

114. If n is a positive integer and if $y=(x-1)^{n-1}\log x$, prove that

$$y_n=(n-1)!\ (1-x^{-n})/(x-1).$$

115. If $\sin^{-1} y=a+b\sin^{-1} x$, prove that, when $x=0$,

$$y_{n+2}=(n^2-b^2)y_n.$$

116. If $y=\sqrt{(1-x^2)}\sin^{-1} x$, prove that

$$(1-x^2)y_2-xy_1+y=-2x$$

and deduce that

$$y=x-\frac{x^3}{3}-\frac{2}{3}\frac{x^5}{5}-\dots\dots-\frac{2.4\ \dots\ (2n-2)}{3.5\ \dots\ (2n-1)}\frac{x^{2n+1}}{2n+1}-\dots.$$

117. If $y\sqrt{(1+x^2)}=\log\{x+\sqrt{(1+x^2)}\}$, prove that

$$(1+x^2)y_1+xy=1$$

and deduce the expansion for y:

$$x-\tfrac{2}{3}x^3+\frac{2.4}{3.5}x^5-\dots\dots+(-1)^n\frac{2.4\ \dots\ 2n}{3.5\ \dots\ (2n+1)}x^{2n+1}+\dots.$$

118. If $\{\text{ch}^{-1}(1+x)\}^2 = \Sigma a_r x^r$, prove that

$$(n+1)(2n+1)a_{n+1} + n^2 a_n = 0$$

and find a_n.

119. Sketch the curve $x^2(x+y) = y^2$.

120. Sketch the curve $(x+y)^2(x-y) = 2xy$.

121. Find the asymptotes of the curve

$$xy(x-y)^2 - x^3 + y^3 + 2xy = 0.$$

122. Find the radii of curvature at the origin of the two branches of the curve given by $x = 1 - t^2$, $y = t - t^3$.

123. Trace the curve $y^2(a+x) = x^2(3a-x)$ and prove that the area of the loop and the area included between the curve and the asymptote are both equal to $3a^2\sqrt{3}$.

124. A solid of mass M and uniform density is formed by the revolution about the x-axis of the area bounded by the positive x-axis and the curve $y = x\sqrt{(1-x)}$. Find its moment of inertia about Ox.

125. A rectangular plate, edges $2a$, $2b$, is of variable thickness kr^2 where r is the distance of the point from a fixed point distant c from the centre of the plate. Prove that the mean thickness of the plate is $\frac{1}{3}k(a^2 + b^2 + 3c^2)$.

ANSWERS TO VOLUME II

(Constants of integration are usually omitted.)

Exercise XI a (p. 243)

1. $(2x+1)(x-3)^6/14$. **2.** $-\frac{1}{8}(4x+1)/(2x+1)^2$.
3. $-\frac{1}{3}(3x^2+6x+4)/(x+2)^3$. **4.** $\frac{2}{3}(x+4)\sqrt{(x-2)}$.
5. $\frac{2}{15}(3x-2)\sqrt{(x+1)^3}$. **6.** $\frac{1}{8}(1+x^2)^4$.
7. $-1/\{16(4x^2+9)^2\}$. **8.** $-\frac{2}{3}\sqrt{(1-x^3)}$. **9.** $-\frac{1}{4}\cos^4 x$.
10. $\frac{1}{4}\tan^4 x$. **11.** $(5x-2)(x+2)^5/30$.
12. $(3x+4)(x-1)^6/21$. **13.** $-\frac{1}{4}(2x^2+4x+3)(x+3)^{-4}$.
14. $-\frac{2}{3}(8+x)\sqrt{(4-x)}$. **15.** $\frac{1}{3}\sin^{-1}(3x/2)$.
16. $\frac{1}{6}\tan^{-1}(2x/3)$. **17.** $\sin^{-1}\sqrt{x}-\sqrt{(x-x^2)}$.
18. $-\frac{2}{5}(5-\cos^2 x)\sqrt{(\cos x)}$.

Exercise XI b (p. 246)

1. 0·4. **2.** $5\frac{1}{6}$. **3.** $32\frac{2}{3}$. **4.** 0·2.
5. $-\sqrt{(1+x^2)^3}\div(3x^3)$. **6.** $-\operatorname{cosec}^{-1} x$ if $x>1$; + if $x<-1$.
7. $\frac{1}{3}\tan^{-1} x^3$. **8.** $\sqrt{2}-\frac{1}{2}\sqrt{5}$. **9.** $\frac{1}{6}\pi$. **10.** $\frac{1}{6}\pi$.
11. $\pi/12$. **12.** 2/99. **13.** $\frac{1}{6}\pi-\frac{1}{4}\sqrt{3}$. **14.** 2/15.
15. $1-\frac{1}{2}\sqrt{3}+\frac{1}{6}\pi$. **16.** The second; $\frac{1}{2}-\frac{3}{4}\pi$, $\frac{1}{2}+\frac{1}{4}\pi$. **17.** $\frac{2}{3}$.
18. Yes, no; $\tan^{-1}\frac{1}{3}$, $\frac{1}{2}\pi$.

Exercise XI c (p. 250)

1. $\frac{1}{9}(\sin 3x-3x\cos 3x)$. **2.** $\frac{1}{4}(\cos 2x+2x\sin 2x)$.
3. $(x^2-2)\sin x+2x\cos x$. **4.** $\frac{1}{8}(2x^2-2x\sin 2x-\cos 2x)$
5. $x\cos^{-1} x-\sqrt{(1-x^2)}$. **6.** $\frac{1}{2}(x^2+1)\tan^{-1} x-\frac{1}{2}x$.
7. $2\pi-4$. **8.** $\pi-2$. **9.** $12-3\pi^2$. **10.** $\sin\theta-\frac{1}{3}\sin^3\theta$.
12. $(\sin 4x-4\sin 2x)/32-x(\cos 4x-2\cos 2x)/8$. **13.** $(\pi-3)/16$.

Exercise XII a (p. 257)

2. 2·08, 2·77, 3·47, − 2·08, − 2·77, − 3·47. **3.** 1·10. **4.** $1/x$.
5. $3/x$. **6.** $-1/x$. **7.** $1/(2x)$. **8.** $\cot x$. **9.** $-1/(1-x)$.
10. $-2\tan x$. **11.** $4/(4x+3)$. **12.** $\sec x\operatorname{cosec} x$.
13. $1+\log x$. **14.** $(1-\log x)/x^2$. **15.** $1/(1-x^2)$. **17.** $1/(2e)$.
18. $\frac{1}{4}\pi$, max.; $\tan^{-1} 2$, min.

Exercise XII b (p. 258)

1. $\frac{1}{5}\log|x|$. 2. $\log|1+x|$. 3. $-\frac{1}{2}\log|1-x^2|$.
4. $\frac{1}{3}\log|1+x^3|$. 5. $\log(3x^2+5x+7)$. 6. $\frac{1}{2}\log|x^2-4x+3|$.
7. $\log|\sin x|$. 8. $\frac{1}{2}\log|\sec 2x|$. 9. $\frac{1}{3}\log|\sin 3x|$.
10. $\log(5/3)$. 11. $\log(4/3)$. 12. $\log 2$. 13. $\frac{1}{2}$. 14. $1-\log 2$.
15. $2-\log 2$. 17. $(1\frac{1}{2}+\log 2)a^2$.

Exercise XII c (p. 259)

1. $4e^{4x}$. 2. $-2e^{-2x}$. 3. $(1+x)e^x$. 4. $2xe^{x^2}$.
5. $e^{2x}(2\sin 3x+3\cos 3x)$. 6. $(3x-1-x^2)e^{-x}$. 7. $\frac{1}{3}e^{3x}$.
8. $-e^{-x}$. 9. $2\sqrt{e^x}$. 10. $\frac{1}{2}e^{x^2}$. 11. $\log(e^x+1)$.
12. $x-\log(e^x+1)$. 13. $2(e^2-1)$. 14. $(e^4-1)/(2e^2)$. 15. $e-1$.
17. $1/e$. 19. sh 1, (1·18). 20. $c(1-e^{-a})$; area $\to c$.
21. 3, max. ; 0, $3\pm\sqrt{3}$, infl. 22. 1, 3. 23. e, 0.
24. $a/(a^2+b^2)$, $-b/(a^2+b^2)$; $e^{ax}(a\sin bx-b\cos bx)/(a^2+b^2)$.

Exercise XII d (p. 262)

1. $(9x^2+34x+29)(x+1)(x+2)^2(x+3)^3$.
2. $(60x^2+13x+32)(x-1)^4(3x+4)/(5x-2)^4$.
3. $(46x^2-59x-90)(x+4)^2/\{6(x-3)^{1/2}(2x+1)^{2/3}\}$.
4. $e^x\{(\sin x+\cos x)\log x+x^{-1}\sin x\}$.
5. $x^2e^x\{(x+3)\sin 2x+2x\cos 2x\}$.
6. $\frac{1}{2}e^{2x}(7\cos 7x+2\sin 7x-\cos x-2\sin x)$. 7. $10^x\log 10$.
8. $x10^x(2+x\log 10)$. 9. $1/(x\log 10)$. 10. $-e^{\cos x}$.
11. $2^x/(\log 2)$. 12. $\frac{1}{2}(\log x)^2$. 13. $e^{1/e}$. 15. $(1-\log x)/x^2$, $(2\log x-3)/x^3$. 16. $y^2/(1-xy)$. 17. $-3\sin t\sin 3t$.

Exercise XII e (p. 264)

1. $\frac{1}{4}x^2(2\log x-1)$. 2. $x\tan^{-1}x-\frac{1}{2}\log(1+x^2)$.
3. $e^{3x}(3x-1)/9$. 4. $e^x(x^2-2x+2)$. 5. $\frac{1}{2}e^x(\sin x-\cos x)$.
6. $\frac{1}{2}e^{-x}(\sin x-\cos x)$. 7. $-(1+\log x)/x$.
8. $-\frac{1}{5}e^{-x}(\sin 2x+2\cos 2x)$. 9. $x\tan x+\log|\cos x|$.
10. $x\tan x+\log|\cos x|-\frac{1}{2}x^2$. 11. $e^x(\sin 2x-2\cos 2x)/10$.
12. $e^x(5+\cos 2x+2\sin 2x)/10$.
13. $e^x\{(\sin 4x-4\cos 4x)/34+(\sin 2x-2\cos 2x)/10\}$.
14. $x^3\{9(\log x)^2-6\log x+2\}/27$. 15. $x^6(\log x^6-1)/36$.
16. 1. 17. $\frac{1}{4}$. 18. $\frac{1}{4}\pi+\frac{1}{2}\log 2$.
19. $(ne^{n+1}+1)/(n+1)^2$, $n\neq-1$; $\frac{1}{2}$, if $n=-1$. 20. $\frac{1}{2}\pi-1$.
21. $\frac{1}{4}\pi^2$. 22. $(-1+\epsilon^5-5\epsilon^5\log\epsilon)/25$; $-1/25$.

Exercise XIII a (p. 265)

1. 0, 1. **7.** e^x. **8.** e^{-x}. **9.** 1. **10.** ch $2x$.
11. sh $2x$. **12.** ch x.

Exercise XIII b (p. 268)

13. sh (A − B) = sh A ch B − ch A sh B ;
sh A − sh B = 2 sh $\frac{1}{2}$(A − B) ch $\frac{1}{2}$(A + B) ;
ch (A − B) = ch A ch B − sh A sh B. **14.** cosech$^2\,\theta$ = coth$^2\,\theta - 1$.
15. 2 sh θ ch ϕ = sh $(\theta+\phi)$ + sh $(\theta-\phi)$.
16. th $(\theta-\phi)$ = (th θ − th ϕ)/(1 − th θ th ϕ).
17. coth $(\theta+\phi)$ = (coth θ coth ϕ + 1)/(coth θ + coth ϕ).
18. − sh$^2\,x$. **19.** − cosech$^2\,x$. **20.** ch $4x$. **21.** 1. **22.** th x.
23. coth$^2\,\frac{1}{2}x$. **24.** $1\frac{1}{4}$. **25.** $\pm 2\sqrt{2}$. **26.** $\frac{2}{3}\sqrt{2}$.
27. 3 and log 3, or $\frac{1}{3}$ and − log 3.

Exercise XIII c (p. 270)

1. 2 ch $2x$. **2.** 3 sh $3x$. **3.** − sh x sech$^2\,x$.
4. $-\frac{1}{2}$ ch $\frac{1}{2}x$ cosech$^2\,\frac{1}{2}x$. **5.** 3 sh $6x$. **6.** coth x.
7. 2 sech$^2\,2x$. **8.** 2 th x sech$^2\,x$. **9.** $-\frac{1}{2}$ cosech$^2\,\frac{1}{2}x$.
10. 2 cosech $2x$. **11.** − 4 sh $2x$ sech$^3\,2x$. **12.** coth$^2\,x$.
13. sh x ch x. **14.** 2 ch $\frac{1}{2}x$. **15.** $\frac{1}{2}$ log ch $2x$. **16.** $\frac{1}{2}$ th $2x$.
17. − 2 coth $\frac{1}{2}x$. **18.** $\frac{1}{3}$ log | sh $3x$ |. **19.** $\frac{1}{2}(x$ + sh x ch $x)$.
20. $\frac{1}{2}$(sh $x - x$). **21.** x − coth x. **22.** $\tan^{-1} e^{2x}$.
23. cosech x, log | th $\frac{1}{2}x$ |.
24. $\frac{1}{8}$(ch $4x$ − 2 ch $2x$), $\frac{1}{8}$(ch $4x$ + 2 ch $2x$).
25. (sh $5x$ + 5 sh x)/10, (sh $5x$ − 5 sh x)/10.
26. $e - e^{-1}$; $(e^4 - 4e^2 - 1)/(8e^2)$. **27.** x ch x − sh x.
28. $\frac{1}{2}x$ sh $2x - \frac{1}{4}$ ch $2x$. **29.** x th x − log ch x.
30. (x^2+2) sh $x - 2x$ ch x.

Exercise XIII d (p. 272)

1. 4. **2.** $\frac{1}{2}c^2(e - e^{-1})$. **4.** $1/e$. **7.** $1 - 1/e$.
8. $x = \frac{1}{4}\pi$, $\alpha = 0$. **9.** 2π th a ; $\to 2\pi$. **10.** k^{-1} log ch (ckt).

Exercise XIII e (p. 277)

1. log 5. **2.** log $(2+\sqrt{3})$. **3.** $\frac{1}{2}$ log (9/5).
5. Always positive and min. at $x = 0$. **6.** $1/(1-x^2)$.
7. $2/\sqrt{(4x^2+9)}$. **8.** $4/\sqrt{(16x^2-25)}$, $x > 1\frac{1}{4}$.

9. $3/(1-9x^2)$, $|x|<\frac{1}{3}$. **10.** $-1/\sqrt{(x^4+x^2)}$, $x\neq 0$.
11. $-1/\{x\sqrt{(1-x^2)}\}$, $0<x<1$. **12.** $-2/(4x^2-1)$, $|x|>\frac{1}{2}$.
13. $\log\{x^{-1}+\sqrt{(1+x^{-2})}\}, x\neq 0$. **14.** $\log\{x^{-1}+\sqrt{(x^{-2}-1)}\}, 0<x<1$.
15. $\frac{1}{2}\log\{(x+1)/(x-1)\}$, $|x|>1$. **17.** $\frac{1}{2}\log(16/27)$.
18. $4\log(1+\sqrt{2})$.

Exercise XIII f (p. 280)

1. $\text{sh}^{-1}\frac{1}{3}x$. **2.** $\text{ch}^{-1}\frac{1}{4}x$. **3.** $\frac{1}{2}\text{ch}^{-1}\frac{2}{3}x$. **4.** $\frac{1}{3}\text{sh}^{-1}3x$.
5. $\frac{1}{3}\sin^{-1}\frac{3}{4}x$. **6.** $\frac{1}{2}\text{sh}^{-1}x^2$. **7.** $\log 3$. **8.** $\log(4-\sqrt{7})$.
9. $\tan^{-1}\frac{7}{24}$. **10.** $\log(5\sqrt{2}+3\sqrt{5}-2\sqrt{10}-6)$.
11. $\text{sh}^{-1}(p/a)-\text{sh}^{-1}(q/a)$. **12.** $\frac{1}{2}\{x\sqrt{(x^2+9)}+9\,\text{sh}^{-1}\frac{1}{3}x\}$.
13. $\frac{1}{2}\{x\sqrt{(x^2-9)}-9\,\text{ch}^{-1}\frac{1}{3}x\}$. **14.** $\frac{1}{2}x\sqrt{(9x^2+16)}+2\frac{2}{3}\text{sh}^{-1}\frac{3}{4}x$.
15. $\frac{1}{2}x\sqrt{(1+x^2)}-\frac{1}{2}\text{sh}^{-1}x$. **16.** $\{\sqrt{(x^2-1)}\}/x$.
17. $\text{ch}^{-1}x-\{\sqrt{(x^2-1)}\}/x$. **18.** $\text{ch}^{-1}(x-1)$.
19. $\frac{1}{2}(x-1)\sqrt{(x^2-2x)}-\frac{1}{2}\text{ch}^{-1}(x-1)$. **20.** $\text{sh}^{-1}(\frac{1}{2}x+\frac{1}{2})$.
21. $x\,\text{sh}^{-1}x-\sqrt{(1+x^2)}$. **22.** $(\frac{1}{2}x^2-\frac{1}{4})\,\text{ch}^{-1}x-\frac{1}{4}x\sqrt{(x^2-1)}$.
23. $x\,\text{th}^{-1}x+\frac{1}{2}\log(1-x^2)$. **24.** $7\frac{1}{2}+8\log 2$.
25. $2-\sqrt{5}+\log\{\frac{1}{2}(\sqrt{5}+1)\}$. **26.** $4\log(2+\sqrt{3})-2\sqrt{3}$.
28. $(\log 2)/k$, $(\log 3)/k$.

Exercise XIV a (p. 284)

1. (i) $x^{2n+1}/(2n+1), n\neq -\frac{1}{2}, \log x$ if $n=-\frac{1}{2}$; (ii) $x^{1-n}/(1-n), n\neq 1$, $\log x$ if $n=1$; (iii) $\log x+1/x$. **2.** $\frac{1}{3}e^{3x}$, $-e^{-x}$, $2\sqrt{e^x}$.
3. $-\frac{1}{2}\cos 2x$, $2\sin\frac{1}{2}x$, $\frac{1}{4}\log|\sec 4x|$.
4. $\frac{1}{3}\tan 3x$, $-\frac{1}{2}\cot 2x$, $2\log|\sin\frac{1}{2}x|$.
5. $\frac{1}{2}\text{ch}(2x+1)$, $\frac{1}{2}\text{sh}(2x-1)$, $\frac{1}{2}\log\text{ch}(2x+3)$.
6. $\frac{1}{2}\log|2x+3|$, $-\frac{1}{4}\log|5-4x|$, $\frac{1}{6}\log(1+3x^2)$.
7. $\sin^{-1}x$; $-\sqrt{(1-x^2)}$; $\text{ch}^{-1}x$, $x>1$, but $-\text{ch}^{-1}(-x)$, $x<-1$.
8. $\sqrt{(x^2+1)}$; $\text{sh}^{-1}x$; $\sqrt{(x^2-1)}+\text{ch}^{-1}x$, $x>1$, but as in No. 7 if $x<-1$.
9. $\sec x$. **10.** $-\text{cosec}\,x$. **11.** $\log|\tan x|$. **12.** $\tan x-x$.
13. $\frac{1}{2}\tan^{-1}\frac{1}{2}x$. **14.** $\text{ch}^{-1}\frac{1}{3}x$, $x>3$, but see No. 7.
15. $\frac{1}{6}\tan^{-1}\frac{2}{3}x$. **16.** $\frac{1}{2}\text{sh}^{-1}\frac{2}{3}x$. **17.** $-\log(1+\cos x)$.
18. $\log(1+\text{ch}\,x)$. **19.** $\frac{1}{2}\sin^{-1}2x$. **20.** $\text{ch}^{-1}\frac{1}{2}x$, $x>2$, but $-\text{ch}^{-1}(-\frac{1}{2}x)$, $x<-2$. **21.** $-\frac{1}{2}\log|4-x^2|$.
22. $\log(1+e^x)$. **23.** $\tan^{-1}e^x$. **24.** $\frac{1}{3}\tan^{-1}3x$.
25. $\frac{1}{2}\log\text{ch}\,2x$. **26.** $-\log|\text{sh}(1-x)|$. **27.** $-\text{sech}\,x$.
28. $-\text{cosech}\,x$. **29.** $\log\text{ch}\,x$. **30.** $\frac{1}{3}\text{ch}^{-1}3x$, $x>\frac{1}{3}$, but $-\frac{1}{3}\text{ch}^{-1}(-3x)$, $x<-\frac{1}{3}$. **31.** $\frac{1}{3}\sin^{-1}3x$. **32.** $\tan^{-1}e^{2x}$.
33. $\frac{1}{3}\sqrt{3}\tan^{-1}\{(x+2)/\sqrt{3}\}$. **34.** $\frac{1}{5}\tan^{-1}\{(x-3)/5\}$.
35. $\frac{1}{6}\tan^{-1}\{(2x+1)/3\}$. **36.** $\sin^{-1}\frac{1}{3}(x-2)$.

37. $\frac{1}{3}\sin^{-1}\{(3x-2)/\sqrt{5}\}$. **38.** $\frac{1}{3}\sin^{-1}\frac{1}{2}(3x+1)$. **39.** $\text{ch}^{-1}\frac{1}{3}(x-2)$, $x>5$; $-\text{ch}^{-1}\frac{1}{3}(2-x)$, $x<-1$. **40.** $\text{sh}^{-1}\frac{1}{3}(x+2)$.
41. $\text{sh}^{-1}\{(x-3)/\sqrt{5}\}$. **42.** $\frac{1}{6}\log|(x-3)/(x+3)|$.
43. $\frac{1}{10}\log|(5+x)/(5-x)|$.
44. $\frac{1}{4}\sqrt{2}\log|(x+3-\sqrt{2})/(x+3+\sqrt{2})|$.

Exercise XIV b (p. 287)

1. $-\frac{3}{5}\log(4-5x)$. **2.** $-\log(1-x)-x$. **3.** $\frac{1}{2}x^2+x+\log(x-1)$.
4. $\frac{1}{3}x^3-\frac{1}{2}x^2+x-\log(x+1)$. **5.** $\frac{1}{2}\log\{(x-1)/(x+1)\}$.
6. $\frac{1}{2}\log(x^2-4)$. **7.** $\log\{(3+x)/(3-x)\}$.
8. $3\log(x-1)-2\log x$. **9.** $2\log(x-1)+3\log(x+4)$.
10. $3\log(x+2)+5/(x+2)$. **11.** $2\log(x-4)-\log(x+3)$.
12. $\frac{1}{9}\log\{(x+3)/x\}-1/(3x)$. **13.** $2\log\{(x-2)/x\}-5/(x-2)$.
14. $\log(x+1)-3\log(x-1)+8\log(x-2)$.
15. $2\log\{x/(x+1)\}-1/x$. **16.** $\log(x-3)-(6x-14)/(x-3)^2$.
17. $\log\{x(x+2)/(x+1)^2\}$. **18** $\frac{1}{2}x^2+3x+8\log(x-2)-\log(x-1)$.
19. $2\log\{(x-1)/(x+2)\}-3/(x-1)$. **20.** $(1+x)/(x-x^2)$.
21. $2\log\{x/(x+1)\}+1/(2x^2)$. **22.** $1/(e^{c-x}-1)$.
23. $x+\log(1+e^x)-2\log(1+2e^x)$.
24. $y=\{b^2/(2a)\}\log\{(x-a)/(x+a)\}+c$.

Exercise XIV c (p. 291)

1. $\frac{1}{2}\sqrt{2}\tan^{-1}(x/\sqrt{2})$. **2.** $\frac{1}{2}\log(x^2+2)$.
3. $\frac{1}{6}\sqrt{3}\log\{(x-\sqrt{3})/(x+\sqrt{3})\}$.
4. $\frac{1}{4}\sqrt{2}\log\{(x\sqrt{2}-1)/(x\sqrt{2}+1)\}$.
5. $(1/\sqrt{10})\tan^{-1}(2x/\sqrt{10})$.
6. $\frac{1}{12}\sqrt{6}\log\{(x\sqrt{3}-\sqrt{2})/(x\sqrt{3}+\sqrt{2})\}$.
7. $x-3\tan^{-1}\frac{1}{3}x$. **8.** $\frac{1}{2}\log(x^2+3)+\sqrt{3}\tan^{-1}(x/\sqrt{3})$.
9. $\frac{1}{4}\log(2x^2+1)+\sqrt{2}\tan^{-1}(x\sqrt{2})$.
10. $\log(2x+3)-\frac{1}{2}\log(2x-3)$. **11.** $x+\log(x^2+1)$.
12. $\frac{1}{2}x^2-\log(x^2+2)$. **13.** $\frac{1}{4}\log(x^4+1)$. **14.** $\frac{1}{2}\tan^{-1}\{\frac{1}{2}(x+1)\}$.
15. $\frac{2}{3}\sqrt{3}\tan^{-1}\{(2x+1)/\sqrt{3}\}$. **16.** $\frac{1}{10}\tan^{-1}\{(2x-3)/5\}$.
17. $\frac{1}{20}\log\{(x-4)/(x+1)\}$.
18. $\frac{1}{6}\sqrt{3}\log\{(\sqrt{3}-2+x)/(\sqrt{3}+2-x)\}$.
19. $\frac{3}{2}\log(x^2+2x+10)-\frac{1}{3}\tan^{-1}\{\frac{1}{3}(x+1)\}$.
20. $x+\log(x^2-2x-2)+\frac{1}{2}\sqrt{3}\log\{(x-1-\sqrt{3})/(x-1+\sqrt{3})\}$.
21. $\frac{1}{5}\sqrt{5}\log\{(\sqrt{5}-1+2x)/(\sqrt{5}+1-2x)\}$. **22.** $\log\{x/\sqrt{(x^2+1)}\}$.
23. $\log\{(x-1)^2/(x^2+1)\}-\tan^{-1}x$.
24. $\frac{1}{2}x^2+3x+\log(x^2+1)-2\tan^{-1}x$.
25. $\frac{1}{4}\log\{(x-1)/(x+1)\}+\frac{1}{2}\tan^{-1}x$.

26. $\frac{1}{4}[\log\{(x+1)^2/(x^2+1)\}-2/(x+1)]$.
27. $\frac{3}{10}\log\{(4+x^2)/(1-x)^2\}-\frac{1}{5}\tan^{-1}\frac{1}{2}x$.
28. $\{\tan^{-1}\frac{1}{3}x+3x/(x^2+9)\}/54$.
29. $-\frac{1}{8}(x+4)/(x^2+4)-\frac{1}{16}\tan^{-1}\frac{1}{2}x$.
30. $\frac{1}{16}\tan^{-1}\{\frac{1}{2}(x+1)\}+(x+1)/\{8(x^2+2x+5)\}$.

Exercise XIV d (p. 295)

1. $\frac{2}{3}(x+3)\sqrt{x}$. 2. $\frac{2}{3}(x-2)\sqrt{(x+1)}$. 3. $-2\sqrt{x}-\log(1-\sqrt{x})^2$.
4. $-\frac{2}{3}(2x+1)\sqrt{(1-x)}$. 5. $\frac{1}{3}\sqrt{3}\log[\{\sqrt{3}-\sqrt{(1-x)}\}^2/(x+2)]$.
6. $\frac{2}{15}(3x^2+8x+32)\sqrt{(x-2)}$. 7. $\text{sh}^{-1}\{\frac{1}{3}(x+1)\}$.
8. $\sin^{-1}\{\frac{1}{3}(x-2)\}$. 9. $\text{ch}^{-1}\{\frac{1}{2}(x+3)\}, x>-1$.
10. $\text{ch}^{-1}(2x+1), x>0$. 11. $\sin^{-1}x-2\sqrt{(1-x^2)}$.
12. $\sqrt{(x^2+4)}+\text{sh}^{-1}\frac{1}{2}x$.
13. $2\sqrt{(x^2+3x+2)}-4\,\text{ch}^{-1}(2x+3), x>-1$.
14. $\frac{1}{2}\text{sh}^{-1}\{(2x-1)/\sqrt{3}\}-\sqrt{(1-x+x^2)}$.
15. $\frac{1}{2}\sin^{-1}\{(2x+1)/\sqrt{5}\}-\sqrt{(1-x-x^2)}$.
16. $\sin^{-1}x+\sqrt{(1-x^2)}$. 17. $\sqrt{(x^2+2x)}-\text{ch}^{-1}(x+1), x>0$.
18. $\frac{1}{8}(x^2-8)\sqrt{(x^2+4)}+\text{sh}^{-1}(\frac{1}{2}x)$. 19. $-\frac{1}{3}\text{sh}^{-1}|3/x|$.
20. $-\frac{1}{3}\text{ch}^{-1}|3/x|, |x|<3, x\neq 0$.
21. $-\frac{1}{3}\sqrt{3}\,\text{ch}^{-1}|(2x-1)/(x-2)|, x>1, x\neq 2$;
$\frac{1}{3}\sqrt{3}\,\text{ch}^{-1}\{(1-2x)/(2-x)\}, x<-1$.
22. $\log\{\sqrt{(1+x^2)}-1\}-\log x+\sqrt{(1+x^2)}$. 23. $\frac{1}{2}\text{sh}^{-1}x^2$.
24. $x(1+x^2)^{-1/2}$. 25. $\text{sh}^{-1}x-x/\sqrt{(1+x^2)}$ if integrand is positive.
26. $\text{ch}^{-1}x+\text{cosec}^{-1}x, x>1$; $-\text{ch}^{-1}(-x)-\text{cosec}^{-1}x, x<-1$.
27. $-\{\sqrt{(1+x^2)}\}/x$. 28. $\frac{1}{2}\text{sh}^{-1}x+\frac{1}{2}x\sqrt{(1+x^2)}$.
29. $\frac{1}{2}x\sqrt{(4-x^2)}+2\sin^{-1}(\frac{1}{2}x)$.
30. $\frac{1}{2}\{x\sqrt{(x^2-9)}-9\,\text{ch}^{-1}(\frac{1}{3}x)\}, x>3$.
31. $\frac{1}{2}x\sqrt{(1+x^2)}-\frac{1}{2}\text{sh}^{-1}x$. 32. $\text{ch}^{-1}x+\sqrt{(1-x^{-2})}, x>1$.
33. $\frac{1}{2}\text{sh}^{-1}(x+1)+\frac{1}{2}(x-3)\sqrt{(x^2+2x+2)}$. 34. $\cos^{-1}(5-2x)$.
35. $\frac{1}{4}(2x-5)\sqrt{\{(x-2)(x-3)\}}-\frac{1}{8}\text{ch}^{-1}(2x-5)$.

Exercise XIV e (p. 299)

1. $\frac{1}{2}\log\tan(x+\frac{1}{4}\pi)$. 2. $2\log\tan\frac{1}{4}x$. 3. $\log\tan x$.
4. $\cos x+\log\tan\frac{1}{2}x$. 5. $2\sin x-\log\tan(\frac{1}{2}x+\frac{1}{4}\pi)$.
6. $4\cos x+3\log\tan\frac{1}{2}x$. 7. $\tan\frac{1}{2}x$. 8. $-1/(1+\sin x)$.
9. $\tan(\frac{1}{2}x-\frac{1}{4}\pi)$. 10. $\frac{1}{2}\tan^{-1}(\frac{1}{2}\tan\frac{1}{2}x)$.
11. $\frac{1}{4}\log\{(2+\tan\frac{1}{2}x)/(2-\tan\frac{1}{2}x)\}$. 12. $\frac{1}{2}\tan^{-1}(2\tan\frac{1}{2}x)$.
13. $\frac{1}{2}x-\frac{1}{2}\log(\cos x+\sin x)$. 14. $\sqrt{2}\tan^{-1}\{(\tan x)/\sqrt{2}\}-x$.
15. $\frac{1}{2}\sqrt{2}\log\tan(\frac{1}{2}x+\frac{1}{8}\pi)$. 16. $\frac{1}{2}\sqrt{2}\tan^{-1}(\sqrt{2}\tan x)$.
17. $x-\log(\sin x+2\cos x)^2$. 18. $\frac{1}{6}\tan^{-1}\{\frac{2}{3}\tan(\frac{1}{2}x-\frac{1}{4}\pi)\}$.

19. $-\log(1+\cot\frac{1}{2}x)$. 20. $\frac{1}{5}\{4x+3\log(\cos x+2\sin x)\}$.
21. $\frac{1}{12}\log\{(5-\tan\frac{1}{2}x)/(5\tan\frac{1}{2}x-1)\}$. 22. $\log\{x+\sqrt{(1+x^2)}\}$.
23. $\log\{x+\sqrt{(x^2-4)}\}$. 24. $-2\,\mathrm{ch}^{-1}\sqrt{(1+\cos x)}$.
25. $\log x-\log\{1+\sqrt{(1-x^2)}\}-(1/x)\sqrt{(1-x^2)}$.

Exercise XIV f (p. 300)

1. $\frac{1}{2}x-\frac{1}{8}\sin 4x$. 2. $\frac{1}{12}(\cos 3x-9\cos x)$.
3. $\frac{1}{32}(\sin 4x+8\sin 2x+12x)$. 4. $\frac{1}{7}\sin^7 x-\frac{1}{9}\sin^9 x$.
5. $\frac{1}{16}(x-\frac{1}{4}\sin 4x+\frac{1}{3}\sin^3 2x)$. 6. $\frac{1}{3}\cos^3 x-\frac{4}{5}\cos^5 x$.
7. $4(\frac{1}{5}\cos^5 x-\frac{1}{3}\cos^3 x)$. 8. $\frac{1}{15}(15\sec x-10\sec^3 x+3\sec^5 x)$.
9. $\operatorname{cosec} x-\frac{1}{3}\operatorname{cosec}^3 x$. 10. $\frac{1}{2}\tan^2 x+\log\cos x$.
11. $\frac{1}{3}\cos^3 x-\frac{2}{5}\cos^5 x$. 12. $\frac{1}{4}\log\tan(x+\frac{1}{4}\pi)-\frac{1}{2}x$.

Exercise XIV g (p. 303)

(In these answers, $s\equiv\sin\theta$, $c\equiv\cos\theta$, $t\equiv\tan\theta$.)

1. $\frac{1}{4}e^{2x}(2x^4-4x^3+6x^2-6x+3)$.
2. $5\theta/16-c(15s+10s^3+8s^5)/48$. 3. $s(5c^6+6c^4+8c^2+16)/35$.
4. $\frac{1}{7}s^7-\frac{2}{9}s^9+\frac{1}{11}s^{11}$.
5. $(15\theta+15sc+10sc^3-40sc^5-80s^3c^5-128s^5c^5)/1280$.
6. $x\{(\log x)^3-3(\log x)^2+6\log x-6\}$.
7. $\frac{1}{5}t^5-\frac{1}{3}t^3+t-\theta$; $\frac{1}{4}t^4-\frac{1}{2}t^2-\log\cos\theta$.
9. $nu_n+(n-1)u_{n-2}=\mathrm{sh}^{n-1}x\,\mathrm{ch}\,x$. 10. $\frac{1}{15}t(15+10t^2+3t^4)$;
$\frac{1}{8}\sec\theta\tan\theta(5+2\tan^2\theta)+\frac{3}{8}\log(\sec\theta+\tan\theta)$.
12. $(x^6+30x^4+360x^2+720)\,\mathrm{sh}\,x-6x(x^4+20x^2+120)\,\mathrm{ch}\,x$.
13. $(x/384)\{32(x^2+2)^{-3}+20(x^2+2)^{-2}+15(x^2+2)^{-1}\}+$
$\{(5\sqrt{2})/256\}\tan^{-1}(x/\sqrt{2})$.
14. $(3/16)\log\{(x+1)/(x-1)\}-\frac{1}{8}(3x^3-5x+12)(x^2-1)^{-2}$.
15. $2(n-1)u_n=(x+1)(x^2+2x+2)^{1-n}+(2n-3)u_{n-1}$;
$\frac{1}{8}\{(x+1)(3x^2+6x+8)(x^2+2x+2)^{-2}+3\tan^{-1}(x+1)\}$.

Exercise XV a (p. 307)

1. $8/15$. 2. $3\pi/16$. 3. $128/315$. 4. $35\pi/256$.
5. $8/35$. 6. $5\pi/16$. 7. $2/35$. 8. $5\pi/256$.
9. $7\pi/2048$. 10. $16/1155$. 11. $5\pi/2$. 12. $21\pi/16$.
13. $5\pi/256$. 14. $3\pi/128$. 15. $(512\sqrt{2})/315$. 16. $1/120$.
17. $\pm 4/35$. 18. $v_n+v_{n-2}=1/(n-1)$; $13/15-\frac{1}{4}\pi$. 19. $\pi-2$.
20. $\frac{1}{4}\pi-\frac{1}{2}$. 21. $\frac{1}{2}\pi-\log 2$. 22. $\frac{1}{5}\pi$. 23. $32/\pi^3$.
24. $\frac{1}{4}\pi-\frac{1}{2}\log 2-\pi^2/32$. 25. $\frac{1}{8}\pi$. 27. $5\pi a^2/4$.
32. $(3\pi+8)/32$. 34. $-\frac{1}{2}\pi$.

Exercise XV b (p. 311)

1. $\frac{1}{2}$. 2. Does not exist. 3. $\pi/6$. 4. 0.
5. Does not exist. 6. 1. 7. $\frac{1}{2}\log 3$. 8. π.
9. $1-\frac{1}{4}\pi$. 10. 1. 11. $\frac{1}{2}\pi$. 12. Does not exist.
13. $\frac{1}{2}\log 2$. 14. 0.

Exercise XV c (p. 314)

1. $2\sqrt{2}$. 2. Does not exist. 3. Does not exist.
4. 6. 5. π. 6. 2. 7. 0. 8. Does not exist.
9. Does not exist. 10. $\log 2$. 11. $\frac{1}{2}\pi$, $a>0$; $-\frac{1}{2}\pi$, $a<0$.
12. $\frac{2}{3}$. 13. $\frac{1}{4}\pi$. 14. $-\frac{1}{4}$. 15. $\log 2-\frac{1}{2}$. 16. π.
17. 2. 18. π. 19. (ii) Does not exist. 21. $n!$
22. $63\pi/512$.

Exercise XV d (p. 317)

1. $\frac{1}{2}$. 2. πa^2. 3. 4π. 7. $4/3$.
10. π. 11. No. 12. π. 13. $\frac{1}{2}\pi$.
15. $\bar{x}=2n/(n+1)$, $\bar{y}=\frac{1}{5}(n^5-1)/(n^5-n^3)$; $(2, \frac{1}{5})$; no. 16. (2, 0).

Exercise XV e (p. 322)

0, $5\pi/16$. 2. 0, $35\pi/128$. 3. $\frac{3}{4}\pi$, 0. 4. 0, $4/35$.
5. 0. 6. 2π. 7. 0. 8. π.
13. $\int_a^b f(x)\,dx$; $\int_a^b f(x)\,dx$. 15. $\frac{1}{3}\sqrt{3}$. 16. $\frac{1}{4}\pi$.
17. $(t-1)/\log t$. 22. π.

Exercise XVI a (p. 328)

1. $c\,\text{sh}\,(x/c)$. 2. $6a$. 5. $\log(\text{sh}\,b/\text{sh}\,a)$, $(b>a)$.
6. $s=8a\sin\frac{1}{3}(\psi-\frac{1}{2}\pi)$, $(-\pi\leqslant\theta\leqslant\pi)$, if $s=0$ when $\theta=0$.
7. $\tan^{-1}(\sqrt{2})-\frac{1}{4}\pi+\sqrt{2}$.

Exercise XVI b (p. 331)

5. $dx:dy:ds=c:s:\sqrt{(c^2+s^2)}$; $x=c\,\text{sh}^{-1}(s/c)$; $y=\sqrt{(c^2+s^2)}$.
7. $(e^{2x/a}-1)^{1/2}$. 9. $x=2a(\cos\psi+\psi\sin\psi)$, $y=2a(\sin\psi-\psi\cos\psi)$.
11. $x=k+a\log\tan\psi$; $s=k'+a(\sec\psi+\log\tan\frac{1}{2}\psi)$.

Exercise XVI c (p. 333)

1. $r^2\sin 2\theta=2c^2$. 2. $r=2a\cos\theta$. 3. $r^2=2ar\,\text{cosec}\,2\theta$.
4. $x\cos\alpha+y\sin\alpha=p$. 5. $x^2-y^2=a^2$.
6. $(x^2+y^2)^2=2a^2xy$. 7. $8a$. 8. $a\alpha$. 11. $8a/3$.
13. $a\int_0^{\theta_1}(\sec 2\theta)^{3/2}\,d\theta$, $a\int_0^{\theta_1}(\sec 2\theta)^{1/2}\,d\theta$; $(-\frac{1}{4}\pi<\theta_1<+\frac{1}{4}\pi)$.
15. $a(15+16\log 2)/32$.

Exercise XVI d (p. 337)

1. $\frac{1}{2}\theta$, $\frac{3}{2}\theta$; $r^3 = 2ap^2$. 2. $\frac{1}{4}\pi$, $\frac{1}{4}\pi + \theta$; $r = p\sqrt{2}$.
3. 2θ, 3θ ; $r^3 = a^2 p$. 6. $\frac{1}{3}\pi$. 7. $r^4 = p^2(a^2 + r^2)$.
8. $a^2 r^2 = p^2(2a^2 - r^2)$. 9. $pr^{n-1} = a^n$.
10. $2l/r = 1 - e^2 + l^2/p^2$.

Exercise XVI e (p. 342)

1. $r = 2a \cos^3 \frac{1}{3}\theta$. 3. $2p = |a \sin 2\psi|$; $r = \frac{1}{2} a \sin 2\theta$.
5. $p = a$, regarded as a system of lines ; ($p^2 = ar$ is a system of parabolas).
6. $p = r \sin \alpha$. 7. $r^5 = a^2 p^3$. 8. $r\sqrt{2} = ae^{\theta + \pi/4}$.
9. $r = a(1 + \cos \theta)$. 10. $r^2 = a^2 \sin 2\theta$. 11. $\phi_1 = \phi$.
12. $x^{2/3} + y^{2/3} = (2a)^{2/3}$. 14. $r = 2a \sin (\theta + \gamma)$.
15. $r = ce^{\theta \cot \alpha}$. 16. $r = a\{1 + \sin (\theta + \gamma)\}$.

Exercise XVII a (p. 348)

1. $1/(4a \cos \alpha)$. 2. $c \tan \alpha$. 3. $s + a = ce^{2\psi}$.
4. $-a \operatorname{cosec} (x/a)$. 7. $\sqrt{(a^2 + b^2)^3}/(2ab\sqrt{2})$. 8. $4a \cos \frac{1}{2}t$.
9. $\frac{3}{2}a \sin 2\psi$. 10. $2a$. 11. $2\sqrt{5}$. 12. $4/3$. 13. $2a$.

Exercise XVII b (p. 351)

1. $2r\sqrt{(r/a)}$. 2. $-\frac{3}{2}r^{8/3}k^{-5/3}$. 3. $1/(4a\sqrt{2})$.
4. $(\sin \alpha)/r$. 6. $p(1 - n^2)$. 11. $s = \frac{1}{2}a\psi^2$.

Exercise XVII c (p. 354)

1. $x^2 + y^2 = a^2$. 2. $x = a \cos^3 t$, $y = a \sin^3 t$ or $x^{2/3} + y^{2/3} = a^{2/3}$.
3. $x^2/a^2 + y^2/b^2 = 1$. 4. $27y^2 = 4x^3$.
5. $x^2 + y^2 = ax$ or $r = a \cos \theta$. 6. $x^{1/2} + y^{1/2} = a^{1/2}$.
7. $4xy = a^2$. 8. As for No. 2. 9. $y = v^2/(2g) - \frac{1}{2}gx^2/v^2$.
10. $x = a(\theta - \sin \theta)$, $y = a(3 + \cos \theta)$.
11. $(x^2 + y^2 - c^2)^2 = 4a^2\{(x - c)^2 + y^2\}$.
12. $(x^2 + y^2)^2 = 4a^2(x^2 - y^2)$ or $r^2 = 4a^2 \cos 2\theta$.

Exercise XVII d (p. 359)

2. $(-\frac{1}{2}, 1\frac{1}{4})$. 3. $(2c, 2c)$. 4. $(\frac{1}{3}\pi - \sqrt{3}, 1 + \log 2)$.
5. $(0, \frac{1}{2})$. 6. $y + tx = 2at + at^3$; $x = 2a + 3at^2$, $y = -2at^3$, that is $27ay^2 = 4(x - 2a)^3$.
8. $4(a^3 - b^3)/(ab)$. 9. $4a$. 10. $c \operatorname{sh}^2 1$. 11. $s = c \cot^2 \psi$.
12. $\{-a(1 + 3\tan^2 t), -2a \tan^3 t\}$; $6a \tan t \sec^3 t$.
16. $p - k = \frac{1}{2}a(9 \sin \frac{1}{3}\psi + \sin \psi)$ by a change of initial line.

Exercise XVIII a (p. 362)

1. $\frac{1}{4}\pi a^2$. 2. $3\pi a^2/8$. 3. $11\pi a^2$. 4. $(e^4-1)/(4e^2)$.
5. $a^2(4-\pi)/8$. 6. $a^2 \log \cot \frac{1}{2}\alpha$. 7. $\frac{1}{2}\pi(a^2+b^2)$.
8. a^2. 9. $\frac{1}{2}\pi a^2$. 10. $a^2(3\pi-8)/8$.
11. $\frac{1}{2}a^2(1-\sqrt{2}+\cot^{-1}\sqrt{2})$. 13. $7\pi a^2/1024$ each.
14. $a^2(2-\frac{1}{2}\pi)$. 15. $\pi(a^2+b^2)/12$.

Exercise XVIII b (p. 365)

1. 6π. 2. $\frac{1}{4}\pi a^2$. 3. $8a^2/15$. 4. $\frac{3}{2}\pi a^2$.
5. $4ab/3$. 6. $\frac{4}{5}$. 7. $1/60$. 8. $\frac{1}{4}\pi a^2$.
9. $6\pi a^2$. 10. $16/35$. 11. $\pi/6$.

Exercise XVIII c (p. 367)

1. $45\pi/16$. 2. $2\pi ac+\pi c^2 \operatorname{sh}(2a/c)$.
3. $\frac{8}{3}\pi\sqrt{\{a(a+b)^3\}}-\frac{8}{3}\pi a^2$. 4. 3π. 5. $6\pi a^2/5$.
6. $64\pi a^2/3$. 7. $2\pi\{\sqrt{2}+\log(\sqrt{2}+1)\}$. 8. $32\pi a^2/5$.
9. $\pi\{2\sqrt{5}-\sqrt{2}+\log(\sqrt{10}-\sqrt{5}+2\sqrt{2}-2)\}$. 10. $3\pi^3-16\pi$.

Exercise XVIII d (p. 370)

1. $(2a\sqrt{2})/\pi$; $(4a\sqrt{2})/(3\pi)$. 2. $\pi a^3\sqrt{2}$, $4\pi a^2\sqrt{2}$.
3. $4\pi^2 ab$; $2\pi^2 a^2 b$. 4. $4\pi^2 a^2$; $2\pi^2 a^3$. 5. $4\pi^2 a^2 b$.
6. $4\pi a^2 b/3$; $[4a/(3\pi), 0]$. 7. $(3\pi a/16, 0)$. 8. $(4a/5, 0)$.
9. $7\pi^2 a^3/2$. 10. $(\pi a, 4a/3)$; $32\pi a^2/3$. 11. $9\pi^2 a^2 b$.

Exercise XVIII e (p. 377)

1. 6. 2. $\frac{1}{4}\pi$. 3. $\frac{1}{4}a^4bc$. 4. $\frac{1}{3}\pi a^3$. 5. $6\frac{7}{15}$.
6. $1/10$. 7. $(6\pi-16\sqrt{2})/9$. 8. $\frac{1}{3}abc(b^2+c^2)$.
9. $\frac{2}{3}$. 10. $(3/7, 15/32)$. 11. $4\pi a^2 c(2-\sqrt{3})$.
12. $a^2(\beta \sim \gamma)(\sin\lambda \sim \sin\mu)$. 13. $3\pi\lambda a^4/2$. 15. $16a^3(3\pi-4)/9$.

Exercise XIX a (p. 383)

1. $1/y$, $-x/y^2$. 2. $2ax+2hy$, $2hx+2by$.
3. $-y/(x^2+y^2)$, $x/(x^2+y^2)$. 4. $2y/(x+y)^2$, $-2x/(x+y)^2$.
5. $1/\sqrt{(y^2-x^2)}$, $-x/\{y\sqrt{(y^2-x^2)}\}$, if $y>0$.
6. $-x/\sqrt{(x^2+y^2)^3}$, $-y/\sqrt{(x^2+y^2)^3}$. 7. 0, 1, 1, 0.
8. $6ax+6by$, $6bx$, $6bx$, $6cy$. 9. $-1/x^2$, 0, 0, $-1/y^2$.
10. Each is e^{x+y}. 11. $-y\cos x$, $-\sin x-\sin y$, $-\sin x-\sin y$, $-x\cos y$.
12. $\operatorname{sh} x \operatorname{ch} 2y$, $2\operatorname{ch} x \operatorname{sh} 2y$, $2\operatorname{ch} x\operatorname{sh} 2y$, $4\operatorname{sh} x\operatorname{ch} 2y$. 13. a/b.

Exercise XIX b (p. 388)

1. $4t^3+6t-2$.
2. $(1/V)dV=(2/x)dx+(3/y)dy$; $(1/V)dV=(1/x)dx-(1/y)dy$; $x^2(x^2+y^2)dV=y^2(ydx-xdy)$.
4. $-x^2/y^2$. 5. y^2/x^2. 6. $(y\cos x-\sin y)/(x\cos y-\sin x)$.
7. $axx_1+byy_1=1$. 8. $xx_1{}^{-1/3}+yy_1{}^{-1/3}=a^{2/3}$.
9. $x(2-y_1/x_1)+y(2-x_1/y_1)=x_1+y_1$.
10. $-yf_y/f_x$, $-yf_x/f_y$. 11. $-3a^4x^2/y^7$; $\frac{1}{2}\operatorname{sh}2y$.
15. $-\cos B\sec C$. 16. 0 ; $-3V/2$.

Exercise XIX c (p. 393)

1. $du=(udx+dy)/(u+v)$, $dv=(dy-vdx)/(u+v)$.
3. $\cos\theta$; $\cos\theta$; $-r\sin\theta$ if $x=f(r,\theta)$; $-(\sin\theta)/r$ if $\theta=g(x,y)$.
4. xV_y-yV_x. 6. $V_r{}^2+(1/r^2)V_\theta{}^2$.
7. $\dot{x}=\dot{r}\cos\theta-\dot{\theta}r\sin\theta$, $\dot{y}=\dot{r}\sin\theta+\dot{\theta}r\cos\theta$.
13. $2f_{uu}+(u+2v/u)(f_{uv}+f_{vu})+(u^2+2v+2v^2/u^2)f_{vv}+2f_v$.
16. $(\pm\sqrt{2},\mp\sqrt{2})$, $(0,0)$.

Exercise XX a (p. 397)

4. $\dfrac{d^2x}{dt^2}=-n^2x$. 5. $xdy=ydx$. 6. $\dfrac{d^2y}{dx^2}=0$.

7. $y\dfrac{d^2y}{dx^2}+\left(\dfrac{dy}{dx}\right)^2=0$; $(1+x^2)\dfrac{d^2y}{dx^2}+2x\dfrac{dy}{dx}=0$.

Exercise XX b (p. 401)

1. $y=\frac{1}{2}x^6+c$. 2. $12x+y^{-6}=c$. 3. $y=x^{-1}-\frac{1}{3}x^{-3}+c$.
4. $x=\log(c\sin y)$. 5. $r\theta+cr+k=0$.
6. $3y=2x^{3/2}+c$. 7. $x^2+2xy=c$. 8. $x^2-2xy-y^2=c$.
9. $x^4-2x^2y^2=c$. 10. $x=\tan(c-1/y)$. 11. $(x-a)^2+y^2=c$.
12. $xy=ax+c$. 13. $x+y=a\log(cx)$. 14. $\tan y+\cot x=c$.
15. $y=\pm\sin(x-c)$. 16. $x^2=2y^2\log(cy)$.
17. $y=\frac{1}{4}x^4+\log(cx)$. 18. $y=ce^{x/k}$. 19. $r=ce^{\theta/k}$.
20. $(1+x-y)/(1-x+y)=ce^{2x}$. 21. $\tan\frac{1}{2}(x+y)=x+c$.
22. $x^{n-1}(xdy+nydx)$; $(n+k)y=x^k+cx^{-n}$, $k\neq-n$; $x^ny=\log(cx)$, $k=-n$.
23. $y=k\operatorname{ch}\{(x-c)/k\}$, a catenary.

Exercise XX c (p. 405)

1. $y+1=ce^x$. **2.** $y=\frac{1}{2}e^x+ce^{-x}$.
3. $y\sin x=c-\frac{1}{4}\cos 2x$. **4.** $y\sec x=\log(c\sec x)+\frac{1}{4}\cos 2x$.
5. $y=ce^x-x-1$. **6.** $y=1+x\tan x+c\sec x$.
7. $y^{-3}=1+ce^{3x}$. **8.** $y^2=4x/(c-x^4)$. **9.** $y^{-3}=ce^x-x-1$.
10. $y=1/(\cos x+c\sin x)$. **12.** (i) $y=cx+1/c$; $y^2=4x$;
(ii) $y=cx\pm\sqrt{(a^2c^2+b^2)}$; $x^2/a^2+y^2/b^2=1$.

Exercise XX d (p. 407)

1. $x+y=k$. **2.** $xy=k$. **3.** $x^2-y^2=k$. **4.** $y^2+2x^2=k$.
5. $5y^2+x^2=k$. **6.** $r=ke^{-\theta^2/2}$. **7.** $r=k(1-\cos\theta)$.
8. $r^2=k\cos 2\theta$. **9.** $r=k\sin^2\theta$.

Exercise XX e (p. 409)

1. $y=-a\log x+\mathsf{B}x+\mathsf{C}$. **2.** $y=-(1/n^2)\sin nx+\mathsf{A}x+\mathsf{B}$.
3. $8y=2ax^2-a\cos 2x+\mathsf{B}x+\mathsf{C}$. **4.** $y=e^{-x}+\mathsf{B}x+\mathsf{C}$.
5. $y=\mathsf{A}\,\mathrm{sh}\,nx+\mathsf{B}\,\mathrm{ch}\,nx$. **6.** $\mathsf{A}(y^2-\mathsf{A})=(kx+\mathsf{B})^2$.
7. $y=\mathsf{B}e^{ax}+\mathsf{C}$. **8.** $y=\mathsf{A}x^3+\mathsf{B}$. **9.** $x=\mathsf{A}+\mathsf{B}y-\frac{1}{3}y^3$.
10. $y=\mathsf{A}\log(\mathsf{B}x)$. **11.** $3y=4c(x+\mathsf{A})^{3/2}+\mathsf{B}$.
12. $y=\mathsf{A}+\log\cos(x+\mathsf{B})$. **13.** $y=\mathsf{A}+\mathsf{B}\tan^{-1}x$.
14. $y^2-\mathsf{B}^2=(\mathsf{B}^4-a^4)^{1/2}\sin(2x+\mathsf{C})$.
15. $\pi(2m\pm\frac{1}{3})/n$; $2\pi(m\pm\frac{1}{3})/n$. **16.** $x=a\,\mathrm{ch}\{t\sqrt{(2g/l)}\}$.
17. The conics in No. 6.

Exercise XX f (p. 415)

1. $y=\mathsf{A}e^x+\mathsf{B}e^{2x}$. **2.** $y=\mathsf{A}e^x+\mathsf{B}e^{-3x}$. **3.** $y=\mathsf{A}e^{2x}+\mathsf{B}e^{-2x}$.
4. $y=\mathsf{A}\sin 3x+\mathsf{B}\cos 3x$. **5.** $y=(\mathsf{A}x+\mathsf{B})e^{3x}$.
6. $y=(\mathsf{A}x+\mathsf{B})e^{-2x}$. **7.** $y=e^{-x}(\mathsf{A}\sin 2x+\mathsf{B}\cos 2x)$.
8. $y=e^x(\mathsf{A}\sin x+\mathsf{B}\cos x)$. **9.** $y=e^{3x}(\mathsf{A}\sin 2x+\mathsf{B}\cos 2x)$.
10. $y=e^{-x/2}(\mathsf{A}\sin\frac{1}{2}x\sqrt{3}+\mathsf{B}\cos\frac{1}{2}x\sqrt{3})$. **11.** $3\frac{1}{2}$. **12.** $4, -6$.
13. $-4\frac{1}{2}, -13\frac{1}{2}, -24\frac{3}{4}$. **14.** $1\frac{1}{4}$. **15.** $-2/17, -11/85$.
16. $\frac{1}{4}, -\frac{1}{8}$. **17.** $-5/9, -5/9, -10/27$. **18.** $1\frac{2}{3}$.
19. $0, -\frac{3}{4}$. **20.** $\frac{1}{2}$. **21.** $y=\mathsf{A}e^x+\mathsf{B}e^{3x}+4$.
22. $y=\mathsf{A}e^{-x}+\mathsf{B}e^{-2x}+2x-3$. **23.** $y=\mathsf{A}e^{3x}+\mathsf{B}+x+4x^2$.
24. $y=\mathsf{A}e^x+\mathsf{B}e^{-x}+\frac{1}{3}e^{2x}$. **25.** $y=\mathsf{A}e^{3x}+\mathsf{B}e^{-3x}+\frac{1}{6}xe^{3x}$.
26. $y=\mathsf{A}e^{3x}+\mathsf{B}e^{-3x}-\frac{1}{10}\sin x$.
27. $y=\mathsf{A}\sin 3x+\mathsf{B}\cos 3x-\frac{1}{6}x\cos 3x$.
28. $y=\mathsf{A}e^x+\mathsf{B}e^{2x}+(\cos x-3\sin x)/10$.

29. $y=(\mathsf{A}+\mathsf{B}x+\frac{1}{2}x^2)e^x$. **30.** $y=(\mathsf{A}-x)e^{2x}+\mathsf{B}e^{3x}$.
31. $y=\mathsf{A}\operatorname{ch}2x+\mathsf{B}\operatorname{sh}2x+\mathsf{C}\cos 2x+\mathsf{D}\sin 2x$.
32. $y=\mathsf{A}e^{2x}+\mathsf{B}e^{-2x}+\frac{1}{5}\sin x-\frac{1}{3}e^x$.
33. $y=\mathsf{A}\cos\log x+\mathsf{B}\sin\log x$; $y=\mathsf{A}x^3\log(\mathsf{B}x)+2x$.
34. $\phi e^{\int \mathsf{P}\,dx}$. **35.** $y=\mathsf{A}x+\mathsf{B}\sqrt{(1+x^2)}$. **36.** $y=\mathsf{A}e^x+\mathsf{B}xe^{-x}$.
38. $y=(\mathsf{A}\sin kx+\mathsf{B}\cos kx)/x$.

Exercise XXI a (p. 423)

(In Nos. 1-6, r takes the values 1, 2, 3, . . .)

1. $1+\Sigma(-2x)^r/r!$; $(-2x)^n e^{-2\theta x}/n!$ **2.** $1+\Sigma(-x^2)^r/(2r)!$; $x^n\cos(\theta x+\frac{1}{2}n\pi)/n!$
3. $\Sigma x^{2r-1}/(2r-1)!$; $\frac{1}{2}x^n\{e^{\theta x}-(-1)^n e^{-\theta x}\}/n!$
4. $-\Sigma x^r/r$; $-x^n/\{n(1-\theta x)^n\}$. **5.** $\cos x+\Sigma h^r\cos(x+\frac{1}{2}r\pi)/r!$
6. $\log a+\Sigma(-1)^{r-1}(1/r)(x/a)^r$. **15.** $\frac{1}{2}$. **16.** $\frac{1}{6}$. **17.** 1.
18. $-\frac{1}{2}$. **19.** $-\frac{1}{2}$. **20.** 2/7. **21.** $y=0$ or 1 according as x is integral or not.

Exercise XXI b (p. 426)

1. $(n+2)!\,x^2/2$. **2.** $(-1)^n n!/x^{n+1}$.
3. $(-1)^{n+1}2^n(n-1)!/(2x+3)^n$.
4. $(-\frac{3}{4})^n(2n)!/\{n!\sqrt{(3x+5)^{2n+1}}\}$. **5.** $3^n\sin(3x+\frac{1}{2}n\pi)$.
6. $-2^{n-1}\cos(2x+\frac{1}{2}n\pi)$. **7.** $(\sqrt{2})^n e^x\sin(x+\frac{1}{4}n\pi)$.
8. $5^n e^{3x}\cos(4x+n\tan^{-1}\frac{4}{3})$. **9.** $\frac{1}{2}\{e^x+(-1)^n e^{-x}\}$.
10. $\frac{1}{4}\{3^n\cos(3x+\frac{1}{2}n\pi)+3\cos(x+\frac{1}{2}n\pi)\}$.
11. $(-1)^n n!\{(x-a)^{-n-1}-(x+a)^{-n-1}\}/(2a)$.
12. $(-1)^n n!\{(x+1)^{-n-1}-(x+2)^{-n-1}\}$.
13. $\frac{1}{4}\{3\sin(x+\frac{1}{2}n\pi)-3^n\sin(3x+\frac{1}{2}n\pi)\}$.
14. $\frac{1}{2}\{5^n e^{5x}+(-1)^n e^{-x}\}$. **15.** $2^{n-1}e^{2x}(2x+n)$.
16. $\{x^3+3nx^2+3n(n-1)x+n(n-1)(n-2)\}e^x$.
17. $(-1)^n(x^2-2nx+n^2-n)e^{-x}$.
18. $3^{n-2}\{(9x^2-n^2+n)\sin(3x+\frac{1}{2}n\pi)-6nx\cos(3x+\frac{1}{2}n\pi)\}$.
19. $-6(-x)^{3-n}(n-4)!$ if $n>3$.
20. $\frac{1}{2}(x^2+2nx+n^2-n+1)e^x-\frac{1}{2}(-1)^n(x^2-2nx+n^2-n+1)e^{-x}$.

(In Nos. 21-25, r takes the values 1, 2, 3, . . .)

21. $(\sqrt{2})^n e^x\sin(x+\frac{1}{4}n\pi)$; $\Sigma(x\sqrt{2})^r(\sin\frac{1}{4}r\pi)/r!$
23. $x+\Sigma 4^r(r!)^2x^{2r+1}/(2r+1)!$
24. $x^2+\Sigma 2^{2r+1}(r!)^2x^{2r+2}/(2r+2)!$
25. $1+\Sigma p^2\{p^2+2^2\}\ldots\{p^2+(2r-2)^2\}x^{2r}/(2r)!$

Exercise XXI c (p. 430)

9. $y = x^3 + x^9$. **10.** $y = x - 2x^3$. **11.** $y = \pm x\sqrt{x} + \frac{1}{2}x^3$.
12. $x = y + 2y^2$, $x = -y^2 + y^3$. **13.** $y = \pm(x\sqrt{x} + \frac{1}{2}x^2\sqrt{x})$.
14. $y = \pm x\sqrt{x} + x^2$. **15.** $y = x^3 - x^5$, $y = -x - 2x^3$.
16. $y = -x + \frac{2}{3}x^2$. **17.** $y = x - \frac{4}{3}x^3$. **18.** $y = \pm x - x^3$.
19. $y = x \pm 2x\sqrt{x}$. **20.** $y = -x + \frac{1}{4}x^2$, $y = x \pm \frac{1}{2}x\sqrt{(2x)}$.
21. $\frac{1}{6}$. **22.** $\frac{1}{2}, -\frac{1}{2}$. **23.** $\frac{5}{2}\sqrt{5}$. **24.** $4\sqrt{2}, -20\sqrt{10}$.
25. (i) No; (ii) only $x = 0$.

Exercise XXI d (p. 436)

1. $x = 1$, $x = -1$, $y = 0$. **2.** $x = a$. **3.** $y = 2x$, $y = 3x$.
4. $y = x - 1$, $y = 2x + 1$. **5.** $y = x$, $y = 2x$, $y = 3x$.
6. $x = 2$, $x = -2$, $y = x$. **7.** $y = \frac{1}{2}$, $y = -\frac{1}{2}$, $y = -4x$.
8. $y + 4x = \frac{1}{16}$. **9.** $y = x + 2$, $y = 2x - 3$, $y = 4x + 1$.
10. $x + y = 2$. **11.** $y = x$, $y = -x$. **12.** $x = 0$, $y = 1$.
13. $x = 0$, $y = x + 2$, $y = x - 2$. **14.** $y + x = 0$, $y = x + 2$, $y = x - 2$.
15. $y = x + 1$, $y = x - 1$. **16.** $x = 0$, $y = x + 1$.
19. $(y + x - 3/32)^2 = \frac{1}{8}(x - y + 13/128)$.
20. $y = 1$, $y = -1$, $y + x = 0$; $x = y^3$.
21. $y = 0$, $y = x + 1$, $y = x - 1$; $x = y^3$.
22. $y = 2x + 4$; $y^2 = 10x$, $x^2 = -5y$.
23. $y = x$, $y = -x$; $y = \frac{1}{2}x^3$, $x = -\frac{1}{2}y^3$.
24. $y + x + 1 = 0$; $y = 2x + 2\frac{1}{4}x^2$, $y = -2x + 1\frac{3}{4}x^2$.
25. $y + x - 1 = 0$, $y + x + 1 = 0$; $y = -\frac{1}{4}x^3$, $x = -\frac{1}{4}y^3$.
26. $x = 1$, $x = -1$, $y = x$, $y + x = 0$; origin is isolated point.
27. $x = 0$, $y = x$; $y = -x \pm x^2\sqrt{2}$.
28. $y = 2x - 3$, $y + 3x = 1$, $y = 5x$; $2y = x - 3$, $y = x - 4$, $y = x + 1$.
30. $y(y^2 - x^2) + xy - 3y^2 + 2x = 0$. **31.** $y = 0$, $(y \simeq 1/x \pm 1/\sqrt{x^3})$.
32. No asymptotes.

Exercise XXII a (p. 440)

1. $\frac{4}{3}Ma^2$. **2.** $\frac{1}{3}Mb^2(3a + b)/(a + b)$. **3.** $\frac{4}{3}Mb^2$.
4. $\frac{1}{2}Ma^2$. **5.** $\frac{1}{2}Mh^2$. **6.** $3Ma^2/10$. **7.** $4Mac/5$.
8. $\frac{4}{3}Mac$. **9.** $7Ma^2/64$. **10.** $\frac{1}{4}Mc^2(2\theta - \sin 2\theta)/\theta$.

Exercise XXII b (p. 445)

1. $\frac{2}{3}M(2a^2 + ab + 2b^2)$. **2.** $\frac{1}{3}M(m^2 + n^2)$. **3.** $5Ma^2/4$.
4. $7Ma^2/5$. **5.** $\frac{1}{4}M(a^2 + b^2)$. **6.** $\frac{2}{5}M(a^5 - b^5)/(a^3 - b^3)$.
7. $\frac{1}{2}Ma^2(3a + 4h)/(a + h)$. **8.** $Mh^2(3\tan^2\alpha + 2)/20$.

9. $\frac{2}{3}Ml^2$; $2Ml^2/75$. 10. $\frac{2}{3}Ma^2$; $\frac{1}{3}Ma^2$; $\frac{4}{3}Ma^2$.
11. $Ma^2(15\pi - 32)/(12\pi)$. 12. $M(3a^2 + 4h^2)/12$; $M(15a^2 + 4h^2)/12$.
13. $M(3a^3 + 6a^2h + 6ah^2 + 4h^3)/(12a + 12h)$. 14. $\frac{1}{2}Mr^2$.
15. $\frac{1}{3}Ma^2$. 16. $10Ma^2/21$.

Exercise XXII c (p. 447)

1. ab^2w, $3ab^2w$; $\frac{1}{2}ab^2w$, $\frac{3}{2}ab^2w$. 2. πa^3w ; $\pi a^2(a + h)w$. $5a/4$; $\frac{1}{4}(4h^2 + 8ah + 5a^2)/(a + h)$.
3. $\frac{1}{2}wa^3\sqrt{3}$. 4. $\frac{3}{4}h$; $\frac{1}{2}(6c^2 + 8ch + 3h^2)/(3c + 2h)$.
5. $\frac{8}{5}wc^2\sqrt{(ac)}$; $5c/7$. 6. $\frac{7}{12}a\sqrt{2}$. 7. $h + a^2/(4h)$.
8. $3\pi a/16$. 10. $\frac{1}{6}(7a^2 + 14ab + 8b^2)/(a + b)$.

ANSWERS

APPENDIX TO VOLUME II

(Constants of integration are usually omitted)

Exercise XI a (p. 449)

51. $(10x+3)(2x-3)^5/120$. **52.** $\frac{1}{6}(1-3x)/(x-1)^3$.
53. $-(2/15)(8+4x+3x^2)\sqrt{(1-x)}$.
54. $(11bx-a)(a+bx)^{11}/(132b^2)$.
55. $(2/15)(3x+5a-2b)\sqrt{(x+b)^3}$. **56.** $\frac{1}{3}\sqrt{(1+x^2)^3}$.
57. $(a+bx^2)^{n+1}/\{2b(n+1)\}$. **58.** $-\frac{1}{3}\operatorname{cosec}^3 x$.
59. $-\frac{1}{3}\cot^3 x$. **60.** $(1/a)\tan^{-1}(x/a)$.
61. $-(1/1215)(2+27x^2)\sqrt{(1-9x^2)^3}$. **62.** $\frac{1}{2}\{\tan^{-1}x - x/(1+x^2)\}$.
63. $\sin^{-1}(x/|a|)$. **64.** $a\sin^{-1}(x/a) - \sqrt{(a^2-x^2)}$. **65.** $\sec^{-1}|x|$.
66. $\{(1-n)bx-a\}/\{(n-1)(n-2)b^2(a+bx)^{n-1}\}$.
67. $(2/15)(8-4x+3x^2)\sqrt{(1+x)}$. **68.** $(2/135)(9x-8)\sqrt{(3x+4)^3}$.
69. $\frac{2}{9}\sqrt{(1+x^3)^3}$. **70.** $\frac{1}{3}\sec^3 x$. **71.** $\frac{1}{3}\tan^3 x$.
72. $x/\{a^2(x^2+a^2)^{1/2}\}$. **73.** $-(1/243)(2+9x^2)\sqrt{(1-9x^2)}$.
74. $-\sqrt{(x^{-2}-1)}$, if $0<x<1$.

Exercise XI b (p. 450)

51. $2\{\sqrt{(a+b)}-\sqrt{a}\}/b$. **52.** $\{(2b-4a)\sqrt{(a+b)}+4a\sqrt{a}\}/(3b^2)$.
53. $3\frac{3}{4}$. **54.** $49/81$. **55.** $\frac{1}{6}$. **56.** $\frac{1}{3}\pi$. **57.** $\pi/12$.
58. $\frac{1}{4}(\sqrt{5}-1)$. **59.** $\pi/24$. **60.** $\frac{1}{4}\sqrt{3}(\sqrt{5}-2)$. **61.** $1/10$.
62. $\frac{1}{2}\sqrt{2}\cot^{-1}2$. **63.** $\frac{1}{3}\pi+\frac{1}{2}\sqrt{3}$. **64.** $\frac{1}{3}\pi$. **65.** $\frac{1}{8}\pi\sqrt{2}$.
66. $-\frac{1}{2}, \frac{1}{2}$. **67.** $\frac{2}{3}$. **68.** $\frac{1}{2}-\frac{1}{4}\sqrt{3}$; $\frac{1}{4}\sqrt{3}-\frac{1}{2}$; -1.

Exercise XI c (p. 451)

51. $(nx\sin nx+\cos nx)/n^2$. **52.** $\frac{1}{4}(\cos 2x+2x\sin 2x-2x^2\cos 2x)$.
53. $(x^3-6x)\sin x+3(x^2-2)\cos x$. **54.** $\frac{1}{2}(x^2+1)\tan^{-1}x-\frac{1}{2}x$.
55. $\frac{1}{2}(x\sec^2 x-\tan x)$. **56.** $\frac{1}{2}(x^2+1)\cot^{-1}x+\frac{1}{2}x$. **57.** $5\pi/32$.
58. $5\pi/32$. **59.** $(x^4-12x^2+24)\sin x+4x(x^2-6)\cos x$.
60. $u_n = -x^n\cos x+nx^{n-1}\sin x-n(n-1)u_{n-2}$;
$4x(x^2-6)\sin x-(x^4-12x^2+24)\cos x$.
61. $\pi^2/16$. **62.** $(3x\sin 3x+\cos 3x)/18 \mp (5x\sin 5x+\cos 5x)/50$.
63. $1/1260$. **66.** $(n-1)u_n=\sec^{n-2}x\tan x+(n-2)u_{n-2}$.

Exercise XII a (p. 452)

51. n/x. **52.** $\tan x$. **53.** $(\cos\log x)/x$.
54. $3(\log x)^2/x$. **55.** $(1-\frac{1}{2}\log x)/(x\sqrt{x})$. **56.** $1/(x\log x)$.
57. $2\cot 2x$. **58.** $\cot x\cos x - \sin x\log\sin x$.
59. $4x/(1-x^4)$. **60.** $2x/(a^2+x^2)$. **61.** $1/\sqrt{(x^2-1)}$.
62. $\sec x$. **63.** $\sec x$. **64.** $1/\sqrt{(a^2+x^2)}$.
65. $x^{m-1}(m\log x+n)(\log x)^{n-1}$. **67.** $-1/e$; $-1/(2e)$.
68. $\pm\sqrt{\sqrt{(4+\sqrt{17})}}$, $\frac{1}{2}\log(2+2\sqrt{17})$. **72.** $\tan^{-1}2$.

Exercise XII b (p. 453)

51. $-2\log(1-x)$. **52.** $-(1/q)\log(p-qx)$.
53. $\frac{1}{2}\log(x^2+2ax+b)$. **54.** $\log\tan x$. **55.** $\frac{1}{2}(\log x)^2$.
56. $\log\log x$. **57.** $\log(b/a)$, $ab>0$. **58.** $\log(b/a)$, $c\neq 0$, $ab>0$.
59. 1. **60.** $-\log 3$. **61.** $-\frac{1}{2}\log 2$. **62.** $\log(3/2)$.
63. $2-\log 2$. **64.** $4-\frac{1}{2}\log 5$. **65.** 1. **66.** $1-2/e$.
67. $\frac{3}{4}+\frac{1}{2}\log(8/27)$. **68.** $(\frac{1}{2}\pi-1+\log 2)/6$.
69. $c^2\log\{(q-1)/(p-1)\}$. **70.** $\frac{1}{2}(1-\log 2)$.
71. $\pi ac^2(\lambda-1-\log\lambda)$. **74.** $\frac{1}{2}k\log\{1+u^2/(gk)\}$.

Exercise XII c (p. 454)

51. 1. **52.** 1. **53.** $x^{n-1}e^x(x+n)$. **54.** $e^{\sqrt{x}}/(2\sqrt{x})$.
55. $(1+2x^2)e^{x^2}$. **56.** $e^{ax}(a\cos bx-b\sin bx)$.
57. $e^{-bx}(a\cos ax-b\sin ax)$. **58.** $2e^x/(e^x+1)^2$.
59. $4/(e^x+e^{-x})^2$. **60.** x^2 ; $2^x/\log 2$; $\frac{1}{3}x^3$. **61.** x^2. **62.** e^{e^x}.
63. $e^x\log x$. **64.** xe^{x^2}.
65. $\frac{1}{2}(e^{2x}-1)\log(1+e^{-x})+\frac{1}{2}(e^x-x)$. **69.** 3, -2.
71. $\pi(3n+1)/(3\sqrt{3})$. **72.** $1/b$. **73.** $\pm\frac{1}{2}$. **74.** $(=c)$.
75. $(n+\frac{1}{4})\pi/a$; $e^{-\pi/4}/\sqrt{2}$. **76.** $\mathrm{T}=\mathrm{T}_0e^{2\mu\pi}$.

Exercise XII d (p. 455)

51. $a^x(2x+x^2\log a)$. **52.** $(\cot x)/\log p$. **53.** $(ex)^x(2+\log x)$.
54. $(1+x)^{m-1}\{1+m\log(1+x)\}$.
55. $x^{n-1}e^x\tan x(n+x+2x\operatorname{cosec}2x)$.
56. $e^{-x}(x+1)^{x-1/2}\{(x+1)\log(x+1)-\frac{1}{2}\}$. **57.** $10^{2x}/\log 100$.
58. $x(\log x-1)/\log p$. **59.** $e^x\tan x$. **65.** $\frac{1}{2}\sin 2t$.

Exercise XII e (p. 456)

51. $\frac{2}{3}(\log x-\frac{2}{3})x\sqrt{x}$. **52.** $x^{n+1}\{(n+1)\log x-1\}/(n+1)^2$.
53. $e^{2x}(2\sin 3x-3\cos 3x)/13$. **54.** $e^{2x}(2\cos 3x+3\sin 3x)/13$.
55. $2\{\log(ax)-2\}\sqrt{x}$. **56.** $\frac{1}{4}x^2\{2(\log x)^2-2\log x+1\}$.

57. $\frac{1}{4}e^x(\cos x + \sin x) - e^x(\cos 5x + 5\sin 5x)/52$.
58. $\frac{1}{4}e^x(\cos x + \sin x) + e^x(\cos 5x + 5\sin 5x)/52$.
59. $x^3\{9(\log x)^2 - 6\log x + 2\}/27$.
60. $x\log\{x + \sqrt{(x^2 - 1)}\} - \sqrt{(x^2 - 1)}$.
61. $x\log\{x + \sqrt{(a^2 + x^2)}\} - \sqrt{(x^2 + a^2)}$. **62.** $x - (\sin^{-1} x)\sqrt{(1 - x^2)}$.
63. $\frac{1}{4}(e^2 - 3)/e^2$. **64.** $7\log 2 - 3$. **65.** $\frac{1}{3}\log 4 - 5/18$.
66. $(\pi + 1 - \log 4)/20$. **67.** $3\log 3 - 2$. **68.** $\pi/8$.
72. $u_n + nu_{n-1} = x^n e^x$.

Exercise XIII b (p. 457)

52. $(3\,\text{th}\,\theta + \text{th}^3\,\theta)/(1 + 3\,\text{th}^2\,\theta)$. **53.** $-2\,\text{sh}^2\,x$. **54.** $2\coth 2x$.
55. $\text{ch}\,(x + y) - \text{ch}\,(x - y)$. **56.** $\frac{1}{2}\,\text{sh}\,(\theta + \phi) + \frac{1}{2}\,\text{sh}\,(\theta - \phi)$.
57. $\frac{1}{2}\,\text{ch}\,(\theta + \phi) + \frac{1}{2}\,\text{ch}\,(\theta - \phi)$. **58.** $-2\,\text{sh}\,A\,\text{sh}\,3A$. **59.** e^x.
60. e^{nx}. **61.** $-\coth\frac{1}{2}x$. **62.** $2/\sqrt{5}$. **63.** $2/\sqrt{3}$.
64. $\pm\sqrt{\{\frac{1}{2}(k - 1)\}}$. **65.** $(1 + t^2)/(1 - t^2)$. **66.** 2.
67. $x^2\,\text{sech}^2\,b + y^2\,\text{cosech}^2\,b = 1$. **68.** $\frac{1}{2}(x^2 - 1)/x$; $\frac{1}{2}(x^2 + 1)/x$.
69. $\sec\theta, \sin\theta$; $\text{ch}\,x = -\sec\theta$, $\text{th}\,x = -\sin\theta$, $x = \log(\tan\theta - \sec\theta)$.
71. $-\coth\frac{1}{2}x$.

Exercise XIII c (p. 458)

51. $\text{sh}\,2x$. **52.** $\text{ch}\,2x$. **53.** $-\text{ch}\,x\,\text{cosech}^2\,x$.
54. $-2\,\text{cosech}^2\,2x$. **55.** $2\,\text{sh}\,2x$. **56.** $4\,\text{ch}^2\,x$. **57.** $\text{th}\,x$.
58. $-\text{sech}\,2x$. **59.** 1. **60.** $na\,\text{sh}^{n-1}\,(ax)\,\text{ch}\,(ax)$.
61. $3\,\text{sech}^4\,x$. **62.** $2\cos x\,\text{sh}\,x$. **63.** $\frac{1}{3}\,\text{ch}\,3x$.
64. $2\log\text{ch}\,\frac{1}{2}x$. **65.** $\frac{1}{2}x + (\text{sh}\,6x)/12$. **66.** $x - \frac{1}{2}\,\text{th}\,2x$.
67. $\frac{1}{2}\log\text{th}\,x$. **68.** $-\frac{1}{2}x^2$. **69.** $\frac{1}{2}(\text{sh}\,x + \frac{1}{5}\,\text{sh}\,5x)$.
70. $\frac{1}{4}(\frac{1}{4}\,\text{sh}\,8x - \text{sh}\,2x)$. **71.** $\frac{1}{2}(\frac{1}{5}\,\text{ch}\,5x - \text{ch}\,x)$. **72.** $2\,\text{sh}\,x\cos x$; $\frac{1}{2}(\text{ch}\,x\cos x + \text{sh}\,x\sin x)$. **73.** $1 + \text{sh}\,(\frac{1}{2}\pi)$.
76. $(x^2 + 2)\,\text{ch}\,x - 2x\,\text{sh}\,x$; $(x^3 + 6x)\,\text{sh}\,x - 3(x^2 + 2)\,\text{ch}\,x$.
77. $x^{\text{sh}\,x}\{\text{ch}\,x\log x + (\text{sh}\,x)/x\}$.
78. $e^{ax}(a\,\text{sh}\,bx - b\,\text{ch}\,bx)/(a^2 - b^2)$. **79.** $e^{-x}\,\text{th}\,x$.

Exercise XIII d (p. 459)

51. $b \geqslant |a|$. **54.** $(=a)$. **55.** $\text{sh}\,(x/c)$.
56. $2\tan^{-1} e^a - \frac{1}{2}\pi$; yes, $(\frac{1}{2}\pi)$. **57.** $\pi/8$.
58. $y = \sqrt{(a^2 + ax)}$; $\frac{2}{3}a^2$.
60. $\pi\{2\log(e^k\,\text{sech}\,k) - \text{th}\,k\}$; $\pi(\log 4 - 1)$.

Exercise XIII e (p. 460)

51. $\log 3$. 52. $\frac{1}{2}\log 2$. 53. $\log(1+\sqrt{2})$.
57. $1/\sqrt{(x^2+x)}$. 58. $\frac{1}{2}\sec x$. 59. $+1/\sqrt{(1+x^2)}$ if $x>0$; $-$ if $x<0$. 60. $-\sqrt{2}/\{|1+x|\sqrt{(1+x^2)}\}$.
61. $(\frac{1}{2}\sqrt{2})/\{(1-x)\sqrt{(1+x)}\}$. 62. $2/(1-x^2)$.
63. (i) $|\sec x|$; (ii) $\sec x$ if $\tan x>0$, $-\sec x$ if $\tan x<0$; (iii) $\sec x$.
64. $2\sqrt{(1+x^2)}$; $\frac{1}{2}\sqrt{2}+\frac{1}{2}\log(1+\sqrt{2})$. 65. 2.
66. $\log(\cot\theta-\operatorname{cosec}\theta)$. 67. $\operatorname{th}^{-1}\{(x+y)/(1+xy)\}$.

Exercise XIII f (p. 461)

51. $\frac{1}{2}\operatorname{sh}^{-1}(\frac{2}{3}x)$. 52. $\frac{1}{5}\operatorname{ch}^{-1}(5x)$.
53. $-\frac{1}{4}\sqrt{(9-4x^2)}+\frac{1}{2}\sin^{-1}(\frac{2}{3}x)$. 54. $\operatorname{sh}^{-1}\frac{1}{2}$.
55. $\operatorname{ch}^{-1}(3/2)-\operatorname{ch}^{-1}2$. 56. $\frac{1}{2}\operatorname{sh}^{-1}1$.
57. $\operatorname{sh}^{-1}x-\{\sqrt{(1+x^2)}\}/x$. 58. $-\operatorname{cosech}^{-1}x$.
59. $-\operatorname{sech}^{-1}x$. 60. $\operatorname{ch}^{-1}(\frac{1}{2}x-1)$.
61. $\frac{1}{2}(x+3)\sqrt{(x^2+6x)}-(9/2)\operatorname{ch}^{-1}\frac{1}{3}(x+3)$.
62. $\frac{1}{2}(x+a)\sqrt{\{(x+a)^2+b^2\}}+\frac{1}{2}b^2\operatorname{sh}^{-1}\{(x+a)/b\}$.
63. $x\operatorname{ch}^{-1}(\frac{1}{2}x)-\sqrt{(x^2-4)}$. 64. $\frac{1}{4}(2x^2+1)\operatorname{sh}^{-1}x-\frac{1}{4}x\sqrt{(1+x^2)}$.
65. $\frac{1}{4}(x^4-1)\operatorname{th}^{-1}x+(3x+x^3)/12$.
66. $x\coth^{-1}(2x)+\frac{1}{4}\log(4x^2-1)$.
67. $\frac{1}{3}x^3\operatorname{ch}^{-1}x-\frac{1}{9}(x^2+2)\sqrt{(x^2-1)}$. 68. $(\operatorname{sh}^{-1}x)\sqrt{(1+x^2)}-x$.
69. $1/\sqrt{(a^2+x^2)}$; $-1/\sqrt{(a^2+x^2)}$. 70. $1/\sqrt{(x^2-a^2)}$. 71. No.
72. $-a/\{x\sqrt{(a^2-x^2)}\}$. 73. $-a/\{x\sqrt{(x^2+a^2)}\}$.
74. $a/(a^2-x^2)$. 75. $2\sqrt{(x^2-a^2)}$.
76. $2\sqrt{(a^2+x^2)}$ if $a>0$; $2x^2/\sqrt{(a^2+x^2)}$ if $a<0$. 77. $1-\frac{1}{2}\sqrt{3}$.
80. $(1/k)\operatorname{ch}^{-1}(b/a)$.

Exercise XIV a (p. 462)

51. $\frac{2}{5}\sqrt{x^5}$. 52. $-2/\sqrt{x}$. 53. $\log(e^x+1)$.
54. $x-\log(e^x+1)$. 55. $-(1/n)\cos nx$.
56. $(1/n)\log|\sin nx|$. 57. $\frac{1}{2}\operatorname{th}2x$. 58. $-2\coth\frac{1}{2}x$.
59. $\log(1-\cos x)$. 60. $-\log(\operatorname{ch}x-1)$. 61. $\frac{1}{2}\log(4+x^2)$.
62. $\frac{1}{3}(\log x)^3$. 63. $\log\log\operatorname{cosec}x$. 64. $\log\tan^2\frac{1}{4}x$.
65. $\sec x$. 66. $-1/(n\sin^n x)$. 67. $\frac{1}{4}\operatorname{sh}^{-1}(4x/5)$.
68. $-(1/16)\sqrt{(25-16x^2)}$. 69. $\frac{1}{4}\sin^{-1}(4x/5)$.
70. $\frac{1}{4}\{\frac{1}{4}\sqrt{(16x^2+25)}+\operatorname{sh}^{-1}(4x/5)\}$. 71. $\log\operatorname{sh}^2\frac{1}{2}x$.
72. $\frac{1}{2}\log\operatorname{ch}(2x+3)$. 73. $\frac{1}{2}\log|\operatorname{th}x|$. 74. $4\tan^{-1}e^{x/2}$.
75. $\frac{1}{2}\tan^{-1}(2x+1)$. 76. $\frac{1}{4}\log|x/(x+1)|$.
77. $\frac{1}{4}\log|(x+2)/(x+1)|$. 78. $\frac{1}{2}\operatorname{sh}^{-1}(2x+1)$.
79. $\frac{1}{2}\operatorname{ch}^{-1}(2x+3)$, $x>-1$; $-\frac{1}{2}\operatorname{ch}^{-1}(-2x-3)$, $x<-2$.

80. $\frac{1}{2}\sin^{-1}(2x-1)$. **81.** $\tan^{-1}(x+1)$. **82.** $\text{ch}^{-1}(x+2)$, $x>-1$; $-\text{ch}^{-1}(-x-2)$, $x<-3$. **83.** $\sin^{-1}(x-2)$.
84. $\text{sh}^{-1}\{(x+k)/|a|\}$. **85.** $\text{ch}^{-1}\{(x+k)/|a|\}$, $x>|a|-k$; $-\text{ch}^{-1}\{-(x+k)/|a|\}$, $x<-|a|-k$.
86. $\sin^{-1}\{(x+k)/|a|\}$. **87.** $\frac{1}{2}(1/a)\log\{(a+k+x)/(a-k-x)\}$.
88. $\cos^{-1}(x/a)$.

Exercise XIV b (p. 463)

51. $-(7/9)\log(8-9x)$. **52.** $-(7/9)\log(9x-8)$.
53. $(5/6)\log(6x-7)$. **54.** $-\frac{3}{4}\log(1-2x)-\frac{3}{2}x$.
55. $x-2\log(x+1)$. **56.** $-3\log(2-x)-\frac{1}{2}x^2-2x$.
57. $\frac{1}{2}ax^2+(b-ap)x+(ap^2-bp+c)\log(x+p)$.
58. $\frac{1}{4}x^4+\frac{1}{3}x^3+\frac{1}{2}x^2+x+\log(x-1)$. **59.** $\frac{1}{6}\log(3x^2-5)$.
60. $\tan^{-1}x^2$. **61.** $\log(x+1)+2/(x+1)$.
62. $\log\{(x-3)/(x+3)\}$. **63.** $3\log(x+1)-(5/4)\log(4x-1)$.
64. $x+\log\{(x-1)/(x+1)\}$. **65.** $\log(2x-5)-(3/7)\log(7x-3)$.
66. $x+2\log(x-2)-\log(x-3)$. **67.** $x+2\log\{(x-2)/(x-3)\}$.
68. $\log(x^2-x)-3/(x-1)$. **69.** $\frac{1}{3}x^3+2x+\log\{(x-1)/(x+1)\}$.
70. $(x^2+1)/x+\log x-2\log(x+1)$. **71.** $\frac{1}{2}\log(x^2-1)-\log(x+2)$.
72. $(\frac{1}{2}-x)/(x-1)^2$. **73.** $\frac{2}{3}\log(x^3-1)$.
74. $\{\log(x-1)+9\log(x+3)-10\log(x+5)\}/24$.
75. $9\log(x-3)-5\log(x-2)-\log(x-1)$.
76. $\frac{1}{2}\log\{(x-1)/(x+1)\}-2/(x-1)$.
77. $\frac{1}{2}\log\{(x-3)/(x-1)\}+\frac{1}{2}(2x-1)/(x-1)^2$.
78. $x+\{a^2\log(x-a)-b^2\log(x-b)\}/(a-b)$.
79. $\{(pa+q)\log(x-a)-(pb+q)\log(x-b)\}/(a-b)$.
80. $\frac{1}{3}x^3-\frac{1}{2}x^2(a+b)+x(a^2+ab+b^2)$
$-\{a^4\log(x+a)-b^4\log(x+b)\}/(a-b)$.

Exercise XIV c (p. 463)

51. $(1/20)\log\{(2x-5)/(2x+5)\}$.
52. $\frac{1}{8}\log(4x^2-3)-(\sqrt{3}/12)\log\{(2x-\sqrt{3})/(2x+\sqrt{3})\}$.
53. $\log(x^2+2x-1)-\sqrt{2}\log\{(x+1-\sqrt{2})/(x+1+\sqrt{2})\}$.
54. $\frac{1}{2}\log(x^2-7)$. **55.** $\frac{1}{2}\log(x^2+7)$.
56. $\log(x^2+7)+(3/\sqrt{7})\tan^{-1}(x/\sqrt{7})$. **57.** $x-2\tan^{-1}(x+1)$.
58. $\frac{1}{2}\log(x^2-2x+6)+(2/\sqrt{5})\tan^{-1}\{(x-1)/\sqrt{5}\}$.
59. $\frac{1}{3}x^3-3x+3\sqrt{3}\tan^{-1}(x/\sqrt{3})$. **60.** $\frac{1}{4}\log\{(x^2-1)/(x^2+1)\}$.
61. $\log(1+x)-\frac{1}{2}\log(1-x+x^2)-(1/\sqrt{3})\tan^{-1}\{(2x-1)/\sqrt{3}\}$.
62. $\frac{1}{4}\sqrt{2}\log\{(x+3-\sqrt{2})/(x+3+\sqrt{2})\}$.
63. $(\sqrt{5}/10)\log\{(x+4-\sqrt{5})/(x+4+\sqrt{5})\}$.
64. $(1/\sqrt{5})\tan^{-1}\{(x+4)/\sqrt{5}\}$.

65. $\frac{1}{2}\log(x^2+8x+11)-(3\sqrt{5}/10)\log\{(x+4-\sqrt{5})/(x+4+\sqrt{5})\}$.
66. $\log(x^2+8x+21)-\sqrt{5}\tan^{-1}\{(x+4)/\sqrt{5}\}$.
67. $\frac{1}{4}\sqrt{2}\log\{(\sqrt{2}+1+x)/(\sqrt{2}-1-x)\}$. 68. $-\frac{1}{2}\log(1-2x-x^2)$.
69. $-\frac{3}{2}\log(1-2x-x^2)+\frac{1}{2}\sqrt{2}\log\{(\sqrt{2}+1+x)/(\sqrt{2}-1-x)\}$.
70. $\log(1+x)+2\tan^{-1}x$. 71. $\frac{1}{2}\log(1+x^2)-2\log(1+x)+\tan^{-1}x$.
72. $\log x-\frac{1}{2}\log(x^2+x+1)-(1/\sqrt{3})\tan^{-1}\{(2x+1)/\sqrt{3}\}$.
73. $\frac{1}{2}\{\log(1+x^{-2})-x^{-2}\}$.
74. $x+\frac{1}{5}\{8\log(x-2)+\log(x^2+1)-\tan^{-1}x\}$.
75. $\frac{1}{2}\log(x^2+1)+(2/\sqrt{3})\tan^{-1}\{(2x+1)/\sqrt{3}\}$.
76. $\sqrt{2}\tan^{-1}(x/\sqrt{2})-\tan^{-1}x$.
77. $\log x-\tan^{-1}x-(x-1)/(x^2+1)$.
78. $\frac{1}{2}a^{-3}\{\tan^{-1}(x/a)+ax/(x^2+a^2)\}$.
79. $\frac{1}{2}(b/a^3)\tan^{-1}(x/a)+\frac{1}{2}(bx-a^2)/(a^4+a^2x^2)$.
80. $\frac{1}{2}x/(x^2+2x+3)+\frac{1}{4}\sqrt{2}\tan^{-1}\{(x+1)/\sqrt{2}\}$.

Exercise XIV d (p. 464)

51. $\log x+2\sqrt{x}$. 52. $\frac{1}{2}\sqrt{2}\log\{(\sqrt{2}+\sqrt{x})/(\sqrt{2}-\sqrt{x})\}$.
53. $\log\{1-\sqrt{(1-x)}\}^2-\log x$. 54. $\log x-3\sqrt[3]{x}$.
55. $-3\log\{1+\sqrt[3]{(1/x)}\}$. 56. $\frac{2}{3}\{\sqrt{(x+1)^3}-\sqrt{x^3}\}$.
57. $\operatorname{ch}^{-1}(x-1)$, $x>2$. 58. $\sin^{-1}(x-1)$. 59. $\operatorname{sh}^{-1}(x-1)$.
60. $\sqrt{(x^2-1)}+2\operatorname{ch}^{-1}x$, $x>1$.
61. $\sqrt{(x^2+2x)}-2\operatorname{ch}^{-1}(x+1)$, $x>0$.
62. $2\sqrt{(x^2+3x+2)}$. 63. $2\sqrt{(x^2+2x+2)}+\operatorname{sh}^{-1}(x+1)$.
64. $\sqrt{(x^2+x+1)}+\frac{1}{2}\operatorname{sh}^{-1}\{(2x+1)/\sqrt{3}\}$.
65. $2a\sin^{-1}\{(x-a)/|a|\}-\sqrt{(2ax-x^2)}$.
66. $\sqrt{(x^2-1)}-\operatorname{ch}^{-1}x$, $x>1$. 67. $\sin^{-1}x-2\sqrt{(1-x^2)}$.
68. $\sqrt{(x^2+1)}+2\operatorname{sh}^{-1}x$. 69. $2\sqrt{(2-1/x)}$, $x>\frac{1}{2}$.
70. $-\frac{1}{2}\sqrt{(1+4/x)}$, $x>0$.
71 $-\operatorname{ch}^{-1}(2+1/x)$, $x>0$; $-\operatorname{ch}^{-1}(-2-1/x)$, $-\frac{1}{3}<x<0$; $\operatorname{ch}^{-1}(2+1/x)$, $x<-1$.
72. $-\frac{1}{2}\sqrt{2}\operatorname{sh}^{-1}\{(1-x)/(1+x)\}$, $x>-1$.
73. $3\operatorname{sh}^{-1}(\frac{1}{2}x)-\sqrt{5}\operatorname{sh}^{-1}\{(x+4)/(2x-2)\}$, $x>1$. 74. $\frac{1}{2}\sec^{-1}x^2$.
75. $\frac{1}{2}\sin^{-1}x-\frac{1}{2}x\sqrt{(1-x^2)}$. 76. $\frac{1}{2}\sec^{-1}x-\frac{1}{2}x^{-2}\sqrt{(x^2-1)}$, $x>1$.
77. $\frac{1}{2}\sec^{-1}x+\frac{1}{2}x^{-2}\sqrt{(x^2-1)}$, $x>1$.
78. $3\sin^{-1}x+(5-x)\sqrt{(1-x^2)}$. 79. $\frac{1}{3}\{\sqrt{(x+1)^3}+\sqrt{(x-1)^3}\}$.
80. $\frac{1}{2}x^2+\frac{1}{2}x\sqrt{(x^2-1)}-\frac{1}{2}\operatorname{ch}^{-1}x$, $x>1$.
81. $\frac{1}{2}\{3a^2\operatorname{sh}^{-1}(x/|a|)+x\sqrt{(x^2+a^2)}\}$.
82. $2\sqrt{(x^2+ax+a^2)}+b\operatorname{sh}^{-1}\{(2x+a)/(a\sqrt{3})\}$.
83. $\frac{1}{2}\operatorname{sh}^{-1}x^2+\frac{1}{2}\operatorname{cosech}^{-1}x^2$. 84. $\cos^{-1}\{(a+b-2x)/(b-a)\}$.
85. $\frac{1}{4}(2x-a-b)\sqrt{\{(x-a)(x-b)\}}-\frac{1}{8}(a-b)^2\operatorname{ch}^{-1}\{(2x-a-b)/(a-b)\}$.
86. $(b-a)\tan^{-1}\sqrt{\{(x-a)/(b-x)\}}-\sqrt{\{(b-x)(x-a)\}}$.

Exercise XIV e (p. 465)

51. $-2\cot 2x$. **52.** $-\cot\frac{1}{2}x$. **53.** $\frac{1}{3}\tan^3 x+\tan x$.
54. $\log(\sec x+\tan x)-\sin x$. **55.** $-\operatorname{cosec} x$.
56. $-\cot x(1+\frac{2}{3}\cot^2 x+\frac{1}{5}\cot^4 x)$. **57.** $\log(\sin x-\cos x)$.
58. $2\sqrt{(1+\tan x)}$. **59.** $(1/b)\log(a+b\sin x)$.
60. $(1/15)\log\{(5\tan\frac{1}{2}x-3)/(5\tan\frac{1}{2}x+3)\}$. **61.** $\frac{1}{2}\tan^{-1}(4\tan\frac{1}{2}x)$.
62. $(2/7)\cot^{-1}\{\frac{1}{7}\tan(\frac{1}{4}\pi-\frac{1}{2}x)\}$.
63. $\{2x-3\log(3\cos x+2\sin x)\}/13$.
64. $\frac{1}{4}\log\{(2+\tan x)/(2-\tan x)\}$. **65.** $-\log(1+\cot\frac{1}{2}x)$.
66. $2x+\log(3+4\sin x+5\cos x)$.
67. $-\frac{1}{2}\{\log(a\cos^2 x-b\sin^2 x)\}/(a+b)$.
68. $x-\frac{3}{2}\tan\frac{1}{2}x+\frac{1}{6}\tan^3\frac{1}{2}x$.
69. $3\log(\sec x+\tan x)+2\tan x-2x$.
70. $\frac{1}{4}\sqrt{2}\log\{(\sec x+\sqrt{2})/(\sec x-\sqrt{2})\}$.
71. $-2\{\log(a+b\cos x)+a/(a+b\cos x)\}/b^2$. **72.** $\frac{1}{2}\cot^{-1}(\cos 2x)$.
73. $\{\tan^{-1}(\sqrt{5}\tan x)+\frac{1}{4}\log[\sqrt{(5-2\cos x)}/(2\cos x+\sqrt{5})]\}/\sqrt{5}$.
74. $b\sin^{-1}\{b\sin x/\sqrt{(a^2+b^2)}\}+a\log\{\sqrt{(a^2\sec^2 x+b^2)}+a\tan x\}$.
75. $\log\{x+\sqrt{(a^2+x^2)}\}$. **76.** $\log\{x+\sqrt{(x^2-a^2)}\}$, $x>|a|$.
77. $\log\{1-\sqrt{(1-x^2)}\}-\log x$, $x>0$.
78. $2\sin^{-1}(\sqrt{2}\sin\frac{1}{2}x)$, $\cos\frac{1}{2}x>0$.
79. $-\frac{1}{2}\sqrt{2}\operatorname{ch}^{-1}\{(5+\sin x)/(3+3\sin x)\}$.
80. $\sqrt{2}\sin^{-1}(\sin x-\cos x)$.

Exercise XIV f (p. 466)

51. $-\cos x(1-\frac{2}{3}\cos^2 x+\frac{1}{5}\cos^4 x)$. **52.** $\frac{1}{2}(x+\frac{1}{6}\sin 6x)$.
53. $\sin^9\theta(\frac{1}{9}-\frac{2}{11}\sin^2\theta+\frac{1}{13}\sin^4\theta)$.
54. $(4x-\sin 2x-\sin 4x+\frac{1}{3}\sin 6x)/64$.
55. $(\sin 6x+9\sin 4x+45\sin 2x+60x)/192$.
56. $(\cos^{n+3}x)/(n+3)-(\cos^{n+1}x)/(n+1)$. **57.** $\frac{1}{3}\sec^3 x-\sec x$.
58. $\log\tan\frac{1}{2}x+\sec x+\frac{1}{3}\sec^3 x$. **59.** $\frac{1}{2}x+\frac{1}{4}\log\tan(\frac{1}{4}\pi+x)$.

Exercise XIV g (p. 467)

(In this exercise, $s\equiv\sin\theta$, $c\equiv\cos\theta$.)

52. $2(n-1)a^2u_n=(2n-3)u_{n-1}+x(x^2+a^2)^{1-n}$.
53. $u_n+u_{n-2}=(\cot^{n-1}x)/(1-n)$. **55.** $-c(5s^6+6s^4+8s^2+16)/35$.
56. $sc(8c^4+10c^2+15)/48+5\theta/16$. **57.** $s^7(7c^2+2)/63$.
58. $sc(384s^8-48s^6-56s^4-70s^2-105)/3840+7\theta/256$.
59. $(5/16)\log\tan\frac{1}{2}\theta-\operatorname{cosec}\theta\cot\theta(8\operatorname{cosec}^4\theta+10\operatorname{cosec}^2\theta+15)/48$.
60. $s(5+6c^2+8c^4+16c^6)/(35c^7)$.
61. $x^m\{m^2(\log x)^2-2m\log x+2\}/m^3$.

62. $a^2u_n = ax^n \sin ax + nx^{n-1} \cos ax - n(n-1)u_{n-2}$;
$\frac{1}{4}(2x^4 - 6x^2 + 3) \sin 2x + \frac{1}{2}x(2x^2 - 3) \cos 2x$.
63. $u_n + u_{n-1} = x^{2n-1}/(2n-1)$. 64. $nu_n = \text{ch}^{n-1} x \,\text{sh}\, x + (n-1)u_{n-2}$.
65. $(n^2+1)u_n = e^x \sin^{n-1} x\,(\sin x - n \cos x) + n(n-1)u_{n-2}$.
66. $u_{n+2} - (n+1)(n+2)u_n = x^{n+2} \,\text{ch}\, x - (n+2)x^{n+1} \,\text{sh}\, x$.
67. $(n-1)u_n = (n-2)u_{n-2} - \text{cosec}^{n-2} x\{x \cot x + 1/(n-2)\}$.
68. $2nu_{n+1} - (2n-1)u_n = x(1-x^2)^{-n}$;
$\frac{1}{8}x(5-3x^2)(1-x^2)^{-2} + (3/16) \log\{(1+x)/(1-x)\}$.

Exercise XV a (p. 468)

51. $128/315$. 52. $35\pi/256$. 53. $\pi/32$.
54. $\pi\{(n-1)!!\}/n!!$, n even ; $2\{(n-1)!!\}/n!!$, n odd ; see p. 307.
55. $2\pi\{(n-1)!!\}/n!!$, n even ; 0, n odd ; see p. 307.
56. $\frac{1}{4}(\frac{3}{4}\pi - 2)$. 57. $8/1287$. 58. $5\pi/2048$. 59. $63\pi/8$.
60. $\frac{1}{5}$. 61. $16/105$. 62. $2|a|^5/15$. 63. $256(a^5\sqrt{a})/3465$.
64. $a^{n+2}/\{(n+1)(n+2)\}$. 65. $28/15$. 66. $\frac{1}{2}\pi$. 67. $\frac{1}{4}$.
68. $3\pi/64$. 69. $2 - \frac{1}{2}\pi$. 70. $(256\sqrt{2})/315$.
71. $5\pi/8$. 72. $3\pi/256$. 73. $(5 - 6 \log 2)/12$.
74. $\frac{1}{4}\pi - 76/105$. 77. $m!\,n!/\{(m+n+1)!\,2^{n+1}\}$.
78. $\frac{1}{4}(3\sqrt{2} - \text{sh}^{-1} 1)$.
79. $v_n = v_{n-2} - (3/5)^{n-1}/(n-1)$; $\log(5/4) - 531/2500$. 85. $32/65$.

Exercise XV b (p. 470)

51. $1/(k-1)$. 52. Does not exist. 53. $\frac{1}{2}\log 3$.
54. $2\log(1+\sqrt{2})$. 55. 1. 56. $\frac{1}{2}$. 57. $\pi/6$.
58. $\frac{1}{3}\log 2$. 59. Does not exist. 60. $\frac{1}{2}\log 2$.
61. $\frac{1}{2}\pi/|a|^3$. 62. $3/(8e^\pi)$.

Exercise XV c (p. 470)

51. $1/(1-k)$. 52. Does not exist. 53. $2\sqrt{a}$.
54. $\frac{1}{2}\pi$, $a > 0$; $-\frac{1}{2}\pi$, $a < 0$. 55. $(14\sqrt{2})/15$. 56. $3\pi/8$.
57. Does not exist. 58. π. 59. 1. 60. Does not exist.
61. $32/33$. 62. $a(\log a - 1)$, $a > 0$. 67. $16/35$; $5\pi/32$.
71. 1 ; no.

Exercise XV d (p. 472)

53. $\frac{1}{2}(4-\pi)a^2$. 54. $b \log 2$. 55. Exists if $k < \frac{1}{2}$; $\pi/(1-2k)$.
56. π. 60. $(0, \frac{1}{4})$. 61. Does not exist. 62. πa^2.

Exercise XV e (p. 473)

51. 0. **52.** $35\pi/128$. **53.** 0. **54.** $63\pi/256$.
55. $\frac{1}{2}\pi a(2a^2 + 3b^2)$. **56.** 0. **57.** n odd.
59. $2/|a|$ if $|a| \geqslant |b|$; $2/|b|$ if $|a| < |b|$. **64.** $3\pi^2/16$.
65. $\frac{1}{4}\pi(a + b)$. **66.** $m!n!(b-a)^{m+n+1}/(m+n+1)!$.
68. $\log\{(e^t - 1)/t\}$.

Exercise XVI a (p. 474)

51. $ae^{-s/a}$. **54.** $(0, 0)$.
59. $as = 4b(a + b)\sin\{a(\psi - \frac{1}{2}\pi)/(a + 2b)\}$. **60.** $(b^2 + c^2)^{1/2}$.

Exercise XVI b (p. 475)

51. $\coth x$; $s = \log(\text{sh}\, x/\text{sh}\, a)$ where $0 < a < x$, if $s = 0$ when $x = a$.
56. $(c/s)ds$, $(1 - c^2s^{-2})^{1/2}ds$; $c\log(s/c)$; $\pm\{\sqrt{(s^2 - c^2)} - c\sec^{-1}(s/c)\}$.

Exercise XVI c (p. 476)

51. $2r(\cos^3\theta + \sin^3\theta) = 3a\sin 2\theta$. **52.** $x^2 + y^2 = (ex - l)^2$.
53. $r = 2a(1 - \cos\theta)$.
54. (i) the part of the hyperbola $y^2 = (3x + 1)(x + 1)$ for which $x \geqslant -\frac{1}{3}$; (ii) the whole of the hyperbola.
55. $2\int_0^\pi \sqrt{(13 + 12\cos\theta)}\,d\theta$. **64.** $t = \sqrt{2} - 1$.

Exercise XVI d (p. 478)

51. $r^2 = ap$, $(a > 0)$. **52.** $pr = 2c^2$. **53.** $p = r\sin\alpha$.
54. $p = |r^{n+1}/a^n|$. **55.** $1/p^2 = (1 - n^2)/r^2 + n^2/a^2$.
56. $1/p^2 = (1 + m^2)/r^2 - m^2/a^2$. **57.** $-m\theta/n$. **58.** $\frac{1}{2}\pi$.

Exercise XVI e (p. 478)

51. $r = ab\sin\alpha\, e^{\theta\cot\alpha}$ where $\log b = (\frac{1}{2}\pi - \alpha)\cot\alpha$.
55. $p = a$, (actually the line $r\cos\theta = a$). **56.** $r^4(r^2 - p^2) = a^2p^4$.
57. $p^{n+1} = |r^{2n+1}/a^n|$.
58. $r^m = a^m\sin m(\frac{1}{2}\pi - \theta)$, where $m = n/(n - 1)$, and $n \neq 1$.
59. $r^2 = a^2\sin^3\frac{2}{3}(\theta - \gamma)$. **60.** $r = f(\theta + \frac{1}{2}\pi)$. **61.** $x^2 + y^2 = ay$.
62. $r^2\sin 2(\theta + \gamma) = a^2$.
63. $x = a + 2\tan^{-1}e^s$, $y = b + \log\text{ch}\, s$. **64.** $a\{\sqrt{2} + \log(1 + \sqrt{2})\}$.
69. k/r^2; kr; k/r^3. **71.** $r = a\sec^2\frac{1}{2}\theta$; $r = a\,\text{cosec}^2\frac{1}{2}\theta$.
72. $3a/2$.

Exercise XVII a (p. 480)

52. $s=a\sin\psi$. **53.** $s=ce^{k\psi}$.
54. The root has the same sign as c. **55.** $\sqrt{2}$.
56. $|a|\{\sqrt{(1+9t^4)^3}\}/(6t)$. **57.** $|a|/b^2$. **58.** $|b|/a^2$.
59. $-\frac{3}{2}|a|(1+t^2)^2$. **62.** 2 ; $\frac{1}{2}\sqrt{2}$. **65.** $\frac{5}{4}|a|\sqrt{10}$.

Exercise XVII b (p. 481)

51. $-mp/(nr^2)$. **52.** $\sqrt{\{k(r^2-a^2)\}}$.
54. $(r^2+2a^2)/\sqrt{(r^2+a^2)^3}$. **55.** $\pm 2/(na)$. **57.** l, $l\sqrt{(1+e^2)^3}$, l.
58. $\rho=0$, point circles $(a, 0)$, $(0, -a)$.
59. $\frac{1}{2}e^{\psi}(3\cos\psi-\sin\psi)$. **60.** $\frac{3}{4}|a|\sin^2\frac{1}{3}\theta$.
61. $|a|\{\sqrt{(1+4\tan^2\theta)^3}\}/(2-3\cos^2\theta)$.

Exercise XVII c (p. 482)

51. $y^2=4a(a-x)$. **52.** $x^{-2}+y^{-2}=a^{-2}$ or $x=a\sec t$, $y=a\operatorname{cosec} t$.
53. $x^2-y^2=a^2$ or $x=a\sec t$, $y=-a\tan t$.
54. $ax:by:c^2=\cos^3 t:-\sin^3 t:1$ or $(ax)^{2/3}+(by)^{2/3}=c^{4/3}$.
55. $x=2eye^{2c/x}$. **56.** $x^{4/3}+y^{4/3}=a^{4/3}$ or $x=a\sqrt{\cos^3 t}$, $y=a\sqrt{\sin^3 t}$. **57.** $(x-2a)^3=27a^2y$. **58.** $r=a\pm b$.
59. $x^{2/3}+y^{2/3}=a^{2/3}$. **60.** $x^{2/5}+y^{2/5}=a^{2/5}$.
61. $x=a(2+3t^2)$, $y=2at^3$ or $4(x-2a)^3=27ay^2$.
62. $x=-\frac{1}{2}at^2(9t^2+2)$, $y=\frac{4}{3}at(3t^2+1)$.
63. (i) the point given by $a_1x+b_1y+c_1=0$, $a_2x+b_2y+c_2=0$;
(ii) $(a_1x+b_1y+c_1)(a_3x+b_3y+c_3)=(a_2x+b_2y+c_2)^2$.

Exercise XVII d (p. 483)

51. $(x_1+\cot x_1, y_1+1)$. **52.** $\{\frac{1}{2}c(3t_1^4+1)/t_1^3, \frac{1}{2}c(t_1^4+3)/t_1\}$.
53. $\{a(\theta-\sin\theta), a(3+\cos\theta)\}$. **54.** $2a(2\sqrt{2}-1)$.
56. $3a\sin t\cos t$. **57.** $p=ck\sin k(\frac{1}{2}\pi-\psi)$.
63. $2r^2=a^2+6p^2+\sqrt{\{(a^2+4p^2)^3/a^2\}}$.

Exercise XVIII a (p. 484)

51. $a^2+\frac{1}{3}b^2$. **52.** $ab\log(1+\sqrt{2})+\pi b^2/8$. **53.** $\frac{1}{4}\pi a^2/n$.
54. $5\pi a^2/16$. **55.** a^2. **56.** $a^2(1-\frac{1}{4}\pi)$.
58. $\pi^3a^2/48$; $61\pi^3a^2/48$.
59. $\frac{1}{4}a^2(e^{b\pi}-1)/b$; $\frac{1}{4}a^2(e^{5b\pi}-e^{4b\pi})/b$. **60.** $\frac{1}{12}\sqrt{(\pi^2-4)^3}$.

Exercise XVIII b (p. 485)

51. $2\frac{2}{3}$. **52.** $3\pi ab/8$. **53.** $8ab/15$. **54.** $\frac{2}{3}a^2$.

Exercise XVIII c (p. 486)

51. $8\pi^2$. 52. $2\pi a^2$. 53. $8\pi/3$. 54. $\frac{1}{3}\pi a^2$.
55. $\pi\{\sqrt{2} + \log(1+\sqrt{2})\}$.
56. $\pi\{c\sqrt{(1+c^2)} - \sqrt{2} + \text{sh}^{-1} c - \text{sh}^{-1} 1\}$.
57. $\pi\{7\sqrt{2} + 3\log(1+\sqrt{2})\}/18$. 58. $\pi c^2\{5 - \frac{1}{2}\sqrt{2}\log(1+\sqrt{2})\}$.
59. $\frac{2}{3}\pi l^2(\sec^3 \frac{1}{2}\alpha - 1)$. 60. $4\pi^2 a^2$. 61. $19\pi/6$.
62. $(\pi b^2/e)\log\{a(1+e)/b\} + \pi a^2$ where $e = \sqrt{(1 - b^2/a^2)}$.

Exercise XVIII d (p. 487)

51. $2\pi^2 a^2 b$. 52. $44\pi^2$. 53. 0·7 %; 0·02 %.
54. $\frac{2}{3}\pi\{f(\alpha)\}^3 \sin\alpha\, \delta\alpha$. 55. $\{3\sqrt{2}\log(1+\sqrt{2}) - 2\}\pi a^3/24$.
56. $\pi^2 a^3/8$. 57. $16\pi a^3/15$; $\frac{1}{2}\pi a^2\{\sqrt{2} + 5\log(\sqrt{2}+1)\}$.
58. $2\pi^2 abc$.

Exercise XVIII e (p. 487)

51. $\frac{2}{3}$. 52. $\frac{1}{2}ab(a+b)$. 53. $\frac{1}{3}c^3(1-\cos\alpha)$. 54. $\frac{1}{6}c^2\sin^3\alpha$.
55. $\frac{1}{4}a^2b^2c$. 56. $1/24$. 57. $1/56$; πa^4. 58. $\pi^2 a^3/24$.
60. $35\pi\lambda a^4/16$. 63. $16a^3/3$. 64. $8a^3(2-\sqrt{2})$.

Exercise XIX a (p. 488)

51. $px^{p-1}y^q,\ qx^py^{q-1}$. 52. $y^2/(x+y)^2,\ x^2/(x+y)^2$.
53. $ae^{ax}\sin by,\ be^{ax}\cos by$. 54. $4xy^2/(x^2+y^2)^2,\ -4x^2y/(x^2+y^2)^2$.
55. $1/\sqrt{(x^2+y^2)},\ -x/\{y\sqrt{(x^2+y^2)}\}$, if $y > 0$.
56. $1/\sqrt{(x^2+y^2)},\ -x/\{y\sqrt{(x^2+y^2)}\}$.
57. $n(n-1)x^{n-2}e^y,\ nx^{n-1}e^y,\ x^ne^y$. 58. $-1/x^2 - 1/(x+1)^2,\ 0,\ -2/y^2$.
59. $-ab^2\sin(bx+cy),\ -abc\sin(bx+cy),\ -ac^2\sin(bx+cy)$.
60. $-y/x^2,\ 1/x,\ 0$. 61. $z\{(2x+y)^2+2\},\ z\{(2x+y)(x+2y)+1\}$, $z\{(x+2y)^2+2\}$ where $z = e^{x^2+xy+y^2}$.
62. $\text{sh}\, x\, \text{ch}\, y,\ \text{ch}\, x\, \text{sh}\, y,\ \text{sh}\, x\, \text{ch}\, y$. 63. $(x^2-y^2)/(x^2+y^2)$.
68. $(n = -\frac{3}{2})$.

Exercise XIX b (p. 490)

54. $(ay - x^2)/(y^2 - ax)$. 55. $y^2/\{x(1-\log y)\}$. 56. y/x.
57. $(ax_1 + by_1)x + (bx_1 + cy_1)y = 1$.
58. $x\sqrt{(a/x_1)} + y\sqrt{(b/y_1)} = 1$. 59. $px/x_1 + qy/y_1 = p + q$.
60. $(x - x_1)(f_y)_1 = (y - y_1)(f_x)_1$. 61. $2a^3xy/(ax - y^2)^3$; $cy(2x+c)/x^4$. 64. $2\tan \mathsf{A} \tan \mathsf{B} \sin \mathsf{C}$.
65. $(\cos \mathsf{B}\, \delta a + \cos \mathsf{A}\, \delta b + 2\mathsf{R}\cos \mathsf{A} \cos \mathsf{B}\, \delta \mathsf{C})/c$.
66. $(1/\mathsf{V})d\mathsf{V} = (y/x)dx + \log x\, dy$.
67. $(2\theta - \tan\theta)/\{(\pi - 2\theta)\tan\theta - 1\}$. 68. $f_xg_x + f_yg_y = 0$.

Exercise XIX c (p. 491)

51. $V_r^2 + r^{-2}V_\theta^2$. 52. $(r^2 + r_1^2)^{3/2}/(r^2 + 2r_1^2 - rr_2)$ where $r_1 = dr/d\theta$, etc. 53. $(3x_2^2 - x_1x_3)/x_1^5$ where $x_1 = dx/dy$, etc.
54. $V_\xi^2 + V_\eta^2$. 55. $V_{\xi\xi} + V_{\eta\eta}$.
61. $(1-u^2)\left\{2uv\dfrac{\partial z}{\partial x} - (u^2+v^2)\dfrac{\partial z}{\partial y}\right\} \Big/ (u^2-v^2)^2$.
64. $2z_{uu} + 2(u + 2v/u)z_{uv} + (u^2 + 2v + 2v^2/u^2)z_{vv} + 2z_v$.
65. 108. 66. $(3\sqrt{3})/8$.

Exercise XX a (p. 493)

54. $y_2^2 = (1 + y_1^2)^3$. 55. $x^2y_2 - 2xy_1 + 2y = 0$; $xy_2 + y_1 = 0$.
57. $y_3 - 3ny_2 + 3n^2y_1 - n^3y = 0$.

Exercise XX b (p. 494)

51. $y = ax^{n+1}/(n+1) + B$ if $n \neq -1$; $y = a \log x + B$ if $n = -1$.
52. $y^{1-n} + a(n-1)x = B$ if $n \neq 1$; $y = Be^{ax}$ if $n = 1$.
53. $ncr = e^{n\theta} + A$ if $n \neq 0$; $cr = \theta + A$ if $n = 0$.
54. $y^2 = A - 2k/x$. 55. $y = Ax^n$. 56. $y = \frac{1}{4}(x + A)^2$.
57. $\sin y = A \sec x$. 58. $x = \frac{1}{2}y + \frac{1}{4}\sin 2y + A$.
59. $y^3 = Ax^3 - 1$. 60. $y^{2a} = B(a+x)/(a-x)$ if $a \neq 0$; $y = Be^{1/x}$ if $a = 0$. 61. $x^2 + 2ax + C = 2\log(y+b)$.
62. $\log y + x \log x = x + A$. 63. $x + y \log(Ax) = 0$.
64. $x = A \sin(y/x)$. 65. $y + x\log(Ay) = 0$. 66. $y = ae^{bx}$.
67. $r^2 = A \sin 2\theta$. 68. $r = k/(A - \theta)$. 69. $y^2 = (1 + y^2)(x + A)^2$.
70. The circles $(x - A)^2 + (y - B)^2 = 1$; $\rho = 1$.
71. The circles $(x - A)^2 + (y - B)^2 = c^2$.
72. $x^2 - 2xy - y^2 + 2x + 6y = A$. 73. $(2y + x^2 + A)(y - Be^x) = 0$.
74. $4y = x^3 + A/x$. 75. $(xdy - ydx)/x^2$; $x(y^2 + A) = 2y$.
76. $y = Ax/(1-x)$. 77. $\log(x^2 + y^2) = 2k \tan^{-1}(y/x) + A$.
78. $(x - A)^2 + y^2 = k^2$. 79. $r = k \sin(\theta + \gamma)$.

Exercise XX c (p. 495)

51. $(n+1)y = x^n + c/x$ if $n \neq -1$; $xy = \log(cx)$ if $n = -1$.
52. $(m+n)y = x^m + c/x^n$ if $m + n \neq 0$; $y = x^m \log(cx)$ if $m + n = 0$.
53. $y(1 - x^2) = x + c$. 54. $16y \cos x = c - \cos 4x$.
55. $y = x^2 + ce^{-x^2}$. 56. $(a^2 + c^2)y = b(a \sin cx - c \cos cx) + Ae^{-ax}$.
57. $y = x \operatorname{th} x + c \operatorname{sech} x - 1$. 58. $1/y = x^2 + 2x + 2 + ce^x$.
59. $x = e^{-y}(y + c)$. 60. $y = be^x + c - x - \frac{1}{2}x^2$.
61. $(y - Ax)(1 + A) + Ac^2 = 0$; $(x + y)^2 - 2c^2(x - y) + c^4 = 0$.
62. $4xy = a^2$.

Exercise XX d (p. 496)

51. $4x - 3y = k$. **52.** $x^2 + y^2 = 2k(x + y)$. **53.** $y^2 + 2 \log \operatorname{sh} x = k$.
54. $2ky = x^2 - k^2$. **55.** $ny^2 + x^2 = k$. **56.** $r = ke^{\theta^2/2}$.
57. $r^2 = k \cos \theta$. **58.** $r^n = k \cos n\theta$. **59.** $r^4(3 - \cos 4\theta) = k$.
60. $r^2 = c \cos (2\theta \pm a)$.

Exercise XX e (p. 496)

51. $y = ax \log x + Bx + C$. **52.** $y = a \log \operatorname{cosec} x + Bx + C$.
53. $ay = B - \log (ax + C)$. **54.** $y^{1-n} = Ax + B$ if $n \neq 1$; $\log y = Ax + B$ if $n = 1$. **55.** $y = A + \cos (x + B)$.
56. $a(n + 2)y = \{a(n + 1)x + B\}^{(n+2)/(n+1)} + C$ if $n \neq -1, n \neq -2$; $y = Be^{ax} + C$ if $n = -1$; $ay = B - \log (ax + C)$ if $n = -2$.
57. $y = Ae^{kx} + Bx + C$. **58.** $y = \frac{1}{2}x + B/x + C$.
59. $y = A$, $y + 1 = 2A \tan (Ax + B)$, $Ax + B = \log\{(y + 1 - A)/(y + 1 + A)\}$.
60. $x = \sin (Ay + B)$, $x = \operatorname{ch} (Ay + B)$. **61.** $y = \log (A + Be^x)$.
62. $y = B \log\{(x - 1)/(x + 1)\} + C$.
63. $y = A$, $x = A \log (y - A) + y + B$. **64.** $x = A + y + By^2$.
65. $x = A \operatorname{ch} my + B \operatorname{sh} my + C$. **67.** $(x + A)^2 + y^2 = B^2$, $y = A \operatorname{ch} (B + x/A)$.

Exercise XX f (p. 497)

51. $y = Ae^{4x} + Be^{-3x}$. **52.** $y = (Ax + B)e^{-5x}$.
53. $y = A \cos 2x + B \sin 2x$. **54.** $y = e^x\{A \operatorname{sh} (x\sqrt{3}) + B \operatorname{ch} (x\sqrt{3})\}$.
55. $y = Ae^{2x} + Be^{-x} + C$. **56.** $y = e^{-4x}(A \cos x + B \sin x)$.
57. $5/3, -20/9$. **58.** $1, -1, 1$. **59.** -2. **60.** 2.
61. $-6/145, -1/290$. **62.** $A = 0, B = -\frac{1}{8}$.
63. $y = x^3 - 3x^2 + 6$. **64.** $y = -\frac{1}{8}e^{5x}$.
65. $50y = \cos x - 3 \sin x$. **66.** $9y = xe^{6x}$.
67. $170y = \cos 5x - 4 \sin 5x$. **68.** $y = e^{2x}$. **69.** $y = x^2 + x$.
70. $y = \frac{1}{2}x^2e^{3x}$.
71. $y = Ae^{4x} + Be^{2x} + \frac{1}{3}e^x + \frac{1}{15}e^{-x} + (8x^2 + 12x + 7)/64$.
72. $y = (A - \frac{1}{2}x)e^{2x} + Be^{4x} + (\sin 2x + 3 \cos 2x)/40$.
73. $y = (A + Bx + \frac{1}{2}x^2)e^{-3x}$.
74. $18y = (A + Bx)e^{-3x} + \sin 3x - \cos 3x$.
75. $y = Ae^{2x} + Be^{-2x} - e^x(2 \sin x + \cos x)/10$.
76. $y = A \sin 2x + B \cos 2x + e^x(2 \cos x + \sin x)/10$.
77. $y\{(b^2 + c^2 - n^2)^2 + 4n^2b^2\}$
$= e^{-bx}(P \cos cx + Q \sin cx) + (b^2 + c^2 - n^2) \cos nx + 2nb \sin nx$.
78. $y = (A + B \log x)/x$. **79.** $y = x^2 + (A + B \log x)/x$.
80. $y = (Ae^x + Be^{-x})/x$.

Exercise XXI a (p. 498)

(In Nos. 51-56, 65, r takes the values, 1, 2, 3, . . .)

51. $1+\Sigma(ax)^r/r!$; $(ax)^n e^{a\theta x}/n!$.
52. $\Sigma(-1)^{r+1}(ax)^{2r-1}/(2r-1)!$; $(ax)^n \sin(a\theta x+\frac{1}{2}n\pi)/n!$.
53. $1+\Sigma(-1)^r(bx)^{2r}/(2r)!$; $(bx)^n \cos(b\theta x+\frac{1}{2}n\pi)/n!$.
54. $1+\Sigma p(p-1)\ldots(p-r+1)x^r/r!$;
$p(p-1)\ldots(p-n+1)x^n(1+\theta x)^{p-n}/n!$.
55. $1+\Sigma(ax)^{2r}/(2r)!$; $\frac{1}{2}(ax)^n\{e^{a\theta x}+(-1)^n e^{-a\theta x}\}/n!$.
56. $\Sigma 2x^{2r-1}/(2r-1)$; $x^n\{(1-\theta x)^{-n}+(-1)^{n-1}(1+\theta x)^{-n}\}/n$.
57. $x+\frac{1}{3}x^3+2x^5/15$. 58. $1+x^2+\frac{2}{3}x^4$. 59. $x-\frac{1}{3}x^3$.
60. $1+\frac{1}{6}x^2+7x^4/360$. 61. $x+x^2+\frac{1}{3}x^3$. 62. $x-\frac{1}{3}x^3+\frac{1}{5}x^5$.
65. $\operatorname{ch} x+\Sigma\frac{1}{2}\{e^x+(-1)^r e^{-x}\}h^r/r!$. 67. $1/90$. 68. $7/360$.
69. $\frac{1}{3}a$. 70. -2.

Exercise XXI b (p. 499)

(In Nos. 71, 72, 76, r takes the values, 1, 2, 3, . . .)

51. $a^n(ax+b)^n(2n)!/n!$. 52. $(-\frac{1}{4}a)^n\{\sqrt{(ax+b)^{-2n-1}}\}(2n)!/n!$
53. $(-1)^{n-1}(n-1)!\{a/(ax+b)\}^n$. 54. $m^n\sin(mx+\frac{1}{2}n\pi)$.
55. $(2n)^n\cos 2n(x+\frac{1}{4}\pi)$. 56. $\frac{1}{2}e^{ax}\{(a+b)^n e^{bx}+(a-b)^n e^{-bx}\}$.
57. $2^{n-2}x\{4x^2-3n(n-1)\}\cos(2x+\frac{1}{2}n\pi)$
$+2^{n-3}n\{12x^2-(n-1)(n-2)\}\sin(2x+\frac{1}{2}n\pi)$.
58. $\{x^4-6n(n-1)x^2+n(n-1)(n-2)(n-3)\}\sin(x+\frac{1}{2}n\pi)$
$-4nx\{x^2-(n-1)(n-2)\}\cos(x+\frac{1}{2}n\pi)$.
59. $2^{n-3}[\{n(n-1)-4x^2\}\cos(2x+\frac{1}{2}n\pi)-4nx\sin(2x+\frac{1}{2}n\pi)]$ if $n>2$.
60. $n!\left(\log x+1+\frac{1}{2}+\frac{1}{3}+\ldots\ldots+\frac{1}{n}\right)$.
61. $\frac{1}{2}(-1)^n n!\{(n+1)(n+2)x^2-2n(n+1)x+n(n-1)\}$.
62. $(\sqrt{2^n})e^x\cos(x+\frac{1}{4}n\pi)$.
63. $\frac{1}{2}e^x\{x^2+(2n+1)x+n^2+1\}+\frac{1}{2}(-1)^n e^{-x}\{x^2-(2n-1)x+(n-1)^2\}$.
64. $\frac{1}{4}\{3\sin(x+\frac{1}{2}n\pi)-3^n\sin(3x+\frac{1}{2}n\pi)\}$.
65. $xf(n)+nf(n-1)$ where $f(n)$ is the answer to No. 64.
66. $2^{n-1}\cos(2x+\frac{1}{2}n\pi)+2^{2n-3}\cos(4x+\frac{1}{2}n\pi)$.
67. $\frac{1}{3}(-1)^n n!\{1/(x-2)^{n+1}-1/(x+1)^{n+1}\}$.
68. $5^n e^{3x}\sin\{4x+n\tan^{-1}\frac{4}{3}\}$. 69. $1+\Sigma(x\sqrt{2})^n(\cos\frac{1}{4}n\pi)/n!$.
71. $\Sigma(-1)^{r-1}1^2.3^2.5^2\ldots(2r-3)^2x^{2r-1}/(2r-1)!$.
72. $mx+\Sigma m(1^2-m^2)(3^2-m^2)\ldots\{(2r-1)^2-m^2\}x^{2r+1}/(2r+1)!$.
74. $\frac{1}{2}\pi-y-\frac{1}{4}\pi y^2+\frac{1}{2}y^3$. 75. $z-2z^2+2z^3$ where $z=x-\frac{1}{4}\pi$.
76. $1+\frac{1}{2}x+\Sigma(2\sqrt{2})\{\sin(\frac{1}{2}r\pi+\frac{1}{4}\pi)\}(2r)!(\frac{1}{4}x)^{r+1}/\{r!(r+1)!\}$.

Exercise XXI c (p. 500)

60. $x=y\pm y\sqrt{y}$. 61. $x=\pm y\sqrt{y}+\frac{1}{2}y^4$, $y=\pm x\sqrt{x}+\frac{1}{2}x^4$.
62. $y=x^3+2x^4$. 63. $y=x^2\pm x^2\sqrt{x}$.

64. $y = x - \frac{3}{2}x^2, y = -x + \frac{1}{2}x^2$. 65. Origin isolated.
66. $\pm y = x\sqrt{x} + x^2\sqrt{x}$. 67. $x = \frac{1}{2}y^3 - \frac{1}{4}y^5$.
68. $2x = \pm y\sqrt{(2y)} - \frac{1}{2}y^2$. 69. $y = -\frac{1}{2}x^2 + \frac{1}{2}x^3$.
70. $x = -\frac{1}{2}y^2 + \frac{1}{2}y^3$. 71. $y = x + \frac{1}{2}x^2$. 72. $2|a|$.
73. 0, (infl.). 74. $\frac{3}{4}\sqrt{2}, \frac{1}{4}\sqrt{2}$. 75. $1/(2|a|)$.
76. $\frac{2}{3}$; $\frac{1}{4}\sqrt{2}$; $1/(30\sqrt{10})$. 77. $\sqrt{2}$.

Exercise XXI d (p. 500)

51. $x = b, x = -b, y = a, y = -a$. 52. $x = -1, x = 2, y = 1, y = x$.
53. $y = 0, y = x, y = -x$. 54. $x = -1, x - y = 1, x + y = 1$.
55. $y = -1, x = 2, y - x = 3$. 56. $x - y = \frac{1}{2}, x + y = \frac{1}{2}$.
57. $x = 0, x = y, x + y = 0$. 58. $y - 2x = \frac{1}{4}$. 59. $y - x = \frac{1}{2}$, $y + x = \frac{1}{2}$. 60. $x = 0, x = -1, y = x - 2$. 61. $y + 2x = 0$, $y = x + \frac{1}{3}\sqrt{6}, y = x - \frac{1}{3}\sqrt{6}$. 62. $y = x - 1, y = x - \frac{1}{2}$.
66. $A = 1, B = -\frac{1}{4}, C = 5/32$. 69. $(x^2 + 6y^2 = 25)$.
70. $2(y-1)(y-x-4)(y+x-4) = 3(y-2x-3)$. 71. $x = 1, y = 0$, $y = x + 2$. 72. $y = 0, x = 0, y = x$. 73. $2y + 2x = 5$, $2y - 2x = 1, y - x = 1$. 74. $y = -2$; $(x^2 = 2y)$.
75. $y - x + 1 = \pm\sqrt{5}$; $(y^2 = x)$. 76. $x = 2, y = 3$; $(y^2 = x)$.
77. None. 78. $(x^4 = ay^3)$. 79. $y + x = 0, 2y + x = 0, x = 0$.
80. $y = 1$; $\{(x-y)^2 = \frac{1}{2}x\}$.

Exercise XXII a (p. 502)

51. $\frac{4}{3}Ma^2 \sin^2 \alpha$. 52. $M(\frac{1}{3}a^2 + c^2)$. 53. $\frac{1}{2}M(a^2 + b^2)$.
54. $29835\pi/256$, (oz., in units). 55. $8Mc^2/3$; $24Mc^2/25$.
56. $5Mb^3/(33a)$. 57. $Mh(20a^2 - 15ah + 3h^2)/\{10(3a - h)\}$.
58. $8Mb^2/35$. 59. $7Ma^6/26$. 60. $M(b^2 + \frac{3}{4}a^2)$.

Exercise XXII b (p. 502)

51. $\frac{1}{2}Ma^2$. 52. $\frac{1}{4}M(a^2 + b^2)$. 53. $129M/4$.
54. $M(\frac{1}{3}a^2 + b^2)$, $M(\frac{4}{3}a^2 + b^2)$, where $PN = b$. 55. $7Ml^2/9$.
56. $\frac{3}{2}Ma^2$. 57. $\frac{2}{5}Ma^2$. 58. $13Ma^2/20$. 59. $\frac{1}{2}Mh^2$, $Mh^2/18$, $\frac{1}{6}Mh^2$. 61. $\frac{2}{3}Ma^2$. 62. $M(\frac{3}{2}a^2 + b^2)$. 64. $4Ma^2/9$, $13Ma^2/9$. 65. $48Ma^2/25$; $288Ma^2/175$.
66. $\frac{1}{3}M(k^2 + 2c^2)$.

Exercise XXII c (p. 504)

51. $\frac{3}{4}$ way down median. 52. mid point of median.
54. $4b/3$. 55. $\frac{1}{2}h \cos \alpha$. 56. $\frac{1}{4}a(15\pi - 32)/(3\pi - 4)$.

Miscellaneous Examples (p. 504)

1. $-3\sin x/(2+\cos x)^2$. **2.** $-2n(1-x)^{n-1}/(1+x)^{n+1}$.
3. $x^{n-1}(x\cos 2x+\frac{1}{2}n\sin 2x)$. **4.** $-1/\{2\sqrt{(1-x^2)}\}, |x|<1$.
5. $1/\{x\sqrt{(x^{-2}-1)}\}, |x|<1$. **6.** $-1, x\neq(n-\frac{1}{4})\pi$.
7. 0 if $x^2<2$ but $x^2\neq 1$, $4(x^2+2)/(x^4+4)$ if $x^2>2$.
$f(x)=0$ if $|x|\leqslant 1$; $f(x)=-\pi$ if $1<x\leqslant\sqrt{2}$;
$f(x)=2\tan^{-1}\{2x/(2-x^2)\}$ if $x>\sqrt{2}$; $f(-x)=-f(x)$.
8. $(2x/a^2)\{1-x^2/\sqrt{(x^4-a^4)}\}$. **9.** $-(b/a)\operatorname{cosec}^3 t$. **10.** $-\frac{1}{4}\sec^4\frac{1}{2}t$.
11. $-\{(a+b)/(4ab)\}\operatorname{cosec}^3\{(a+b)t/(2b)\}\operatorname{cosec}\{(a-b)t/(2b)\}$.
14. 1, −1. **15.** 1 if $x<0$, 0 if $0<x\leqslant 1$, $4(x-1)$ if $x>1$.
16. Yes, no. **19.** $c^2+\sqrt{(c^4+1)}$, max.; $c^2-\sqrt{(c^4+1)}$, min.
20. 0, min.; 6/5, max. **21.** Least, $x=-1$; greatest, $x=1$.
22. (i) $(3a+2b)/5$, max.; a, min.; (ii) none; (iii) a, max.; $(3a+2b)/5$, min. **23.** $k\pi/6(\text{mod } 2\pi)$; $k=1, 3, 5, 8, 10$, max.; $k=2, 4, 7, 9, 11$, min. **25.** 1, 2. **26.** 2, $\frac{2}{3}$; max. 2, min. $\frac{2}{3}$ and $\sqrt{3}$. **33.** $3(a+x)^2/(2b)$.
34. $x=2t/(t^2-1)$; inflexion at $x=\pm\sqrt{3}$. **37.** $y(x^2+a^2)/(2ax)$; $y(x^2-a^2)/(2ax)$. **38.** $(1+x)^{m-1}\{m\log(1+x)+1\}$.
39. $(1-\log x)/x^2$. **40.** $c-1/x-\log(1+1/x)^2$.
41. $\frac{1}{2}\log\{(x^2+1)/(x+1)^2\}+\tan^{-1}x+c$.
42. $\frac{1}{2}(1/a^2)\log\{x^2/(x^2+a^2)\}+c$. **43.** $(2\sqrt{x})(\log x-2)+c$.
44. $c-\sin^{-1}(x/|a|)-(1/x)\sqrt{(a^2-x^2)}$.
45. $\frac{1}{4}\sqrt{2}\log\tan^2(\frac{1}{2}x+\frac{1}{8}\pi)+c$. **47.** 1; $\frac{2}{3}$.
48. $(8\log 2-4\frac{1}{2}\log 3+1)$. **49.** $x\sqrt{2}+1-\sqrt{2}$.
56. $\frac{1}{2}\pi-\pi/2^{n+1}$. **57.** $4a^2/105$. **58.** $\frac{3}{4}\pi$.
59. $2\pi\sqrt{2}$. **62.** $(|c|)$. **65.** $|a|\tan^2\theta/\sqrt{(1+4\tan^2\theta)}$.
72. $\{|a|\sqrt{(9-56\lambda^2+144\lambda^4)^3}\}/\{12(1+4\lambda^2)\}$. **75.** $\frac{1}{5}$.
76. $(5\sqrt{5})/12$, $(5\sqrt{5})/16$. **78.** $x(x^2+y^2)=2ay^2$; $-\frac{1}{3}a\lambda^2(6+\lambda^2)$, $8a\lambda/3$. **80.** $r=1$. **85.** $\frac{1}{2}\pi(\pi-2)a^2c$. **86.** $-1/9$.
87. $\pi a^4/32$. **88.** $13\frac{1}{2}$. **89.** $\pi^2c^5/10$. **94.** $(dl)/l-2(dt)/t$.
99. $2r/a=1+\cos(\theta-\gamma)$. **100.** $r^n=c\sin n\theta$.
101. $\log(y-1)=\frac{1}{2}x^2+x+\mathsf{A}$. **102.** $y^2=\mathsf{A}-x^2-2\cos x$.
103. $y=\cos x(\mathsf{A}+\sin x)$. **104.** $y=\mathsf{A}x+\mathsf{B}+\log\sec x$.
105. $y=\mathsf{A}+\mathsf{B}\log\tan\frac{1}{2}x$. **106.** $y=\mathsf{A}e^{4x}+\mathsf{B}e^{-2x}$.
107. $y=(\mathsf{A}x+\mathsf{B})e^{-6x}$. **108.** $y=\mathsf{A}e^{x}+\mathsf{B}e^{-4x}+\frac{1}{6}e^{2x}$.
109. $y=(\mathsf{A}+\frac{1}{5}x)e^{x}+\mathsf{B}e^{-4x}$. **110.** $y=(\frac{1}{2}x^2+\mathsf{A}x+\mathsf{B})e^{6x}$.
111. $\frac{1}{2}$; 0. **112.** $\frac{1}{2}(\sin x-x\cos x)$. **113.** 120, 60, 180.
118. $(-2)^{n+1}\{(n-1)!\}^2/(2n)!$. **121.** $x=-1$, $y=1$, $x-y=1$, $x-y=2$. **122.** $2\sqrt{2}$ for each. **124.** $2\mathsf{M}/35$.

INDEX TO VOLUMES I AND II

(Pages 1 to 240 are in Volume I)